Barron's Review Course Series

Let's Review:

Algebra 2/Trigonometry

Bruce C. Waldner, M.A.
Coordinator of Mathematics, K-12
Syosset Central School District
Syosset, New York

Adjunct Professor of Mathematics
Suffolk County Community College

Barron's Educational Series, Inc.

About the Author

Bruce C. Waldner is past President of the New York State Association of Mathematics Supervisors, the current President of the Nassau County Association of Mathematics Supervisors, and serves on the Long Island Conference Board, where he has been Program Chair for the Long Island Mathematics Conference. In addition, he is the current Coordinator of Mathematics, K-12, for the Syosset Central School District—a position he has held for the past nine years—and an adjunct full professor at Suffolk County Community College. He has also served as Mathematics Chairperson of the Malverne Union Free School District, K-12 and Mathematics Chairperson at Huntington High School.

Dedication

To my loving wife Lois
and our family . . .
For your love and caring
and understanding.
You have sacrificed much
to make this possible.
All my love.

All inquiries should be addressed to:
Barron's Educational Series, Inc.
250 Wireless Boulevard
Hauppauge, New York 11788
www.barronseduc.com

Library of Congress Control Number: 2009010408

ISBN-13: 978-0-7641-4186-7
ISBN-10: 0-7641-4186-4

Library of Congress Cataloging-in-Publication Data

Waldner, Bruce.
Algebra 2/Trigonometry / Bruce Waldner.
p. cm.—(Let's review)
ISBN-13: 978-0-7641-4186-7
ISBN-10: 0-7641-4186-4
1. Trigonometrical functions—Outlines, syllabi, etc. 2. Functions—Outlines, syllabi, etc. 3. Algebra—Examinations—Study guides. 4. High schools—New York (State)—Examinations—Study guides. I. Title. II. Title: Algebra two/trigonometry.

QA342.W35 2009
512□13—dc22

2009010408

PRINTED IN THE UNITED STATES OF AMERICA
9 8 7 6 5 4 3 2 1

TABLE OF CONTENTS

PREFACE

To Students, Teachers, and Parents:

Let's Review: Algebra 2/Trigonometry provides complete coverage of both the content strands and process strands included in the New York State Mathematics Standard 3. Graphing calculator skills are included so as to help you get ready for the NYS Regents Examination in Algebra 2/Trigonometry. In order to qualify for the Advanced Regents Diploma, students entering high school in September 2008 or thereafter will be required to pass the Algebra 2/Trigonometry Regents Examination.

The Regents Examination in Algebra 2/Trigonometry consists of 39 questions, including four different types of questions as indicated in the table below, worth a total of 88 credit points.

Question Type	Number of Questions
Multiple choice	27
2-credit open ended	8
4-credit open ended	3
6-credit open ended	1
Total	39

Key Features of This Review Book

- **A design for self-study and rapid learning**
  - Written in a dialogue style to involve students in the concepts
  - Key ideas to motivate and introduce sections
  - Math facts to highlight key formulas or ideas and to summarize concepts
  - Helpful diagrams and calculator instructions
  - Numerous examples worked out in full detail
  - Easy to read
  - Practice problems presented in Regents style format and an actual Regents exam with answers

- **Graphing calculator skills**
  - Step-by-step calculator solutions to problems
  - Infused into topics as needed
  - Graphing calculator used on the Algebra 2/Trigonometry Regents Examination
  - Used in algebraic topics, linear and nonlinear regression, and statistics
  - Draw connections between graphical, analytical, and numerical solutions
- **Regents-type questions**
  - Practice exercises at end of each section
  - Includes an actual Regents exam
  - Includes questions designed to strengthen understanding
  - Answers provided to all practice exercises at the end of the book

Who Should Use This Book?

Students
Students enrolled in a high school course preparing to take the Algebra 2/Trigonometry Regents Examination will find this book helpful if they need additional explanation and/or practice on specific topics, if they have been absent and missed instruction on a specific topic, or when they are preparing for the Regents Examination in Algebra 2/Trigonometry.

Teachers
Teachers will find this a helpful source of classroom exercises, homework, and test problems. The topics conform to the Mathematics Learning Standards that support the Algebra 2/Trigonometry Regents Examination. *Let's Review: Algebra 2/Trigonometry* belongs in department, school, and personal libraries. This book can be used as a valuable lesson planning aid for teachers. The book is designed to be compatible with all instructional styles and is organized in a logical curricular order.

School and School District Mathematics Departments
This book is a cost-effective way for school districts to provide supplementary resources for instruction in the topics of Algebra 2/Trigonometry. It includes detailed keystrokes for use with the TI-83, TI-84, and INSPIRE calculators. It represents a comprehensive review for the Algebra 2/Trigonometry Regents Examination in a single volume.

Chapter One ALGEBRAIC OPERATIONS

1.1 OPERATIONS WITH POLYNOMIALS

Special note: Parts of this section are a review from Integrated Algebra and are not part of the Algebra 2/Trigonometry curriculum. However, the information should be reviewed before studying the part of this chapter that covers the Algebra 2/Trigonometry curriculum. The new content in this section involves using rational coefficients for polynomials.

KEY IDEAS

Do you remember studying polynomials in Integrated Algebra? If so, remember that a polynomial is an algebraic expression that contains any number of terms where each term is a monomial separated by either addition or subtraction.

What Is a Polynomial?

A monomial is an algebraic expression that is the product of constants and variables. Some examples of monomials are $3x^5$, $-32p^2q^4$, $\frac{-4}{7}t^6$, and $\frac{3}{5}r^3s^2$. Algebraic expressions that are the sum or difference of expressions that are the product of constants and variables are polynomials. The standard form of a polynomial of a single variable is $a_nx^n + a_{n-1}x^{n-1} + a_{n-2}x^{n-2} + \ldots + a_2x^2 + a_1x + a_0$, where $a_0, a_1, a_2, \ldots, a_{n-2}, a_{n-1}, a_n$ are constants, $a_n \neq 0$, and x is a variable.

In a polynomial of a single variable, the degree of the nonzero term of highest degree (the term with the largest exponent) is the degree of the polynomial itself. For instance, the degree of the polynomial $\frac{2}{3}x^3 - 9x^2 + \frac{1}{2}x - 12$ is 3 because the highest exponent is a 3.

Example 1

Determine the degree of the polynomial:

$$4y^3 - 7y^4 + 16y^8 - 3y^5 + 2y + 7 + 19y^2$$

Solution: Notice that the largest exponent is 8, so the degree of this polynomial is 8.

Example 1 also shows that it is better to reorganize the terms of a polynomial and put them in ascending order of exponents, as in the standard form described earlier. In this example, the terms can be rearranged in ascending order of exponents. The resulting polynomial will be $16y^8 - 3y^5 - 7y^4 + 4y^3 + 19y^2 + 2y + 7$. Notice that it is important to maintain the signs of each individual term. Also notice that the sign for the first term in the original form of this polynomial, $4y^3$, is positive. Therefore the $4y^3$ is understood to have an addition sign before it. Similarly, the $16y^8$ in the original form of this polynomial does not require the addition sign preceding it in the standard form since the $16y^8$ is now the lead term of this expression. When a polynomial is put into standard form, the first term is called the *lead term* and its coefficient is referred to as the *lead coefficient* of the polynomial.

Also recall that a polynomial can be categorized by the number of terms in the polynomial. A monomial has one term, a binomial two terms, a trinomial three terms, and so on.

Terms of a polynomial are monomials that are either like terms or unlike terms. Like terms have the same variable(s) raised to the same exponent(s). Here are some examples of like and unlike terms:

First Monomial	Second Monomial	Like or Unlike
$3x^3y$	$\frac{6}{7}a^3b$	Unlike—the variables are not the same but the exponents are the same
$6r^2s^3$	$-3r^2s^3$	Like—same variables with the same exponents
$\frac{1}{8}pt^3$	$\frac{4}{5}p^3t$	Unlike—the exponents on the same variables are not the same
$100x^5y^3$	$100x^3y^5$	Unlike—the exponents on the same variables are not the same
$17k^3f^2$	$22k^3f^2$	Like

Addition and Subtraction of Polynomial Expressions

To add or subtract polynomials, simply add or subtract the coefficients of like terms. When doing so, it is best to write one polynomial beneath the other, aligning the like terms in a column.

Example 2

Determine the sum: $4x^2 + 7x - 4$ and $8x^2 - 12x + 13$

Solution: First write the polynomials one under the other, with like terms lined up:

$$\begin{array}{r} 4x^2 + 7x - 4 \\ 8x^2 - 12x + 13 \\ \hline 12x^2 - 5x + 9 \end{array}$$

Example 3

Determine the sum: $3t^2 - 7t^3 - 6t + 15$ and $7t^3 + 5t - 2t^4 + 8$

Solution: First rearrange the terms so that they are in descending order of exponents. Then write the polynomials one under the other, with like terms lined up:

$$\begin{array}{rrrrr} & -7t^3 & +3t^2 & -6t & +15 \\ -2t^4 & +7t^3 & & +5t & +8 \\ \hline -2t^4 & & +3t^2 & -t & +23 \end{array} \text{ or } -2t^4 + 3t^2 - t + 23$$

Example 4

Determine the sum: $\left(\frac{3}{4}x^2 - \frac{2}{3}x + \frac{1}{5}\right)$ and $\left(\frac{1}{2}x^2 + \frac{5}{6}x - \frac{3}{10}\right)$

Solution: First write the polynomials one under the other, with like terms lined up:

$$\begin{array}{l} \frac{3}{4}x^2 - \frac{2}{3}x + \frac{1}{5} \\ \frac{1}{2}x^2 + \frac{5}{6}x - \frac{3}{10} \\ \hline \frac{5}{4}x^2 + \frac{1}{6}x - \frac{1}{10} \end{array}$$

and

Add the coefficients of the like terms:

$$\frac{3}{4} + \frac{1}{2} = \frac{3}{4} + \frac{2}{4} = \frac{5}{4}$$

$$-\frac{2}{3} + \frac{5}{6} = -\frac{4}{6} + \frac{5}{6} = \frac{1}{6}$$

$$\frac{1}{5} - \frac{3}{10} = \frac{2}{10} - \frac{3}{10} = -\frac{1}{10}$$

If need be, a graphing calculator can be used to add the numerical coefficients of the like terms. For instance, you can enter **3** **÷** **4** **ENTER** and the answer .75 will be displayed on the screen of the calculator. To change this to fractional form, enter **MATH** **1** **ENTER** and that provides the display of $\frac{3}{4}$.

Example 5

Determine the difference between $6y^3 + 4y^2 - 5y + 7$ and $2y^3 - 9y^2 + 3y - 4$

Solution: First write the polynomials one under the other with like terms lined up. Then change the signs of the second polynomial and add:

$$\begin{array}{r} 6y^3 + \ 4y^2 - 5y + 7 \\ -2y^3 + \ 9y^2 - 3y + 4 \\ \hline 4y^3 + 13y^2 - 8y + 11 \end{array}$$

Example 6

Subtract $8m^3 + 3m - 12$ from $7m^4 - 2m^3 + 6m^2 + 9$

Solution: First write the polynomials one under the other with like terms lined up. Notice that it is necessary to begin with the $7m^4 - 2m^3 + 6m^2 + 9$ on the top line since the $8m^3 + 3m - 12$ is being subtracted from that. Also remember to change the signs of the second polynomial:

$$\begin{array}{rrrrr} 7m^4 & - 2m^3 & + 6m^2 & & + 9 \\ & - 8m^3 & & - 3m & + 12 \\ \hline 7m^4 & - 10m^3 & + 6m^2 & - 3m & + 21 \end{array}$$

Example 7

Subtract $\frac{4}{9}y^3 + \frac{1}{4}y - \frac{7}{8}$ from $\frac{1}{3}y^3 - \frac{1}{2}y^2 + \frac{1}{4}$

Solution: First write the polynomials one under the other with like terms lined up. Notice that you must begin with the $\frac{1}{3}y^3 - \frac{1}{2}y^2 + \frac{1}{4}$ on the top line since the $\frac{4}{9}y^3 + \frac{1}{4}y - \frac{7}{8}$ is being subtracted from that:

$$
\begin{array}{r}
\frac{1}{3}y^3 - \frac{1}{2}y^2 \qquad\quad + \frac{1}{4} \\
-\frac{4}{9}y^3 + \qquad -\frac{1}{4}y + \frac{7}{8} \\
\hline
-\frac{1}{9}y^3 - \frac{1}{2}y^2 + \frac{1}{4}y + \frac{9}{8}
\end{array}
$$

Add the coefficients of the like terms:

$$\frac{1}{3} - \frac{4}{9} = \frac{3}{9} - \frac{4}{9} = -\frac{1}{9}$$

$$\frac{1}{4} + \frac{7}{8} = \frac{2}{8} + \frac{7}{8} = \frac{9}{8}$$

Notice that there is only one y^2 term and one y term.

Multiplication of Polynomial Expressions

Before multiplying polynomials in the same variable, it is necessary to recall how to multiply monomials.

MATH FACTS

In order to multiply monomials, simply multiply the coefficients and add the exponents for that variable.

Example 8

Multiply $5a^3$ by $4a^6$

Solution: Multiply the 5 and the 4 and add the 3 and the 6 to get $20a^9$.

What if the variables are not the same or if there is more than one variable? In this case, simply multiply the coefficients and add the exponents only on like variables.

Example 9

Multiply $\frac{4}{5}p^2q^5$ by $\frac{5}{7}q^2r^4$

Solution: Multiply the coefficients together: $\frac{4}{5}\cdot\frac{5}{7}=\frac{4}{7}$. Add the exponents 5 and 2 on the variable q to get 7. The resulting answer is $\frac{4}{7}p^2q^7r^4$.

Now that it is known how to multiply monomials together, it is possible to work with polynomials. Many students recall multiplication through a process called FOIL. When multiplying binomials, multiply the first terms from each binomial, then the outer terms (first term from the first binomial and last term from the second binomial), the inner terms (last term from the first binomial and first term from the second binomial), and finally the last terms from each binomial. This process is easy. However, the problem with the **FOIL** method is that it works for only binomials and not for any other type of polynomial.

Two other approaches work for all polynomials: the distributive method and the box method. In the table below are several examples. Binomial multiplied by binomial is worked out in each of the three methods. Binomial multiplied by trinomial is worked out using the distributive and the box methods.

Example	FOIL Method	Distributive Method	Box Method
Example 10 $(x + 5)(x - 3)$	First terms x^2 Outer terms $-3x$ Inner terms $+5x$ Last terms -15 $x^2 - 3x + 5x - 15$ $= x^2 + 2x - 15$	Distribute x over $(x - 3)$ $x^2 - 3x$ Distribute $+5$ over $(x - 3) + 5x - 15$ $x^2 - 3x + 5x - 15 =$ $x^2 + 2x - 15$	$x - 5$ $x + 3$ $+3x - 15$ $x^2 - 5x$ $x^2 - 2x - 15$
Example 11 $(a - 4)(a - 7)$	First terms a^2 Outer terms $-7a$ Inner terms $-4a$ Last terms $+28$ $a^2 - 7a - 4a + 28$ $= a^2 - 11a + 28$	Distribute a over $(a - 7)$ $a^2 - 7a$ Distribute -4 over $(a - 7) - 4a + 28$ $a^2 - 7a - 4a + 28 =$ $a^2 - 11a + 28$	$a - 4$ $a - 7$ $-7a + 28$ $a^2 - 4a$ $a^2 - 11a + 28$

Example	FOIL Method	Distributive Method	Box Method
Example 12 $\left(t-\frac{1}{3}\right)\left(t+\frac{1}{3}\right)$	First terms t^2 Outer terms $+\frac{1}{3}t$ Inner terms $-\frac{1}{3}t$ Last terms: $-\frac{1}{9}$ (Notice that the outer and inner terms cancel) $t^2+\frac{1}{3}t-\frac{1}{3}t-\frac{1}{9}$ $=t^2-\frac{1}{9}$	Distribute t over $\left(t+\frac{1}{3}\right)$ $t^2+\frac{1}{3}t$ Distribute $-\frac{1}{3}$ over $\left(t+\frac{1}{3}\right)$ $-\frac{1}{3}t-\frac{1}{9}$ $t^2+\frac{1}{3}t-\frac{1}{3}t-\frac{1}{9}=$ $t^2-\frac{1}{9}$	$t-\frac{1}{3}$ $t+\frac{1}{3}$ $+\frac{1}{3}t-\frac{1}{9}$ $t^2-\frac{1}{3}t$ $r^2 \quad -\frac{1}{9}$
Example 13 $(x+2)$ (x^2-3x+5)	Cannot do because x^2-3x+5 is not a binomial	Distribute x over (x^2-3x+5) x^3-3x^2+5x Distribute +2 over $(x^2-3x+5)+2x^2-6x+10$ $x^3-3x^2+5x+2x^2-6x+10=$ x^3-x^2-x+10	$x^2-3x+\ 5$ $x+\ 2$ $2x^2-6x+10$ x^3-3x^2+5x $x^3-\ x^2-\ x+10$

Division of Polynomial Expressions

MATH FACTS

In order to divide a polynomial by a monomial, it is necessary to divide each term of the polynomial individually by that monomial.

When dividing $12x + 15y$ by 3, it is necessary to divide the $12x$ by 3 to get $4x$ and the $15y$ by 3 to get $5y$. The result is the binomial $4x + 5y$. This can also be written in the following way: $\frac{12x+15y}{3}=\frac{12x}{3}+\frac{15y}{3}=4x+3y.$

Example 14

Divide $54a^3b^5c^2 - 30a^4b^4 + 42a^6b^7c^5$ by $6ab^3$

Solution: $$\frac{54a^3b^5c^2 - 30a^4b^4 + 42a^6b^7c^5}{6ab^3}$$

$$= \frac{54a^3b^5c^2}{6ab^3} - \frac{30a^4b^4}{6ab^3} + \frac{42a^6b^7c^5}{6ab^3}$$

$$= 9a^2b^2c^2 - 5a^3b + 7a^5b^4c^5$$

Example 15

Divide $\frac{5}{8}x^3y^5 - \frac{3}{4}x^2y^6$ by $\frac{5}{16}x^2y^5$

Solution: $$\frac{\frac{5}{8}x^3y^5 - \frac{3}{4}x^2y^6}{\frac{5}{16}x^2y^5} = \frac{\frac{5}{8}x^3y^5}{\frac{5}{16}x^2y^5} - \frac{\frac{3}{4}x^2y^6}{\frac{5}{16}x^2y^5} = 2x - \frac{12}{5}y$$

The approach to division of polynomial expressions is similar to the approach used in long division.

The division of 156 by 12 is performed below. Notice the similarity to $(x^2 + 3x - 10) \div (x - 2)$.

$156 \div 12$	$(x^2 + 3x - 10) \div (x - 2)$
$\begin{array}{r} 13 \\ 12\overline{)156} \\ -\underline{12} \\ 36 \\ \underline{36} \\ 0 \end{array}$	$\begin{array}{r} x+5 \\ x-2\overline{)x^2+3x-10} \\ \underline{-(x^2-2x)} \\ 5x-10 \\ \underline{5x-10} \\ 0 \end{array}$

In both cases, first find how many times the divisor divides into the first part of the dividend, multiply that by the divisor, subtract this amount from the dividend, and then repeat that process with the remaining expression.

Check Your Understanding of Section 1.1

Reminder: Part of this section reviews work from Integrated Algebra that is not part of the Algebra 2/Trigonometry curriculum but still needs to be reviewed here. Questions 1–3 below reflect these review topics that will not be part of the Algebra 2/Trigonometry Regents Examination. The rest of the questions listed here reflect content on the Algebra 2/Trigonometry Regents Examination.

1. The degree of the polynomial $12y^5 + 7y^8 - 4y^3 - 9y^7 + 3y^2 - 18 + 4y$ is
(1) 7 (2) 18 (3) 8 (4) 12

2. Subtract $4a^3 + 5a^2 - 3a + 8$ from $2a^3 - 4a^2 + 3a - 17$.
(1) $2a^3 + 9a^2 - 6a + 25$
(2) $2a^3 + a^2 + 25$
(3) $-2a^3 - 9a^2 + 6a - 25$
(4) $-2a^3 - a^2 - 25$

3. What is the product of $2x + 5$ and $3x - 4$?
(1) $6x^2 - 20$
(2) $6x^2 + 7x - 20$
(3) $5x + 1$
(4) $5x^2 - 7x + 1$

4. What is the quotient when $p^2 - 7p + 6$ is divided by $p - 6$?
(1) $p + 1$
(2) $p - 1$
(3) $p + 6$
(4) $p - 7$

5. From the sum of $\frac{3}{4}s^3 - 5s^2 + \frac{7}{3}s + \frac{1}{2}$ and $6s^2 - \frac{2}{3}s + \frac{1}{2}$ subtract $\frac{1}{4}s^3 - \frac{5}{3}s + \frac{5}{4}$.
(1) $\frac{1}{2}s^3 + s^2 + \frac{10}{3}s - \frac{1}{4}$
(2) $-\frac{1}{2}s^3 - s^2 + \frac{1}{4}$
(3) $\frac{1}{2}s^3 + s^2 + \frac{1}{4}$
(4) $\frac{1}{2}s^3 + \frac{10}{3}s + 2$

6. Determine the sum of the product of $\frac{2}{3}y+\frac{5}{9}$ and $\frac{3}{5}y-\frac{1}{3}$ and the product of $\frac{1}{3}y+\frac{5}{27}$ and $\frac{3}{5}y+\frac{2}{3}$.

(1) $\frac{1}{5}y^2-\frac{2}{9}y-\frac{25}{81}$

(2) $-\frac{3}{5}y^2-\frac{4}{9}y+\frac{5}{81}$

(3) $-\frac{2}{5}y^2-\frac{1}{9}y+\frac{5}{27}$

(4) $\frac{3}{5}y^2+\frac{4}{9}y-\frac{5}{81}$

7. Determine the product of $\frac{7}{2}c-7$ and $3c^2-\frac{1}{2}c+\frac{1}{7}$.

(1) $\frac{21}{2}c^3-\frac{91}{4}c^2+4c-1$

(2) $3c^2-3c-\frac{50}{7}$

(3) $-\frac{21}{2}c^3+\frac{91}{4}c^2-4c+1$

(4) $\frac{21}{2}c^3-7c^2+3c-1$

8. Subtract the product of $\frac{1}{2}x+\frac{1}{3}y$ and $2x-\frac{3}{5}y$ from $\frac{7}{3}x^2+\frac{5}{6}xy-\frac{4}{5}y^2$.

(1) $x^2+\frac{11}{30}xy-\frac{1}{5}y^2$

(2) $-\frac{4}{3}x^2-\frac{7}{15}xy+y^2$

(3) $\frac{4}{3}x^2+\frac{7}{15}xy-\frac{3}{5}y^2$

(4) $\frac{10}{3}x^2+\frac{6}{5}xy-\frac{1}{5}y^2$

9. Kimberly wants to multiply $\frac{2}{3}x+\frac{1}{5}$ by a binomial to get a product equal to $\frac{4}{9}x^2-\frac{1}{25}$. By what binomial should she multiply the $\frac{2}{3}x+\frac{1}{5}$ to do this?

(1) $\frac{2}{3}x+\frac{1}{5}$

(2) $\frac{-2}{9}x-\frac{1}{25}$

(3) $2x-4$

(4) $\frac{2}{3}x-\frac{1}{5}$

10. Divide $\frac{1}{3}a^3b^4$ by $\frac{2}{3}a^2b$

(1) $2ab^3$

(2) $\frac{2}{9}ab^3$

(3) $\frac{2}{9}a^5b^5$

(4) $\frac{1}{2}ab^3$

11. Divide $\frac{8}{15}r^5 + \frac{4}{9}r^3s - \frac{20}{27}r^4t^3$ by $\frac{2}{3}r^2$

(1) $\frac{16}{45}r^3 + \frac{8}{27}rs - \frac{40}{81}r^2t^3$ (3) $\frac{4}{5}r^7 + \frac{2}{3}r^5s - \frac{10}{9}r^6t^3$

(2) $\frac{4}{5}r^3 + \frac{2}{3}rs - \frac{10}{9}r^2t^3$ (4) $\frac{4}{5}r^{\frac{5}{2}} + \frac{2}{3}r^{\frac{3}{2}}s - \frac{10}{9}rt^3$

12. Determine the product: $\frac{4}{5}y^2 - \frac{1}{3}y + \frac{3}{2}$ and $5y^2 + \frac{3}{2}y - \frac{1}{4}$

13. The area of a square is $\frac{16}{25}s^2$ square cm. If one side of the square is increased by $\frac{2}{3}$ cm and the other side is increased by $\frac{3}{8}$ cm, what is the difference in their areas?

1.2 FACTORING METHODS

In Integrated Algebra, you learned some methods to factor polynomial expressions. Remember that factoring reverses the process of multiplication. When multiplying 5 by $(x - 2)$, the answer $5x - 10$ is obtained by the distributive law. This same distributive law can be reversed, yielding $5x - 10 = 5(x - 2)$.

Factoring Polynomials with Common Factors

The first factoring method is called *greatest common factoring*. What common factor do the terms of the polynomial $4x^2y + 6xy^2$ have? Notice that there is a common numerical factor of 2. Both also have an x and a y in them. Because of this, it is possible to factor out $2xy$ from this binomial by division to get $2x + 3y$. Therefore, the expression $4x^2y + 6xy^2$ can be factored as $2xy(2x + 3y)$ or $4x^2y + 6xy^2 = 2xy(2x + 3y)$. Notice that these factors can be checked by distributing the $2xy$ over the $2x + 3y$ to see that it is equal to $4x^2y + 6xy^2$.

Another way to find the greatest common factor between the different terms of a polynomial would be to factor each term into its prime factors.

For instance, rewrite the expression $4x^2y + 6xy^2$ as $\underline{2} \cdot 2 \cdot \underline{x} \cdot x \cdot \underline{y} + \underline{2} \cdot 3 \cdot \underline{x} \cdot \underline{y} \cdot y$, underline the common factors, write the product of those factors outside the parentheses, and put the factors from each term that are not underlined inside the parentheses, $2xy(2x + 3y)$. This second method is rather cumbersome. It is much more reasonable to examine the different terms to decide what the greatest common factor is and factor it out.

Just in case the greatest common factor is not selected, it is necessary to examine the expression inside the parentheses to see if there is still a common factor between the terms. When comparing the different monomial expressions or terms in the polynomial, it is important to find the greatest common divisor of the coefficients and the smallest exponent for common variables.

Example 1

Factor: $3p^3q^7 - 12p^5q^4$

Solution: The greatest common factor between the $3p^3q^7$ and $12p^5q^4$ is $3p^3q^4$. Dividing $3p^3q^7 - 12p^5q^4$ by the common factor $3p^3q^4$ yields $q^3 - 4p^2$. Therefore, $3p^3q^7 - 12p^5q^4 = 3p^3q^4(q^3 - 4p^2)$.

Example 2

Factor: $18a^2b^4 + 27b^5c^4 - 36a^4b^3c^2$

Solution: The greatest common factor of the coefficients is 9. The only variable in each term is b, where the lowest exponent on the b is 3. Therefore, the greatest common factor is $9b^3$. Divide each term by the greatest common factor, $9b^3$. That yields $2a^2b + 3b^2c^4 - 4a^4c^2$. So $18a^2b^4 + 27b^5c^4 - 36a^4b^3c^2 = 9b^3(2a^2b + 3b^2c^4 - 4a^4c^2)$.

PLEASE NOTE

Always look to see if there is a greatest common factor between the terms of a polynomial as the first step when factoring, even if another method can be applied.

Examine the polynomial $5x^3 - 25x^2 + 3x + 15$. Like the previous examples, there are some common factors. However, they occur in only two pairs of the terms. In other words, $5x^3 - 25x^2$ has a common factor of $5x^2$ while $3x + 15$ has a common factor of 3. If the terms of the polynomial are grouped into these two binomials and each is factored separately, the result is $5x^2(x + 5) + 3(x + 5)$. Now notice that there is a common binomial factor of $(x + 5)$. Factor that out to get the final result of $(x + 5)(5x^2 + 3)$. This method of factoring by grouping and factoring out a common binomial will be used later in this section.

The Difference of Two Perfect Squares

Remember that sometimes when multiplying two binomials together, the product itself is a binomial. Specifically, this product is in the form $a^2 - b^2$. This type of binomial is called the difference between two perfect squares because both terms of the binomial are squares of a monomial, i.e., $(a)^2 - (b)^2$. When did this happen in multiplication? It happened when $(a + b)$ was multiplied by $(a - b)$. Two binomials that have the same two terms, but one is the sum of these terms and the other is their difference, are called *conjugates*. So the difference of two perfect squares factors as the product of these conjugate pairs, or the sum and difference of the square roots of the original terms. Symbolically, it can be written as $a^2 - b^2 = (a + b)(a - b)$.

MATH FACTS

To factor a binomial that is a difference of two perfect squares, set up two parentheses with the sum and difference of the square roots of the original terms.

Symbolically, you can write $a^2 - b^2 = (a + b)(a - b)$.

	Factor	Written as the Difference of Two Perfect Squares	Solution
Example 3	$d^2 - k^2$	$(d)^2 - (k)^2$	$(d + k)(d - k)$
Example 4	$16x^2 - 25y^2$	$(4x)^2 - (5y)^2$	$(4x + 5y)(4x - 5y)$
Example 5	$400p^2 - 81q^4$	$(20p)^2 - (9q^2)^2$	$(20p + 9q^2)(20p - 9q^2)$
Example 6	$\frac{x^2}{49} - \frac{100y^2}{121}$	$\left(\frac{x}{7}\right)^2 - \left(\frac{10y}{11}\right)^2$	$\left(\frac{x}{7} + \frac{10y}{11}\right)\left(\frac{x}{7} - \frac{10y}{11}\right)$

Factoring Trinomials

In order to factor trinomials, it is important to recognize how two binomials multiply together to result in a trinomial. In the last section, it was established that $(x + 5)(x - 3) = x^2 - 3x + 5x - 15 = x^2 + 2x - 15$. Notice that there is a relationship between the coefficients of the two factors whose product is $x^2 + 2x - 15$ and the coefficients of this trinomial. The x^2 is the product of the first terms in both parentheses. The -15 is the product of the last terms found in both parentheses. The $2x$ is the sum of the product of the two outer terms and the product of the two inner terms.

When multiplying $(ax + b)(cx + d)$, the resulting polynomial is $acx^2 + (ad + bc)x + bd$. This starts to suggest a way to factor a trinomial. In order to factor the trinomial $a^2 + 5a + 6$, look for two numbers whose product is $+6$ and whose sum is $+5$. Since $a \cdot a = a^2$ and $+2 \cdot +3 = +6$ and $(+2) + (+3) = +5$, the factors $(a + 3)(a + 2)$ can be determined.

MATH FACTS

To factor a trinomial whose lead coefficient is 1, look for the factors of the last term whose sum is equal to the coefficient of the middle term. In two sets of parentheses, place the variable from the first term. Into each set, place one of the two factors whose product is the last term and whose sum is the coefficient of the middle term.

Example 7

Factor: $x^2 - 6x + 8$

Solution: Since the lead coefficient is 1, set up two sets of parentheses each, starting with an x. The factors of 8 that add up to -6 are -2 and -4. Therefore, the answer is $(x - 2)(x - 4)$.

Example 8

Factor: $y^2 + 3y - 10$

Solution: Look for two factors of -10 that add up to $+3$. These factors are $+5$ and -2. Therefore, set up the two parentheses as $(y + 5)(y - 2)$.

Example	Supporting Work	Solution
Example 9 $a^2 - 2a - 15$	Find two numbers whose product is -15 and whose sum is -2. $-5 \cdot +3 = -15$ and $-5 + (+3) = -2$	$(a - 5)(a + 3)$
Example 10 $r^2 - 7r + 12$	$-4 \cdot -3 = +12$ and $(-4) + (-3) = -7$	$(r - 4)(r - 3)$
Example 11 $x^2 - 5x + 6$	$-3 \cdot -2 = +6$ and $(-3) + (-2) = -5$	$(x - 3)(x - 2)$
Example 12 $p^2 + 13p + 12$	$+12 \cdot +1 = +12$ and $(+12) + (+1) = +13$	$(p + 12)(p + 1)$
Example 13 $x^2 + 5xy - 24y^2$	$+8y \cdot -3y = -24y^2$ and $(+8) + (-3) = +5$	$(x + 8y)(x - 3y)$

Not all quadratic expressions have a lead coefficient of 1. You can use two methods to factor a trinomial of the form $ax^2 + bx + c$ where $a > 1$. One method involves trial and error. The other uses factoring out a common binomial. To understand these, let's examine how to factor the trinomial $2x^2 + 7x - 4$. Similar to the approach used to factor trinomials with lead coefficients of 1, set up two sets up parentheses with factors of the first term in the first part of the parentheses and the factor of the last term in the second part of each parentheses. In this problem, the factors of the $2x^2$ are $2x$ and x. The factors of the last term are -4 and 1, or $+4$ and -1, or $+2$ and -2. So the possible factors are:

$$(2x + 1)(x - 4) = 2x^2 - 8x + x - 4 = 2x^2 - 7x - 4$$
$$(2x - 4)(x + 1) = 2x^2 + 2x - 4x - 4 = 2x^2 - 2x - 4$$
$$(2x - 1)(x + 4) = 2x^2 + 8x - x - 4 = 2x^2 + 7x - 4$$
$$(2x + 4)(x - 1) = 2x^2 - 2x + 4x - 4 = 2x^2 + 2x - 4$$
$$(2x + 2)(x - 2) = 2x^2 - 4x + 2x - 4 = 2x^2 - 2x - 4$$
$$(2x - 2)(x + 2) = 2x^2 + 4x - 2x - 4 = 2x^2 + 2x - 4$$

Obviously, the correct set of factors is $(2x + 1)(x - 4)$. Be careful to notice a few things from the six possible sets of factors that will make this process a little simpler. First of all, recognize that in each of the last three possible sets of factors, when multiplying these, there is a common factor of 2 in their products. This should be no surprise since there was a common factor of 2 in one of the two parentheses that were multiplied together. When using trial and error, never put two monomials with a common factor inside one of the parentheses unless the original expression also has that common factor. (If that occurs, there is a better method that could be used.)

One other thing to notice is that the answer found by multiplying the first and third sets of possible factors differ only by the sign of the middle term. However, further examination shows that these factors themselves differ only by the signs between the terms inside each set of parentheses. So if one set of factors is tried and they have a result that is different from the desired original trinomial by only the sign of the middle term, then just change the signs in each parentheses.

The second method that can be used to factor $2x^2 + 7x - 4$ is the grouping/common binomial method. Find two numbers whose product is the same as the product of the lead coefficient and the last term and whose sum is the coefficient of the middle term. In other words, multiply together the a and the c from the quadratic trinomial in the form of $ax^2 + bx + c$. Then look for two numbers whose product is that number and whose sum is the coefficient of the middle term. For $2x^2 + 7x - 4$, first multiply the 2 by the -4 to get -8. Now look for two factors of -8 that add up to $+7$. These would be $+8$ and -1. Separate the $+7x$ into $+8x$ and $-x$. The resulting polynomial would now be $2x^2 + 8x - x - 4$. Notice that the first two terms of this polynomial have a common factor of $2x$ and that the last two terms of that same polynomial have a common factor of -1. This gives $2x^2 + 7x - 4 = 2x^2 + 8x - x - 4 = 2x(x + 4) - 1(x + 4)$. In this expression, there is a common binomial factor of $x + 4$. Therefore, $2x(x + 4) - 1(x + 4) = (x + 4)(2x - 1)$.

Example	Trial and Error Method	Grouping/Common Binomial Method
Example 14 $3x^2 - 4x - 4$	Set up two parentheses with the factors of $3x^2$ inside them: $(3x\ )(x\ )$ The factors of -4 are either $+4 \cdot -1$, $-4 \cdot 1$, or $+2 \cdot 2$. $(3x + 4)(x - 1)$ has a middle term of x. $(3x - 4)(x + 1)$ has a middle term of $-x$. $(3x - 2)(x + 2)$ has a middle term of $+4x$, instead of $-4x$, so the correct factoring is $(3x + 2)(x - 2)$.	$3 \cdot -4 = -12$ and factors of -12 that add up to -4 are -6 and $+2$. Rewrite the polynomial as $3x^2 - 6x + 2x - 4$. Factor common factors in two groupings to get $3x(x - 2) + 2(x - 2)$. Factor out the common binomial, $(x - 2)$. The result is $(x - 2)(3x + 2)$.

Example	Trial and Error Method	Grouping/Common Binomial Method
Example 15 $6x^2 + 13x - 5$	Set up two parentheses with the factors of $6x^2$ inside them: $(3x\ )(2x\ )$ or $(6x\ )(x\ )$ The factors of -5 are either $+5 \cdot -1$, or $-5 \cdot 1$. $(6x + 5)(x - 1)$ has a middle term of $-x$. Don't try $(6x - 5)(x + 1)$. $(6x - 1)(x + 5)$ will have a middle term of $+29x$. Don't try $(6x + 1)(x - 5)$. $(3x - 5)(2x + 1)$ has a middle term of $-7x$. Don't try $(3x + 5)$ $(2x - 1)$. $(3x + 1)(2x - 5)$ has a middle term of $-13x$, instead of $+13x$, so the correct factoring is $(3x - 1)(2x + 5)$.	$6 \cdot -5 = -30$ and factors of -30 that add up to $+13$ are -2 and $+15$. Rewrite the polynomial as $6x^2 - 2x + 15x - 5$. Factor common factors in two groupings to get $2x(3x - 1) + 5(3x - 1)$. Factor out the common binomial, $(3x - 1)$. The result is: $(3x - 1)(2x + 5)$.

Complete Factoring

There are also factoring problems for which once one correct factoring method is used, one or more of the factors can themselves be factored. Recall the note mentioned previously. Always look to see if there is a greatest common factor between the terms of a polynomial as the first step when factoring even if another method can be applied. For instance, factoring $3x^2 - 75$ looks hard until the fact that there is a common factor of 3 between the two terms of this binomial is observed. So $3x^2 - 75 = 3(x^2 - 25) = 3(x + 5)(x - 5)$ because the factor $x^2 - 25$ is a difference between two perfect squares.

Not all complete factoring situations involve a common factor. For instance, when factoring $x^4 - 16$, the result may appear to be $x^4 - 16 = (x^2 + 4)(x^2 - 4)$. However, that is not complete because the $x^2 - 4$ is still factorable as it is also a difference between two perfect squares, yielding a final set of factors equal to $(x^2 + 4)(x + 2)(x - 2)$.

Example	First Step	Additional Step(s)	Final Answer
Example 16 $4ay^2 - 100a$	$4a$ is a common factor. $4a(y^2 - 25)$	$y^2 - 25 =$ $(y + 5)(y - 5)$	$4a(y + 5)$ $(y - 5)$
Example 17 $5x^4 + 30x^3 + 45x^2$	$5x^2$ is a common factor. $5x^2(x^2 + 6x + 9)$	$x^2 + 6x + 9 =$ $(x + 3)(x + 3)$	$5x^2(x + 3)$ $(x + 3)$
Example 18 $x^4y - 81y$	y is a common factor. $y(x^4 - 81)$	$x^4 - 81 =$ $(x^2 + 9)(x^2 - 9) =$ $(x^2 + 9)(x + 3)$ $(x - 3)$	$y(x^2 + 9)$ $(x + 3)$ $(x - 3)$
Example 19 $6x^3 + 24x^2 - 126x$	$6x$ is a common factor. $6x(x^2 + 4x - 21)$	$x^2 + 4x - 21 =$ $(x + 7)(x - 3)$	$6x(x + 7)$ $(x - 3)$
Example 20 $y^4 - 26y^2 + 25$	$y^4 - 26y^2 + 25 =$ $(y^2 - 25)(y^2 - 1)$	$y^2 - 25 =$ $(y + 5)(y - 5)$ $y^2 - 1 =$ $(y + 1)(y - 1)$	$(y + 5)$ $(y - 5)$ $(y + 1)$ $(y - 1)$
Example 21 $15x^3 - 5x^2 - 10x$	$5x$ is a common factor. $5x(3x^2 - x - 2)$	$3x^2 - x - 2 = 3x^2$ $- 3x + 2x - 2 =$ $3x(x - 1) + 2(x - 1)$ $= (x - 1)(3x + 2)$	$5x(x - 1)$ $(3x + 2)$

Check Your Understanding of Section 1.2

1. Factor $24x^3y^5 + 30x^4y^4z^2 - 48x^5y^6z$ completely.
(1) $3x^3y^4(8y + 10xz^2 - 16x^2y^2z)$
(2) $6x^3y^4z(4y + 5xz^2 - 8x^2y^2)$
(3) $6x^3y^4(8y + 10xz^2 - 16x^2y^2z)$
(4) $6x^3y^4(4y + 5xz^2 - 8x^2y^2z)$

2. Factor $24x^3y - 54xy^3$ completely.
(1) $6xy(4x^2 - 9y^2)$
(2) $12(2x - 4y^2)$
(3) $6xy(2x + 3y)(2x - 3y)$
(4) $4xy(6x^2 - 16y^2)$

3. Factor $256a^8 - 1$ completely.
(1) $(16a^4 + 1)(16a^4 - 1)$
(2) $(256a^4 + 1)(a^4 - 1)$
(3) $(8a^4 + 1)(32a^4 - 1)$
(4) $(16a^4 + 1)(4a^4 + 1)(2a + 1)(2a - 1)$

4. Factor $10y^2 - 29y - 21$ completely.
(1) $(5y + 3)(2y - 7)$
(2) $(10y + 7)(y - 3)$
(3) $(5y - 3)(2y + 7)$
(4) $(5y + 7)(2y - 3)$

5. Factor $\frac{4}{5}s^2 - 5t^2$ completely.
(1) $\frac{4}{5}(s+t)(s-t)$
(2) $\frac{4}{5}(s+5t)(s-5t)$
(3) $\frac{(2s+t)(2s-t)}{5}$
(4) $\frac{1}{5}(2s+5t)(2s-5t)$

6. Factor $4a^2 + 6ab + 2b^2$ completely.
(1) $(2a + 2b)(2a + b)$
(2) $(4a + b)(a + 2b)$
(3) $2(a + b)(2a + b)$
(4) $2(a - b)(2a - b)$

7. Factor $0.36d^4 - 0.25g^2$ completely.
(1) $(0.6d + 0.5g)(0.6d - 0.5g)$
(2) $(0.6d^2 + 0.5g)(0.6d^2 - 0.5g)$
(3) $(6d^2 + g)(6d^2 - g)$
(4) $(0.9d^2 + 0.25g)(0.4d^2 - g)$

8. Factor $18t^5 - 42t^4 - 120t^3$ completely.
(1) $6t^3(t - 4)(3t + 5)$
(2) $2t(3t^3 - 12t^2)(3t - 5)$
(3) $6t^3(t + 4)(3t - 5)$
(4) $6t^3(3t^2 - 7t - 20)$

9. Factor $10j^3k^2 + 15j^2k^2 - 45jk^2$ completely.
(1) $5jk^2(2j^2 + 3j - 15)$
(2) $5jk^2(2j + 3)(j - 3)$
(3) $5jk^2(2j - 9)(j + 1)$
(4) $5jk^2(2j - 3)(j + 3)$

10. If the area of a square is $4x^2 + 20x + 25$, express the length of a side of the square as a binomial in x.
(1) $2x + 3$
(2) $2x + 5$
(3) $(2x + 5)^2$
(4) $x + 5$

11. The area of a square is $x^2 - 10x + 25$. If one side of the square is increased by 4 and the other side is decreased by 4, determine the area of the resulting rectangle.
(1) $x - 5$
(2) $x^2 - 10x + 9$
(3) $x^2 + 10x + 9$
(4) $x^2 - 16$

12. Which of the following binomials is a factor of $15x^2 + 11x - 14$?

(1) $5x - 7$
(2) $3x + 2$
(3) $3x + 7$
(4) $3x - 2$

Express as the product of two binomials:

13. $a^2 - 11a + 28$

14. $5x^2 + 26xy + 5y^2$

15. $3r^2 - 2rs - 8s^2$

16. $p^2 + 7p - 8$

17. $t^4 - 25$

18. $y^2 - 12y - 28$

19. $20a^2 + 47ab + 21b^2$

20. $\frac{4}{9}x^2 - \frac{16}{25}y^2$

21. $r^4x^2 - 8r^2sx + 15s^2$

Factor completely:

22. $3p^3 + 18p^2 + 24p$

23. $81j^4 - 625k^4$

24. $10x^2y - 25xy - 60y$

25. $75a^4b - 12a^2b^3$

26. $6m^3n - 108m^2n + 270mn$

27. $3ax^3 + 60ax^2 + 288ax$

1.3 RATIONAL EXPRESSIONS

Do you remember working with numerical fractions? The same skills learned with numerical fractions need to be applied when performing operations on algebraic fractions. For instance, if $\frac{1}{3}+\frac{1}{4}=\frac{4}{12}+\frac{3}{12}=\frac{7}{12}$, then it is easy to see why $\frac{1}{a}+\frac{1}{b}=\frac{b}{ab}+\frac{a}{ab}=\frac{b+a}{ab}$. This section will review and teach how to simplify, multiply, divide, add, and subtract algebraic fractions.

What Is a Rational Expression?

A rational expression is the ratio of two polynomial expressions in which the denominator is not zero. Some examples of rational expression are $\frac{x+2}{x^2-9}$, $\frac{x^2-5}{2x^2+3x+1}$, and $\frac{3x^5-2x^4+7x^3+x^2-6x+15}{x^4-5x^3+8x^2+3x-7}$. Notice that the degree of the numerator is independent of the degree of the denominator. In other words, the degree of the numerator can be <, =, or > the degree of the denominator.

Simplifying Rational Expressions

To simplify a rational expression, first factor both the numerator and the denominator and then cancel the same factors from each. That is exactly what was done to simplify fractions. For example, to simplify the fraction $\frac{36}{60}$, factors of both the numerator and the denominator are determined and then canceled if any factor is found in both of them. The work for this could be written as $\frac{36}{60}=\frac{\cancel{12}\cdot 3}{\cancel{12}\cdot 5}=\frac{3}{5}$. Similarly, when simplifying $\frac{16x^5y^2}{20x^3y^4}$, factor and cancel as follows: $\frac{16x^5y^2}{20x^3y^4}=\frac{\cancel{4x^3y^2}\cdot 4x^2}{\cancel{4x^3y^2}\cdot 5y^2}=\frac{4x^2}{5y^2}$. In both of these examples, the greatest common factor of the numerator and the denominator was factored out. Some students find it easier to break each into prime factors and cancel individually. Here is how that is done: $\frac{36}{60}=\frac{2\cdot 2\cdot 3\cdot 3}{2\cdot 2\cdot 3\cdot 5}=\frac{3}{5}$ and

$$\frac{16x^5y^2}{20x^3y^4}=\frac{\cancel{2}\cdot\cancel{2}\cdot 2\cdot 2\cdot \cancel{x}\cdot\cancel{x}\cdot\cancel{x}\cdot x\cdot x\cdot \cancel{y}\cdot\cancel{y}}{\cancel{2}\cdot\cancel{2}\cdot 5\cdot \cancel{x}\cdot\cancel{x}\cdot\cancel{x}\cdot y\cdot y\cdot \cancel{y}\cdot\cancel{y}}=\frac{4x^2}{5y^2}.$$

MATH FACTS

To simplify a rational expression, first factor both the numerator and the denominator. Then cancel (or divide out) the same factors from each.

Example 1

Simplify: $\frac{45a^5b^7}{81a^8b^3}$

Solution: $\frac{45a^5b^7}{81a^8b^3}=\frac{\cancel{9}\,\cancel{a^5}\,\cancel{b^3}\cdot 5b^4}{\cancel{9}\,\cancel{a^5}\,\cancel{b^3}\cdot 9a^3}=\frac{5b^4}{9a^3}$

Example 2

Simplify: $\frac{5x-20y}{8x-32y}$

Solution: $\frac{5x-20y}{8x-32y}=\frac{5(\cancel{x-4y})}{8(\cancel{x-4y})}=\frac{5}{8}$

Example 3

Simplify: $\frac{x^2-9y^2}{x^2+2xy-15y^2}$

Solution: $\frac{x^2-9y^2}{x^2+2xy-15y^2}=\frac{(x+3y)(\cancel{x-3y})}{(\cancel{x-3y})(x+5y)}=\frac{x+3y}{x+5y}$

Caution: Be careful to notice that one cannot cancel out the x's or the y's in Example 3 because they are not factors of the numerator or denominator. A factor indicates multiplication, which are not separated by either addition or subtraction. The x's and y's are terms of the polynomial.

Example	Solution
Example 4 $\dfrac{a^2+3a-18}{2a^2-13a+15}$	$\dfrac{a^2+3a-18}{2a^2-13a+15}=\dfrac{(a+6)(a-3)}{(2a-3)(a-5)}=\dfrac{a^2+3a-18}{2a^2-13a+15}$ Because there were no common factors, the original expression is in simplest form.
Example 5 $\dfrac{b^2+3b-18}{2b^2+9b-18}$	$\dfrac{b^2+3b-18}{2b^2+9b-18}=\dfrac{(\cancel{b+6})(b-3)}{(2b-3)(\cancel{b+6})}=\dfrac{b-3}{2b-3}$ Notice that neither the b or the 3 can cancel.
Example 6 $\dfrac{4y^2+4y-24}{6y^2-30y+36}$	$\dfrac{4y^2+4y-24}{6y^2-30y+36}=\dfrac{4(y^2+y-6)}{6(y^2-5y+6)}$ $=\dfrac{2\cdot\cancel{2}(y+3)(\cancel{y-2})}{3\cdot\cancel{2}(\cancel{y-2})(y-3)}=\dfrac{2(y+3)}{3(y-3)}=\dfrac{2y+6}{3y-9}$
Example 7 $\dfrac{x^2+5x}{2x^2+9x-5}$	$\dfrac{x^2+5x}{2x^2+9x-5}=\dfrac{x(\cancel{x+5})}{(2x-1)(\cancel{x+5})}=\dfrac{x}{2x-1}$
Example 8 $\dfrac{6a^2b^2-3a^2b-45a^2}{24ab^2+84ab+60a}$	$\dfrac{6a^2b^2-3a^2b-45a^2}{24ab^2+84ab+60a}=\dfrac{3a^2(2b^2-b-15)}{12a(2b^2+7b+5)}$ $=\dfrac{\cancel{3}\cancel{a}\cdot a(\cancel{2b+5})(b-3)}{\cancel{3}\cdot 4\cancel{a}(\cancel{2b+5})(b+1)}=\dfrac{a(b-3)}{4(b+1)}=\dfrac{ab-3a}{4b+4}$

Example 9

Simplify: $\dfrac{x^2+2x-15}{9-x^2}$

Solution: $$\frac{x^2+2x-15}{9-x^2}=\frac{(x+5)(\overset{-1}{\cancel{x-3}})}{(3+x)(\cancel{3-x})}=\frac{-1\cdot(x+5)}{(3+x)}=\frac{-x-5}{x+3}$$

Notice that in this example, a factor of $x - 3$ cancels with a factor of $3 - x$. Why is that so? This works because $-1(x - 3) = -x + 3$ or $3 - x$.

Multiplication of Rational Expressions

Remember how you multiplied fractions together. For example, $\frac{15}{27}\cdot\frac{18}{25}=$ $\frac{\cancel{5}\cdot\cancel{3}}{\cancel{3}\cdot\cancel{9}}\cdot\frac{\cancel{9}\cdot 2}{\cancel{5}\cdot 5}=\frac{2}{5}$. Notice that a factor in either numerator can cancel with the same factor found in either denominator. Also recall that it is inefficient to find the product of the numerators first and put that over the product of the denominators. To multiply rational expressions, the same process will be applied.

MATH FACTS

To multiply rational expressions, first factor the numerators of all the fractions and the denominators of all the fractions. Then cancel factors found in any numerator with that same factor in any denominator. Finally, multiply the remaining fractions.

Example 10

Multiply: $\frac{6r^4s^3}{25s^5t^2}\cdot\frac{40rt^5}{21r^5st^4}$

Solution: $\frac{6r^4s^3}{25s^5t^2}\cdot\frac{40rt^5}{21r^5st^4}=\frac{2\cdot\cancel{3}r^4s^3}{\cancel{5}\cdot 5s^5t^2}\cdot\frac{\cancel{5}\cdot 8rt^5}{7\cdot\cancel{3}r^5st^4}=\frac{2\cdot 8\cancel{r^5}\cancel{s^3}\cancel{t^5}}{5\cdot 7\cancel{r^5}s^{\cancel{6}3}t^{\cancel{6}1}}=\frac{16}{35s^3t}$

Example 11

Multiply: $\frac{2y-4}{3y+12}\cdot\frac{9y^2-144}{10y^2-40}$

Solution: $\frac{2y-4}{3y+12}\cdot\frac{9y^2-144}{10y^2-40}=\frac{2(y-2)}{3(y+4)}\cdot\frac{9(y^2-16)}{10(y^2-4)}$

$=\frac{\cancel{2}\cancel{(y-2)}}{\cancel{3}\cancel{(y+4)}}\cdot\frac{\cancel{3}\cdot 3\cancel{(y+4)}(y-4)}{\cancel{2}\cdot 5(y+2)\cancel{(y-2)}}=\frac{3(y-4)}{5(y+2)}=\frac{3y-12}{5y+10}$

Example 12

Multiply: $\frac{x^2+7x+12}{x^2-8x+7}\cdot\frac{x^2-5x-14}{x^2+6x+8}$

Solution:

$$\frac{x^2+7x+12}{x^2-8x+7}\cdot\frac{x^2-5x-14}{x^2+6x+8}=\frac{(x+3)\cancel{(x+4)}}{(x-1)\cancel{(x-7)}}\cdot\frac{\cancel{(x-7)}\cancel{(x+2)}}{\cancel{(x+2)}\cancel{(x+4)}}=\frac{x+3}{x-1}$$

Example 13

Multiply: $\frac{a^2-6a+8}{a^2+6a+5}\cdot\frac{a^2-2a-35}{a^2+a-6}\cdot\frac{a^2+4a+3}{a^2-12a+35}$

Solution: $\frac{a^2-6a+8}{a^2+6a+5}\cdot\frac{a^2-2a-35}{a^2+a-6}\cdot\frac{a^2+4a+3}{a^2-12a+35}=$

$$\frac{(a-4)\cancel{(a-2)}}{\cancel{(a+1)}\cancel{(a+5)}}\cdot\frac{\cancel{(a+5)}\cancel{(a-7)}}{\cancel{(a-2)}\cancel{(a+3)}}\cdot\frac{\cancel{(a+3)}\cancel{(a+1)}}{(a-5)\cancel{(a-7)}}=\frac{a-4}{a-5}$$

Example 14

Multiply: $\frac{2a^2-10a}{5ab+20b}\cdot\frac{ab^3+3b^3}{40a^2-8a^3}$

Solution: $$\frac{2a^2-10a}{5ab+20b}\cdot\frac{ab^3+3b^3}{40a^2-8a^3}=\frac{\cancel{2a}\,\overset{-1}{\cancel{(a-5)}}}{5\cancel{b}(a+4)}\cdot\frac{\cancel{b}\cdot b^2(a+3)}{4\cdot\cancel{2}\,\cancel{a}\cdot a\cancel{(5-a)}}$$

$$=\frac{-b^2(a+3)}{20a(a+4)}=\frac{-ab^2-3b^2}{20a^2+80a}$$

Division of Rational Expressions

Remember that in order to divide fractions, it is necessary to invert the second fraction or divisor and then multiply. For example, $\frac{12}{25}\div\frac{8}{35}=\frac{12}{25}\cdot\frac{35}{8}$ $=\frac{\cancel{4}\cdot 3}{5\cdot\cancel{5}}\cdot\frac{7\cdot\cancel{5}}{\cancel{4}\cdot 2}=\frac{21}{10}$. For division of rational expressions, the same routine is followed. To divide $\frac{3x}{x^2-4}$ by $\frac{2x}{x+2}$, first invert the second fraction and multiply.

$$\frac{3x}{x^2-4} \div \frac{2x}{x+2} = \frac{3x}{x^2-4} \cdot \frac{x+2}{2x} = \frac{3\cancel{x}}{(\cancel{x+2})(x-2)} \cdot \frac{\cancel{x+2}}{2\cancel{x}} = \frac{3}{2(x-2)} = \frac{3}{2x-4}$$

MATH FACTS

To divide rational expressions, first invert the divisor. Then factor the numerators and the denominators of all the fractions. Cancel factors found in any numerator with that same factor in any denominator. Finally, multiply the remaining fractions.

Example 15

Divide: $\frac{3x+15}{7x+7} \div \frac{2x+10}{35x-70}$

Solution:

$$\frac{3x+15}{7x+7} \div \frac{2x+10}{35x-70} = \frac{3x+15}{7x+7} \cdot \frac{35x-70}{2x+10} = \frac{3(\cancel{x+5})}{\cancel{7}(x+1)} \cdot \frac{5 \cdot \cancel{7}(x-2)}{2(\cancel{x+5})}$$

$$= \frac{15(x-2)}{2(x+1)} = \frac{15x-30}{2x+2}$$

Example 16

Divide: $\frac{z^2-7z+12}{z^2+6z+8}$ by $\frac{z^2-4z+3}{z^2+3z+2}$

Solution: $$\frac{z^2-7z+12}{z^2+6z+8} \div \frac{z^2-4z+3}{z^2+3z+2} = \frac{(z-3)(z-4)}{(z+4)(z+2)} \div \frac{(z-3)(z-1)}{(z+2)(z+1)}$$

$$= \frac{(\cancel{z-3})(z-4)}{(z+4)(\cancel{z+2})} \cdot \frac{(\cancel{z+2})(z+1)}{(\cancel{z-3})(z-1)} = \frac{(z-4)(z+1)}{(z+4)(z-1)} = \frac{z^2-3z-4}{z^2+3z-4}$$

Be careful to notice that there are no common factors left between numerator and denominator. Therefore, the fraction with a trinomial in both the numerator and denominator is the final answer! It is not possible to cancel the z^2 or the -4 because they are not factors.

Example 17

Determine the quotient of $\dfrac{3x^3-15x^2}{2x^2+13x-7}$ and $\dfrac{9x-45}{x^2-49}$

Solution: $$\frac{3x^3-15x^2}{2x^2+13x-7}\div\frac{9x-45}{x^2-49}=\frac{3x^3-15x^2}{2x^2+13x-7}\cdot\frac{x^2-49}{9x-45}$$

$$=\frac{\cancel{3}x^2(\cancel{x-5})}{(2x-1)(\cancel{x+7})}\cdot\frac{(\cancel{x+7})(x-7)}{\cancel{3}\cdot 3(\cancel{x-5})}=\frac{x^2(x-7)}{3(2x-1)}=\frac{x^3-7x^2}{6x-3}$$

Example 18

Perform the indicated operations and simplify the answer:

$$\frac{5x^2-45}{2x^2+3x-2}\div\frac{x^2-4x-21}{6x^2-5x-1}\cdot\frac{2x^2-10x-28}{3x^2-8x-3}$$

Solution: $$\frac{5x^2-45}{2x^2+3x-2}\div\frac{x^2-4x-21}{6x^2-5x-1}\cdot\frac{2x^2-10x-28}{3x^2-8x-3}$$

$$=\frac{5x^2-45}{2x^2+3x-1}\cdot\frac{6x^2-5x-1}{x^2-4x-21}\cdot\frac{2x^2-10x-28}{3x^2-8x-3}$$

$$=\frac{5(\cancel{x+3})(\cancel{x-3})}{(\cancel{2x-1})(\cancel{x+2})}\cdot\frac{(\cancel{2x-1})(3x-1)}{(\cancel{x+3})(\cancel{x-7})}\cdot\frac{2(\cancel{x+2})(\cancel{x-7})}{(3x+1)(\cancel{x-3})}$$

$$=\frac{10(3x-1)}{(3x+1)}=\frac{30x-10}{3x+1}$$

Addition of Rational Expressions

In order to add fractions, remember that it is necessary to have a common denominator. Why is this so? That is because a person is able to add only like quantities. Is it possible to add five oranges and three bananas? Not really. One can say that there are a total of eight pieces of fruit but not simply bananas or oranges. How do common denominators make fractions like quantities? What is really important to understand is the concept of unit fractions. A unit fraction is a fraction whose numerator is 1. Converting each fraction to multiples of unit fractions creates like quantities that can be added. Everyone knows that it is possible to add $\dfrac{2}{7}+\dfrac{3}{7}$ because they have a common denominator of 7. This is because what really is being done is $2\left(\dfrac{1}{7}\right)+3\left(\dfrac{1}{7}\right)$. No one ever writes addition of fractions in this way or wants anyone else to write it this way. However, this is the basis upon which one can see that if there are 2 of one quantity and 3 of the same quantity, there is a total of 5 of that quantity. So the answer is $\frac{5}{7}$.

In order to discuss addition of rational expressions, let's first start with the easy fractions—those that already have a common denominator.

Example 19

Add: $\frac{2}{3x}+\frac{4}{3x}$

Solution: $\frac{2}{3x}+\frac{4}{3x}=\frac{6}{3x}=\frac{2\cdot\cancel{3}}{\cancel{3}x}=\frac{2}{x}$

Notice that although the fractions added together have a common denominator of $3x$, the sum was simplified or reduced so that the answer does not have that same denominator. Go back to fraction arithmetic a moment and recall doing that same thing in a problem like $\frac{1}{8}+\frac{3}{8}=\frac{4}{8}=\frac{1}{2}$.

Of course, not all fractions and, in fact, most fractions do not have a common denominator. When fractions do not have a common denominator, equivalent fractions that have the same denominator must be created in order to add the fractions. This has been done many times with numerical fractions. For instance, when adding $\frac{3}{8}+\frac{7}{12}$, first determine the lowest common denominator (the least common multiple of the denominators). In arithmetic, this task is relatively simple. It's easy to examine the multiples of 12 to find a multiple that is also divisible by the 8, which is 24. Equivalent fractions are fractions that are equal in value. They are fractions that can be simplified to the same fraction. The task now is to create two fractions equivalent to the ones being added that have denominators of 24. Since 8 divides into 24 a total of 3 times, multiply the fraction $\frac{3}{8}$ by $\frac{3}{3}$. Remember, $\frac{3}{3}$ equals 1. The multiplicative property of 1 says $a \cdot 1 = 1 \cdot a = a$. Similarly, it is possible to multiply the $\frac{7}{12}$ by $\frac{2}{2}$. So the addition now looks like $\frac{9}{24}+\frac{14}{24}=\frac{23}{24}$.

Once again, the same process can be applied to rational expressions. However, the task of computing the lowest common denominator is much harder. In some examples, it may be possible just to look at the denominators and tell what the common denominator will be. In other cases, a new tool is needed. To find the lowest common denominator, simply factor each denominator into simplest factors (prime factors for numbers). The lowest common denominator or LCD must contain all the factors of the first denominator and any additional factors found in the other denominators that are not present in the first denominator. In the problem just examined, the denominators are 8 and 12. The prime factors of 8 are $2 \cdot 2 \cdot 2$, and the prime factors of 12 are $2 \cdot 2 \cdot 3$. So the LCD must contain the $2 \cdot 2 \cdot 2$ and since the other denominator has only two factors of 2 (not more than the three factors determined from the 8), it is not necessary to add an additional 2 to the LCD. However, the second denominator also has a factor of 3 that is not already in the LCD, so that factor needs to be included in the LCD. The LCD = $2 \cdot 2 \cdot 2 \cdot 3 = 24$. Yes, that method does look more complicated than when the LCD was previously determined to be 24.

MATH FACTS

To add rational expressions, it is necessary first to determine the lowest common denominator and then change each fraction to an equivalent fraction with that lowest common denominator. Once this is accomplished, simply add the numerators of these equivalent fractions and put that sum over the lowest common denominator. A final step that must be done is to simplify this new fraction if it can be simplified.

Example 20

Determine the sum: $\frac{3}{2x} + \frac{5}{7y}$

Solution: $$\frac{3}{2x} + \frac{5}{7y} = \frac{3}{2x} \cdot \frac{7y}{7y} + \frac{5}{7y} \cdot \frac{2x}{2x} = \frac{21y}{14xy} + \frac{10x}{14xy} = \frac{21y + 10x}{14xy}$$

The LCD was easy to find this time because none of the factors of $2x$ are repeated in $7y$. So the LCD is the product of $2x$ and $7y$, which is $14xy$.

Example 21

Determine the sum: $\frac{x}{x+1} + \frac{1}{x^2 - 1}$

Solution: To find the LCD, first factor $x^2 - 1$ as $(x + 1)(x - 1)$. Now it should be clear that $x^2 - 1$ is the LCD and that the first fraction must be multiplied by the missing factor of $x - 1$ in the form of $\frac{(x-1)}{(x-1)}$ to begin the problem.

$$\frac{x}{x+1}+\frac{1}{x^2-1}=\frac{x}{(x+1)}\cdot\frac{(x-1)}{(x-1)}+\frac{1}{x^2-1}$$

$$=\frac{x^2-x}{x^2-1}+\frac{1}{x^2-1}=\frac{x^2-x+1}{x^2-1}$$

Since $x^2 - x + 1$ is not factorable, the fraction cannot be simplified.

Example 22

$$\frac{a+b}{a^2-a}+\frac{b-c}{ab-b}$$

Solution: $a^2 - a = a(a - 1)$ and $ab - b = b(a - 1)$, so the LCD = $ab(a - 1)$.

$$\frac{a+b}{a^2-a}+\frac{b-c}{ab-b}=\frac{a+b}{a(a-1)}+\frac{b-c}{b(a-1)}=\frac{(a+b)}{a(a-1)}\cdot\frac{b}{b}+\frac{(b-c)}{b(a-1)}\cdot\frac{a}{a}$$

$$=\frac{ab+b^2}{ab(a-1)}+\frac{ab-ac}{ab(a-1)}=\frac{2ab+b^2-ac}{ab(a-1)}$$

Example 23

$$\frac{x}{x+3}+\frac{2}{x-1}$$

Solution: LCD = $(x + 3)(x - 1)$

$$\frac{x}{x+3}+\frac{2}{x-1}=\frac{x}{(x+3)}\cdot\frac{(x-1)}{(x-1)}+\frac{2}{(x-1)}\cdot\frac{(x+3)}{(x+3)}$$

$$=\frac{x^2-x}{(x+3)(x-1)}+\frac{2x+6}{(x+3)(x-1)}=\frac{x^2+x+6}{(x+3)(x-1)}=\frac{x^2+x+6}{x^2+2x-3}$$

Example 24

$$\frac{2}{x^2+x-2}+\frac{16}{x^2+2x-3}$$

Solution: $x^2 + x - 2 = (x - 1)(x + 2)$

$x^2 + 2x - 3 = (x + 3)(x - 1)$

$\text{LCD} = (x-1)(x + 2)(x + 3)$

$$\frac{2}{x^2+x-2}+\frac{16}{x^2+2x-3}=\frac{2}{(x+2)(x-1)}+\frac{16}{(x+3)(x-1)}$$

$$=\frac{2}{(x-1)(x+2)}\cdot\frac{(x+3)}{(x+3)}+\frac{16}{(x+3)(x-1)}\cdot\frac{(x+2)}{(x+2)}$$

$$=\frac{2x+6}{(x-1)(x+2)(x+3)}+\frac{16x+32}{(x+3)(x-1)(x+2)}$$

$$=\frac{18x+36}{(x-1)(x+2)(x+3)}=\frac{18\cancel{(x+2)}}{(x-1)\cancel{(x+2)}(x+3)}=\frac{18}{x^2+2x-3}$$

Subtraction of Rational Expressions

As always, subtraction means to add the opposite of an expression, or $a - b = a + (-b)$. In Example 20, $\frac{3}{2x}+\frac{5}{7y}$ was examined. Change this problem to $\frac{3}{2x}-\frac{5}{7y}$. Now this can be done by $\frac{3}{2x}-\frac{5}{7y}=\frac{3}{2x}+\frac{-5}{7y}=\frac{-2}{14xy}$ $=\frac{-1}{7xy}$.

Example 25

From $\frac{x^2}{x^2+7x+12}$ subtract $\frac{2x}{x+4}$.

Solution: First factor the denominators: $x^2 + 7x + 12 = (x + 4)(x + 3)$ and $x + 4$.

$\text{LCD} = (x + 4)(x + 3)$

$$\frac{x^2}{x^2+7x+12} - \frac{2x}{x+4} = \frac{x^2}{(x+4)(x+3)} - \frac{2x}{(x+4)} \cdot \frac{(x+3)}{(x+3)}$$

$$= \frac{x^2}{(x+4)(x+3)} - \frac{2x^2+6x}{(x+4)(x+3)}$$

$$\frac{x^2}{(x+4)(x+3)} + \frac{-2x^2-6x}{(x+4)(x+3)} = \frac{-x^2-6x}{(x+4)(x+3)} = \frac{-x^2-6x}{x^2+7x+12}$$

Example 26

Subtract $\frac{a+b}{a-b}$ from $\frac{b^2}{a^2-ab}$.

Solution: Factoring $a^2 - ab$ yields $a(a - b)$, which is the LCD for this problem.

$$\frac{b^2}{a^2-ab} - \frac{a+b}{a-b} = \frac{b^2}{a^2-ab} - \frac{(a+b)}{(a-b)} \cdot \frac{a}{a} = \frac{b^2}{a^2-ab} - \frac{a^2+ab}{a^2-ab}$$

$$= \frac{b^2}{a^2-ab} + \frac{-a^2-ab}{a^2-ab} = \frac{b^2-ab-a^2}{a^2-ab}$$

Complex Fractions

A complex fraction is simply a fraction that contains fractions in the numerator or the denominator or both. Let's examine a complex arithmetic fraction such as $\frac{\frac{1}{2}+\frac{1}{3}}{\frac{1}{2}-\frac{1}{3}}$. There are two approaches to simplifying this complex fraction. The first method is to add the fractions in the numerator, subtract the fractions in the denominator, and perform the resulting division of one fraction by another by inverting the denominator and multiplying it by the numerator. A second method is to multiply each term in both the numerator and denominator by the lowest common denominator for all the fractional terms.

First Method	Second Method
$\dfrac{\frac{1}{2}+\frac{1}{3}}{\frac{1}{2}-\frac{1}{3}}=\dfrac{\frac{3}{6}+\frac{2}{6}}{\frac{3}{6}-\frac{2}{6}}=\dfrac{\frac{5}{6}}{\frac{1}{6}}=\frac{5}{6}\cdot\frac{6}{1}=5$	$\dfrac{\frac{1}{2}+\frac{1}{3}}{\frac{1}{2}-\frac{1}{3}}=\dfrac{6\cdot\frac{1}{2}+6\cdot\frac{1}{3}}{6\cdot\frac{1}{2}-6\cdot\frac{1}{3}}=\dfrac{3+2}{3-2}=\frac{5}{1}=5$

The same procedures can be followed for complex algebraic fractions. Let's examine the complex fraction $\dfrac{\frac{1}{x}+\frac{2}{y}}{\frac{3}{x}-\frac{1}{y}}$.

First Method	Second Method
$\dfrac{\frac{1}{x}+\frac{2}{y}}{\frac{3}{x}-\frac{1}{y}}=\dfrac{\frac{1}{x}\cdot\frac{y}{y}+\frac{2}{y}\cdot\frac{x}{x}}{\frac{3}{x}\cdot\frac{y}{y}-\frac{1}{y}\cdot\frac{x}{x}}=\dfrac{\frac{y}{xy}+\frac{2x}{xy}}{\frac{3y}{xy}-\frac{x}{xy}}=$ $\dfrac{\frac{y+2x}{xy}}{\frac{3y-x}{xy}}=\frac{y+2x}{xy}\cdot\frac{xy}{3y-x}=\frac{y+2x}{3y-x}$	$\dfrac{\frac{1}{x}+\frac{2}{y}}{\frac{3}{x}-\frac{1}{y}}=\dfrac{xy\cdot\frac{1}{x}+xy\cdot\frac{2}{y}}{xy\cdot\frac{3}{x}-xy\cdot\frac{1}{y}}=\frac{y+2x}{3y-x}$

The same procedures are used when working with an algebraic complex fraction as when working with a numerical complex fraction.

Example 27

Simplify: $\dfrac{\frac{a}{a-b}+b}{a-\frac{b}{a-b}}$

Solution:

First Method	Second Method
$\dfrac{\dfrac{a}{a-b}+b}{a-\dfrac{b}{a-b}}=\dfrac{\dfrac{a}{a-b}+\dfrac{b}{1}\cdot\dfrac{(a-b)}{(a-b)}}{\dfrac{a}{1}\cdot\dfrac{(a-b)}{(a-b)}-\dfrac{b}{a-b}}$ $=\dfrac{\dfrac{a}{a-b}+\dfrac{ab-b^2}{a-b}}{\dfrac{a^2-ab}{a-b}-\dfrac{b}{a-b}}$ $=\dfrac{\dfrac{a+ab-b^2}{a-b}}{\dfrac{a^2-ab-b}{a-b}}=\dfrac{(a+ab-b^2)}{\cancel{(a-b)}}\cdot\dfrac{\cancel{(a-b)}}{(a^2-ab-b)}$ $=\dfrac{a+ab-b^2}{a^2-ab-b}$	$\dfrac{\dfrac{a}{a-b}+b}{a-\dfrac{b}{a-b}}$ $=\dfrac{\dfrac{a}{\cancel{(a-b)}}\cancel{(a-b)}+\dfrac{b}{1}\cdot(a-b)}{\dfrac{a}{1}\cdot(a-b)-\dfrac{b}{\cancel{(a-b)}}\cdot\cancel{(a-b)}}$ $=\dfrac{a+b(a-b)}{a(a-b)-b}$ $=\dfrac{a+ab-b^2}{a^2-ab-b}$

Example 28

Simplify: $\dfrac{\dfrac{x}{x-3}+\dfrac{x-3}{x+3}}{\dfrac{x}{3}-\dfrac{2x}{x^2-9}}$

Solution: $\dfrac{\dfrac{x}{x-3}+\dfrac{x-3}{x+3}}{\dfrac{x}{3}-\dfrac{2x}{x^2-9}}=\dfrac{\dfrac{x}{(x-3)}\cdot\dfrac{3(x+3)}{3(x+3)}+\dfrac{(x-3)}{(x+3)}\cdot\dfrac{3(x-3)}{3(x-3)}}{\dfrac{x}{3}\cdot\dfrac{(x^2-9)}{(x^2-9)}-\dfrac{2x}{(x^2-9)}\cdot\dfrac{3}{3}}$

$$=\frac{\dfrac{3x^2+9x}{3x^2-27}+\dfrac{3x^2-18x+27}{3x^2-27}}{\dfrac{x^3-9x}{3x^2-27}-\dfrac{6x}{3x^2-27}}=\frac{\dfrac{6x^2-9x+27}{3x^2-27}}{\dfrac{x^3-15x}{3x^2-27}}$$

$$=\frac{6x^2-9x+27}{\cancel{3x^2-27}}\cdot\frac{\cancel{3x^2-27}}{x^3-15x}=\frac{6x^2-9x+27}{x^3-15x}$$

or

$$\frac{\frac{x}{x-3}+\frac{x-3}{x+3}}{\frac{x}{3}-\frac{2x}{x^2-9}}=\frac{3(x+3)(x-3)\cdot\frac{x}{x-3}+3(x+3)(x-3)\cdot\frac{x-3}{x+3}}{3(x+3)(x-3)\cdot\frac{x}{3}-3(x+3)(x-3)\cdot\frac{2x}{x^2-9}}$$

$$=\frac{3x(x+3)+3(x-3)(x-3)}{x(x+3)(x-3)-3\cdot 2x}=\frac{3x^2+9x+(3(x^2-6x+9))}{x(x^2-9)-6x}$$

$$=\frac{3x^2+9x+3x^2-18x+27}{x^3-9x-6x}=\frac{6x^2-9x+27}{x^3-15x}$$

Check Your Understanding of Section 1.3

1. Simplify: $\dfrac{\frac{1}{r}-\frac{1}{s}}{\frac{r}{s}-\frac{s}{r}}$

(1) $\dfrac{s-r}{r^2-s^2}$

(2) $\dfrac{-1}{r-s}$

(3) $-\dfrac{1}{r+s}$

(4) $\dfrac{1}{r+s}$

2. The length of a rectangular garden is represented by $\dfrac{a^2-9}{3a+15}$ and its width is represented by $\dfrac{a^2+7a+10}{3a-9}$. Express its area as a rational expression in simplest form.

(1) $\dfrac{a^2+5a+6}{9}$

(2) $\dfrac{a^3+2a^2-7a-18}{9a-17}$

(3) $\dfrac{a^2+5a+2}{3}$

(4) $\dfrac{a^2+7a+10}{18}$

3. What is the sum of $\frac{x}{x+2}$ and $\frac{1}{x-2}$?

(1) $\frac{x-2}{4}$

(2) $\frac{x^2-x+2}{x^2-4}$

(3) $\frac{x^2+x+2}{x^2-4}$

(4) $\frac{x^2+x}{x^2-4}$

4. Subtract $\frac{z^2}{z^2-1}$ from $\frac{2z}{z-1}$ and express this difference in simplest form.

(1) $\frac{2-z^2}{z^2-1}$

(2) $-\frac{z^2+2z}{z^2-1}$

(3) $\frac{z^2+2z}{z^2-1}$

(4) $\frac{z^2-2}{z^2-1}$

5. When written in simplest form, what is the expression $\frac{t^2-s^2}{s^2+st-2t^2}$ equivalent to?

(1) $\frac{s+t}{s+2t}$

(2) $\frac{-s-t}{s+2t}$

(3) $\frac{-t+s}{s+2t}$

(4) $\frac{s+t}{s+2t}$

6. What is the sum of $(3y + 5)$ and $\frac{2y}{y+3}$?

(1) $\frac{3y^2+10y+15}{y+3}$

(2) $13y + 5$

(3) $\frac{5y+5}{y+3}$

(4) $\frac{3y^2+16y+15}{y+3}$

7. $\frac{x^2-2x-35}{x^2+7x+12} \div \frac{x^2-4x-21}{x^2+9x+20} =$

(1) $\frac{10x+25}{6x+9}$

(2) $\frac{-14x+49}{8x+16}$

(3) $\frac{x^2+25}{x^2+9}$

(4) $\frac{x^2+10x+25}{x^2+6x+9}$

8. Determine the lowest common denominator for the rational expressions $\frac{x}{x^2-16}$, $\frac{4x-16}{3x-12}$, and $\frac{20}{x^2+4x}$.

(1) $3x^3 - 48x$
(2) $80x^2 - 320x$
(3) $3x$
(4) $x^2 - 16$

9. When Mr. Berger divided $\frac{3st-3s}{5t+10}$ by $\frac{9s^2}{25t}$, he wrote his answer on an overhead transparency. By the time he went to display his work on the overhead projector, he found that part of the final answer was missing. The transparency showed: $\frac{5t^2}{3st+6s}$. What was missing in the numerator?

(1) $+5t$
(2) $-5t$
(3) $+5s^2$
(4) $-15t$

Perform the indicated operations, and express your answer in lowest terms:

10. Simplify: $\frac{36-k^2}{5k+30}$

11. $\frac{5}{x+5}-\frac{3}{x-3}$

12. $\frac{x^2+6x+5}{x^2-5x+6}\cdot\frac{x^2+2x-8}{x^2+9x+20}+\frac{x^2+3x+2}{x^2-x-6}\cdot\frac{x^2-x-20}{x^2-4x-5}$

13. $\frac{x^2-xy-6y^2}{x^2+2xy-3y^2}\div\frac{x^2-2xy-3y^2}{x^2+4xy+3y^2}-\frac{6x^2-5xy+y^2}{2x^2+xy-y^2}\div\frac{3x^2-4xy+y^2}{3x^2+4xy+y^2}$

14. $\frac{16-x^2}{x^2-x-12}-\frac{x^2+6x+5}{x^2+8x+15}$

15. $\frac{3a^3b^2}{5a^4c^5}\div\frac{12b^3c^3}{20abc^4}$

16. Simplify: $\dfrac{\frac{1}{z-7}+\frac{7}{z}}{\frac{z}{z^2-7z}-1}$

17. Simplify: $\dfrac{4-\frac{a}{a+5}}{\frac{2a}{a+3}+6}$

18. $\frac{9r^2-25}{6r^2-5r-4}\cdot\frac{6r^2+r-12}{3r^2-7r-20}\div\frac{6r^2-r-15}{2r^2-4r-16}$

19. $x-\frac{x+5}{x-3}+\frac{6}{x+3}$

20. $\frac{5b}{12a}\cdot\frac{3}{25ab^3}-b$

1.4 RATIONAL EXPONENTS

Procedures to work with rational numbers and rational expressions have been mastered. What is the meaning of rational exponents? All the rules learned about working with whole-number exponents also apply to working with rational exponents.

Exponentiation Rules

Listed below are several rules learned in Integrated Algebra. These rules apply to rational exponents also.

Rules	Examples
To multiply like bases raised to exponents, simply add their exponents: $x^a \cdot x^b = x^{a+b}$	$r^5 \cdot r^3 = r^{5+3} = r^8$ $c^{\frac{2}{3}} \cdot c^{\frac{1}{6}} = c^{\frac{2}{3}+\frac{1}{6}} = c^{\frac{5}{6}}$
To divide like bases raised to exponents, simply subtract their exponents: $x^a \div x^b = x^{a-b}$	$z^6 \div z^2 = z^{6-2} = z^4$ $s^{\frac{4}{5}} \div s^{\frac{1}{4}} = s^{\frac{4}{5}-\frac{1}{4}} = s^{\frac{11}{20}}$
To raise a base that is raised to an exponent to another exponent, simply multiply their exponents: $(x^a)^b = x^{x \cdot b}$	$(p^3)^4 = p^{3 \cdot 4} = p^{12}$ $\left(a^{\frac{2}{6}}\right)^{\frac{7}{9}} = a^{\frac{2}{7} \cdot \frac{7}{9}} = a^{\frac{2}{9}}$
To raise the product of two bases to the same exponent, raise each base to that exponent: $(x \cdot y)^a = x^a \cdot y^a$	$(r \cdot s)^4 = r^4 \cdot s^4$ $(a \cdot b)^{\frac{1}{5}} = a^{\frac{1}{5}} \cdot b^{\frac{1}{5}}$
To raise the quotient of two bases to the same exponent, raise each base to that exponent: $\left(\frac{x}{y}\right)^a = \frac{x^a}{y^a}$	$\left(\frac{r}{t}\right)^3 = \frac{r^3}{t^3}$ $\left(\frac{x}{y}\right)^{\frac{3}{7}} = \frac{x^{\frac{3}{7}}}{y^{\frac{3}{7}}}$

Simplifying Exponential Expressions

Example 1

Express $9^4 \cdot 3^3$ as a power of base 3 only.

Solution: Since $9 = 3^2$, $9^4 = (3^2)^4 = 3^8$. Therefore, $9^4 \cdot 3^3 = 3^8 \cdot 3^3 = 3^{11}$.

Example 2

Simplify: $(4xy^2)^3$

Solution: $(4xy^2)^3 = 4^3 \cdot x^3 \cdot (y^2)^3 = 64x^3y^6$

Example 3

Simplify: $(2x^2y^4)^3 + (3x^3y^6)^2$

Solution: $(2x^2y^4)^3 + (3x^3y^6)^2$

$$= 2^3(x^2)^3(y^4)^3 + 3^2(x^3)^2(y^6)^2 = 8x^6y^{12} + 9x^6y^{12} = 17x^6y^{12}$$

Example 4

Simplify: $(5a^3b^2)^2(2a^4b^3)^3$

Solution: $(5a^3b^2)^2(2a^4b^3)^3 = (25a^6b^4)(8a^{12}b^9) = 200a^{18}b^{13}$

Calculators and Exponents

Do you remember how to use the calculator to square a number? You type **2** **x²** **ENTER**. To evaluate 5^6, you must type **5** **∧** **6** **ENTER**.

Algebraic Expressions that Contain Negative Exponents

The previous chart reviewed the rules of exponents. There are two addition exponent rules that we can derive relative to negative and zero exponents.

Rules	Examples
When $a = b$, $\frac{x^a}{x^b} = x^{a-b} = x^0 = 1$, provided $x \neq 0$.	$\frac{3^7}{3^7} = 3^{7-7} = 3^0 = 1$
When $a < b$, $c > 0$ so that $\frac{x^a}{x^b} = x^{-c} = \frac{1}{x^c}$.	$\frac{y^2}{y^7} = y^{-5} = \frac{1}{y^5}$ $d^{-\frac{1}{3}} = \frac{1}{d^{\frac{1}{3}}}$

Example 5

Evaluate $s^{-3} + s^0$ if $s = 2$.

$$\textit{Solution: } s^{-3} + s^0 = \frac{1}{s^3} + s^0 = \frac{1}{2^3} + 2^0 = \frac{1}{8} + 1 = \frac{9}{8}$$

Example 6

Simplify and express the answer with only positive exponents: $(s^5t^3)^{-4}$

$$\textit{Solution: } (s^5t^3)^{-4} = \frac{1}{(s^5t^3)^4} = \frac{1}{s^{20}t^{12}}$$

Example 7

Simplify and express the answer with only positive exponents: $(a^{-2}b^3)^{-2}$

$$\textit{Solution: } (a^{-2}b^3)^{-2} = (a^4b^{-6}) = a^4 \cdot b^{-6} = a^4 \cdot \frac{1}{b^6} = \frac{a^4}{1} \cdot \frac{1}{b^6} = \frac{a^4}{b^6}$$

Note: There is more than one way to think out the solution. It can also be accomplished in the following way:

$$(a^{-2}b^3)^{-2} = \left(\frac{1}{a^2} \cdot b^3\right)^{-2} = \left(\frac{b^3}{a^2}\right)^{-2} = \left(\frac{1}{\frac{b^3}{a^2}}\right)^2 = \left(1 \cdot \frac{a^2}{b^3}\right)^2 = \left(\frac{a^2}{b^3}\right)^2 = \frac{a^4}{b^6}$$

Example 8

Simplify and express your answer with only positive exponents: $\left(r^{\frac{3}{5}}\right)^{-5} \cdot \left(r^{-\frac{2}{3}}\right)^3$

$$\textit{Solution: } \left(r^{\frac{3}{5}}\right)^{-5} \cdot \left(r^{-\frac{2}{3}}\right)^3 = r^{-3} \cdot r^{-2} = r^{-5} = \frac{1}{r^5}$$

Example 9

Simplify and express the answer with only positive exponents: $\left(2x^{-3}y^{\frac{1}{2}}\right)^{-2} \cdot \left(8x^{\frac{2}{3}}z^{-1}\right)^3$

$$Solution: \left(2x^{-3}y^{\frac{1}{2}}\right)^{-2}\cdot\left(8x^{\frac{2}{3}}z^{-1}\right)^{3}=\left(2^{-2}(x^{-3})^{-2}\left(y^{\frac{1}{2}}\right)^{-2}\right)\cdot\left(8^{3}(x^{\frac{2}{3}})^{3}(z^{-1})^{3}\right)$$

$$\left(\frac{1}{2^{2}}x^{6}y^{-1}\right)(512x^{2}z^{-3})=\left(\frac{1}{4}\cdot\frac{x^{6}}{1}\cdot\frac{1}{y}\right)\left(\frac{512x^{2}}{1}\cdot\frac{1}{z^{3}}\right)=\frac{x^{6}}{4y}\cdot\frac{512x^{2}}{z^{3}}$$

$$=\frac{x^{6}}{y}\cdot\frac{128x^{2}}{z^{3}}=\frac{128x^{8}}{yz^{3}}$$

Example 10

Simplify and express the answer with only positive exponents: $\frac{z^{-2}-1}{1-z^{-1}}$.

$$Solution: \frac{z^{-2}-1}{1-z^{-1}}=\frac{\frac{1}{z^{2}}-1}{1-\frac{1}{z}}=\frac{\frac{1}{z^{2}}\cdot z^{2}-1\cdot z^{2}}{1\cdot z^{2}-\frac{1}{z}\cdot z^{2}}=\frac{1-z^{2}}{z^{2}-z}=\frac{(1+z)(1-z)}{z(z-1)}$$

$$=\frac{(1+z)(-1)}{z(1)}=\frac{-1-z}{z}$$

Check Your Understanding of Section 1.4

1. Evaluate $a^2 + 3a^{-4} - a^0$ if $a = 2$.

(1) $3\frac{3}{8}$

(2) $3\frac{3}{16}$

(3) $2\frac{3}{8}$

(4) $3\frac{1}{1296}$

2. Simplify: $\left(2a^{\frac{3}{5}}b^{\frac{7}{10}}c\right)^{5}$

(1) $2a^{3}b^{\frac{7}{2}}c^{6}$

(2) $10a^{3}b^{\frac{7}{2}}c^{5}$

(3) $10a^{3}b^{\frac{7}{2}}c$

(4) $32a^{3}b^{\frac{7}{2}}c^{5}$

3. $\dfrac{36p^{\frac{1}{3}}q^{\frac{4}{5}}}{\left(3p^{\frac{2}{3}}q\right)^2} =$

(1) $\dfrac{4}{pq^{\frac{6}{5}}}$

(2) $\dfrac{12p^{\frac{5}{3}}}{q^{\frac{8}{5}}}$

(3) $\dfrac{12p^{\frac{5}{3}}}{q^{\frac{6}{5}}}$

(4) $\dfrac{12}{pq^{\frac{6}{5}}}$

4. If $3^x = 5$, then $3^{x+4} =$

(1) 14 (2) 86 (3) 405 (4) 50,000

5. If $c \neq 0$, simplify $(2c^4 - c^0)^{-2}$

(1) $\dfrac{1}{4c^8 - 4c^4 + 1}$

(2) $\dfrac{1}{4c^8} - 1$

(3) $\dfrac{1}{4c^8 + 1}$

(4) $\dfrac{1}{4c^8 - 2c^4 + 1}$

6. Simplify: $(3y^{-2} + y^3)^{-1}$

(1) $\dfrac{1}{3}y^2 + \dfrac{1}{y^3}$

(2) $\dfrac{y^2}{3 + y^5}$

(3) $\dfrac{y^3}{3y^2 + 1}$

(4) $\dfrac{3 + y^5}{y^2}$

7. Simplify: $\dfrac{a^{-1} + b^{-1}}{a^{-2} + b^{-2}}$

(1) $\dfrac{ab}{b + a}$

(2) $\dfrac{b + a}{ab}$

(3) $\dfrac{ab^2 + a^2b}{b^2 + a^2}$

(4) $a + b$

8. $\left(3x^{\frac{1}{4}}y^{\frac{5}{6}}\right)^2\left(2x^4y^{\frac{1}{2}}\right)^3 =$

(1) $6x^{\frac{9}{2}}y^{\frac{19}{6}}$

(2) $72x^{\frac{57}{4}}y^{\frac{19}{3}}$

(3) $6x^{\frac{25}{2}}y^{\frac{19}{6}}$

(4) $72x^{\frac{25}{2}}y^{\frac{19}{6}}$

9. Evaluate $(y^0 + 3y^{-2})^3$ when $y = 2$.

10. Using only positive exponents, rewrite and simplify $\frac{12r^{-3}s^5}{20r^8s}$.

11. Using only positive exponents, rewrite and simplify $\frac{1-x^{-2}}{(x-1)^{-2}}$.

12. Using only positive exponents, rewrite and simplify $\frac{y^{-2}-1}{1+y}$.

13. Using only positive exponents, rewrite and simplify $\frac{6p^{-2}q^4}{2^{-3}p^5q^{-1}}$.

1.5 OPERATIONS WITH RADICALS

Do you remember what radicals are? The symbol for a radical is $\sqrt{\ }$ and the number inside this symbol is called the radicand. If a radical has a number raised in front of the radical, that is called the index of the radical. For $\sqrt[5]{32}$, 5 is the index and 32 is the radicand. The expression $x^{\frac{a}{b}}$ is equal to $\sqrt[b]{x^a}$ and vice versa. This rule will provide extra tools to use when working with radical expressions.

Square and Cube Roots

What is the meaning of a square root? Remember that $\sqrt{4}=2$. This is because $2^2=4$. It also true that $(-2)^2=4$. The positive square root of a number is referred to as the principal square root of that number. The equation $x^2 = 4$ has two solutions, 2 and −2. However, the square root has only the **principal square root** as its answer. In order to have a negative answer for a square root, it is necessary to have a negative sign in front of the radical sign ($\sqrt{\ }$) itself so that $-\sqrt{4}=-2$. The number inside the radical is called the radicand. Think of it as $\sqrt{4}$ asks for what positive number multiplied by itself yields the radicand 4.

The symbol for a cube root is $\sqrt[3]{\ }$. The 3 is called the index of the radical. Notice the 3 is written in the front corner of this symbol. Obviously, the index stands for the fact that this is a cube root rather than a square root. A 2 is understood to be the index for a square root even though the 2 is not written. Since $2^3 = 8$, $\sqrt[3]{8}=2$. There is no issue about positive and negative cube roots as there was with square roots. Only one number when cubed equals 8, the number 2. The only number whose cube root is –2 is –8 since $(-2)^3 = -8$.

To be able to work with square roots and cube roots, it is best to be familiar with the table below that lists the perfect squares and perfect cubes of some numbers.

Number	Number Squared	Number Cubed
1	1	1
2	4	8
3	9	27
4	16	64
5	25	125
6	36	216
7	49	343
8	64	512
9	81	729
10	100	1000
20	400	8000

Simplifying Radical Expressions

To simplify a radical expression, factor the radicand and find the greatest factor that occurs the number of times indicated by the index. In a square root, look for the greatest factor that occurs twice. For a cube root, the factor must occur three times, so on. The repeated factor comes out of the radicand and becomes a one-time factor that is multiplied by the radical in which the rest of its factors remain in the radicand. For instance, $\sqrt{24} = \sqrt{\underline{2 \cdot 2} \cdot 2 \cdot 3} = 2\sqrt{6}$ and $\sqrt[3]{24} = \sqrt[3]{\underline{2 \cdot 2 \cdot 2} \cdot 3} = 2\sqrt[3]{3}$.

Example 1

Simplify: $\sqrt{150}$

Solution: $\sqrt{150} = \sqrt{5 \cdot 30} = \sqrt{5 \cdot 5 \cdot 6} = \sqrt{\underline{5 \cdot 5} \cdot 3 \cdot 2} = 5\sqrt{6}$

Note: Another way to simplify this would be to look for the largest perfect square factor of 150 and take the square root of that out of the radicand. $\sqrt{50} = \sqrt{25 \cdot 6} = \sqrt{25} \cdot \sqrt{6} = 5\sqrt{6}$

Example 2

Simplify: $\sqrt[4]{972}$

Solution: $972 = 2 \cdot 486 = 2 \cdot 2 \cdot 243 = 2 \cdot 2 \cdot 3 \cdot 81 = 2 \cdot 2 \cdot 3 \cdot 9 \cdot 9 = 2 \cdot 2 \cdot 3 \cdot 3 \cdot 3 \cdot 3 \cdot 3$

$$\sqrt[4]{972} = \sqrt[4]{2 \cdot 2 \cdot \underline{3 \cdot 3 \cdot 3 \cdot 3} \cdot 3} = 3\sqrt[4]{2 \cdot 2 \cdot 3} = 3\sqrt[4]{12}$$

Rational Exponents and Radical Expressions

There is a relationship between rational exponents and radical expressions.

MATH FACTS

$$x^{\frac{a}{b}} = \sqrt[b]{x^a} \text{ or } \sqrt[b]{x}^{\,a}$$

In other words, the *b*th root of *x* to the *a* power is equal to *x* raised to the *a*/*b* power.

The math fact shown above can be used to evaluate radical expressions. It is possible to evaluate such things as $8^{\frac{2}{3}}$ by rewriting it as $\sqrt[3]{8^2}$ or even better as $\sqrt[3]{8}^{\,2}$, which is equal to 2^2 or 4.

Example 3

Evaluate: $4^{\frac{5}{2}}$

Solution: $4^{\frac{5}{2}} = \sqrt{4}^{\,5} = 2^5 = 32$

Example 4

Evaluate $\left(x^{\frac{1}{3}} - x^0 + 2x^{\frac{3}{2}}\right)$ if $x = 64$

Solution:

$$\left(x^{\frac{1}{3}} - x^0 + 2x^{\frac{3}{2}}\right) = \left(64^{\frac{1}{3}} - 64^0 + 2(64)^{\frac{3}{2}}\right) = \left(\sqrt[3]{64} - 1 + 2(\sqrt{64}^{\,3})\right)$$

$$= 4 - 1 + 2 \cdot 8^3 = 4 - 1 + 2 \cdot 512 = 4 - 1 + 1024 = 3 + 1024 = 1027$$

Example 5

Evaluate: $125^{-\frac{4}{3}}$

Solution: $125^{-\frac{4}{3}} = \frac{1}{125^{\frac{4}{3}}} = \frac{1}{\sqrt[3]{125}^{4}} = \frac{1}{5^4} = \frac{1}{625}$

On the graphing calculator, press **1** **2** **5** **∧** **(** **(–)** **4** **÷** **3** **)** **ENTER**. The displayed result is .0016. Now press **MATH** **1** **ENTER**. The calculator now displays $\frac{1}{625}$. Note, this calculator exercise was done as a check. It is really necessary to know how to perform the arithmetic operations above.

Example 6

Rewrite the expressions $a^{\frac{3}{5}}$, $3r^{-\frac{2}{7}}$, and $\frac{4}{y^{-\frac{6}{5}}}$ using radicals and positive exponents.

Solution: $\sqrt[5]{a^3}$, $\frac{3}{r^{\frac{2}{7}}} = \frac{3}{\sqrt[7]{r^2}}$, and $\frac{4}{\frac{1}{y^{\frac{6}{5}}}} = 4y^{\frac{6}{5}} = 4\sqrt[5]{y^6}$

Example 7

Rewrite the expressions $\sqrt{r^3 s}$, $\sqrt[4]{81a^2 b}$, and $7\sqrt[3]{d^5}$ with fractional exponents.

Solution: $(r^3 s)^{\frac{1}{2}} = r^{\frac{3}{2}} s^{\frac{1}{2}}$, $(81a^2 b)^{\frac{1}{4}} = 3a^{\frac{1}{2}} b^{\frac{1}{4}}$, and $7d^{\frac{5}{3}}$

Multiplication and Division of Radical Expressions

To multiply radical expressions that have the same index, multiply the radicands and simplify if possible. As this is being done, be careful to notice if there are any factors common to both radicands that will help simplify the final product. For example, $\sqrt{15} \cdot \sqrt{6} = \sqrt{90} = \sqrt{9} \cdot \sqrt{10} = 3\sqrt{10}$.

Example 8

Determine the product: $\sqrt[3]{2ab^2} \cdot \sqrt[3]{16a^4b}$

Solution: $\sqrt[3]{2ab^2} \cdot \sqrt[3]{16a^4b} = \sqrt[3]{2 \cdot 2 \cdot 8a^3a^2b^3} = \sqrt[3]{8a^3b^3 \cdot 4a^2} = 2ab\sqrt[3]{4a^2}$

Example 9

Multiply: $3\sqrt{5} \cdot 2\sqrt[4]{5}$

Solution: $3\sqrt{5} \cdot 2\sqrt[4]{5} = 3 \cdot 5^{\frac{1}{2}} \cdot 2 \cdot 5^{\frac{1}{4}} = 6 \cdot 5^{\frac{3}{4}} = 6\sqrt[4]{5^3}$ or $6\sqrt[4]{125}$

To divide radical expressions that have the same index, divide the radicands and simplify if possible. For instance, $\frac{\sqrt[3]{32}}{\sqrt[3]{2}} = \sqrt[3]{\frac{32}{2}} = \sqrt[3]{16} = \sqrt[3]{8 \cdot 2} = 2\sqrt[3]{2}$.

Example 10

Determine the quotient: $\frac{8\sqrt{45x^5}}{2\sqrt{5x}}$

Solution: $\frac{8\sqrt{45x^5}}{2\sqrt{5x}} = \frac{8}{2}\sqrt{\frac{45x^5}{5x}} = 4\sqrt{9x^4} = 4 \cdot 3x^2 = 12x^2$

Addition and Subtraction of Radical Expressions

In order to add or subtract radical expressions, it is necessary to have like terms. Like radicals have the same index and the same radicand. Be careful to see if different radicands can be simplified to the same radicand. For instance, look at $\sqrt{18}$ and $\sqrt{50}$. They clearly have different radicands. When each is simplified, they have the same radicand. In order to add $\sqrt{18} + \sqrt{50}$, simplify and get $\sqrt{9 \cdot 2} + \sqrt{25 \cdot 2} = 3\sqrt{2} + 5\sqrt{2}$. To add like radical expressions, simply add their coefficients, just as was done for polynomials. Therefore, $\sqrt{18} + \sqrt{50} = 8\sqrt{2}$.

Example 11

Determine the sum: $\sqrt[3]{81} + \sqrt[3]{24}$

Solution: $\sqrt[3]{81} + \sqrt[3]{24} = \sqrt[3]{27 \cdot 3} + \sqrt[3]{8 \cdot 3} = 3\sqrt[3]{3} + 2\sqrt[3]{3} = 5\sqrt[3]{3}$

The process of subtraction is the same since subtraction always means add the opposite.

Example 12

Determine the sum: $2\sqrt{98} - 3\sqrt{32}$

Solution: $2\sqrt{98} - 3\sqrt{32} = 2\sqrt{49 \cdot 2} - 3\sqrt{16 \cdot 2} = 2 \cdot 7\sqrt{2} - 3 \cdot 4\sqrt{2}$
$= 14\sqrt{2} - 12\sqrt{2} = 2\sqrt{2}$

Multiplication of Expressions Containing Radicals and Conjugates

What method is used to perform operations on expressions that contain radical expressions within them? The answer is simply to perform the indicated operations and incorporate the rules for operations on radical expressions. The chart below will help illustrate this by comparing operations on polynomials with similar expressions that contain radicals.

Multiplication of Polynomials	Compared to Multiplication of Expressions Containing Radicals
$3x(2x - 5) = 6x^2 - 15$	$3\sqrt{2}\left(4\sqrt{6} - 7\right) = 12\sqrt{12} - 21\sqrt{2}$ $= 12\sqrt{4 \cdot 3} - 21\sqrt{2} = 12 \cdot 2\sqrt{3} - 21\sqrt{2}$ $= 24\sqrt{3} - 21\sqrt{2}$
$(x - 3)(x + 4) = x^2 + 4x - 3x$ $- 12 = x^2 + x - 12$	$\left(2 + \sqrt{5}\right)\left(3 - \sqrt{3}\right) = 6 - 2\sqrt{3} + 3\sqrt{5} - \sqrt{15}$ In this case, there are no like terms.
$\frac{12a - 15b}{3} = \frac{12a}{3} - \frac{15b}{3} = 4a - 5b$	$\frac{8\sqrt[3]{12} + 20\sqrt[3]{16}}{2\sqrt[3]{4}} = \frac{8\sqrt[3]{12}}{2\sqrt[3]{4}} + \frac{20\sqrt[3]{16}}{2\sqrt[3]{4}}$ $= \frac{8}{2}\sqrt[3]{\frac{12}{4}} + \frac{20}{2}\sqrt[3]{\frac{16}{4}} = 4\sqrt[3]{3} + 10\sqrt[3]{4}$
$(x + y)(x - y) = x^2 - xy + xy - y^2$ $= x^2 - y^2$ **Note**: When binomial factors are multiplied that differ by only the sign between them, the resulting product is the difference of two perfect squares and there is no middle term. Two such binomial factors are called conjugates.	$\left(2\sqrt{5} + 3\right)\left(2\sqrt{5} - 3\right) = \left(2\sqrt{5}\right)^2 - 9 = 4 \cdot 5 - 9$ $= 20 - 9 = 11$ **Note**: These are also binomial factors that differ by only the sign between them. When multiplying numerical conjugates, the product is always a single number.

Example 13

Perform the indicated operations: $\sqrt{5}\left(3+\sqrt{2}\right)-4\sqrt{2}\left(2\sqrt{5}-1\right)$

$$\textit{Solution: } \sqrt{5}\left(3+\sqrt{2}\right)-4\sqrt{2}\left(2\sqrt{5}-1\right)=3\sqrt{5}+\sqrt{10}-8\sqrt{10}+4\sqrt{2}$$
$$=3\sqrt{5}-7\sqrt{10}+4\sqrt{2}$$

Example 14

Perform the indicated operations: $\left(2+\sqrt{3}\right)\left(3-2\sqrt{3}\right)+4\left(5\sqrt{3}\right)$

$$\textit{Solution: } \left(2+\sqrt{3}\right)\left(3-2\sqrt{3}\right)+4\left(5\sqrt{3}\right)=6-4\sqrt{3}+3\sqrt{3}-2\cdot 3+20\sqrt{3}$$
$$=6-\sqrt{3}-6+20\sqrt{3}=19\sqrt{3}$$

Rationalizing the Denominator: Division of Expressions Containing Radicals

Does division of expressions containing radicals simply follow the procedure for division of polynomials? No. It is not that simple. Before developing this concept, it is necessary to examine division of monomial expressions containing radicals. Dividing the expression $\frac{12\sqrt{6}}{6\sqrt{30}}$ is not the same as dividing the radicals seen in Example 9. Here, $\frac{12\sqrt{6}}{6\sqrt{30}}$ simplifies to the fraction $\frac{2}{\sqrt{5}}$. Unfortunately, the answer is in an awkward form. It has an irrational denominator. Remember, to be a rational number, the number must be able to be expressed as the ratio of two integers. Of course, using a graphing calculator to divide 2 by $\sqrt{5}$ will give .894427191. However, this result is an approximation. It is not an exact answer.

Mathematicians have developed a standard way to deal with a radical in the denominator. It is called rationalizing the denominator. This process is based on the fact that when a radical with index 2 is squared, the result is an integer represented by the radicand itself. That is, $\sqrt{a}\cdot\sqrt{a}=a$. So to rationalize the denominator, multiply by the radical over itself (which equals 1). For instance, $\frac{2}{\sqrt{5}}=\frac{2}{\sqrt{5}}\cdot\frac{\sqrt{5}}{\sqrt{5}}=\frac{2\sqrt{5}}{5}$.

Example 15

Rationalize the denominator: $\frac{9}{4\sqrt{3}}$

Solution: To rationalize this denominator, just multiply both the numerator and denominator by the irrational part of the denominator, $\sqrt{3}$.
$$\frac{9}{4\sqrt{3}} = \frac{9}{4\sqrt{3}} \cdot \frac{\sqrt{3}}{\sqrt{3}} = \frac{9\sqrt{3}}{4 \cdot 3} = \frac{3\sqrt{3}}{4}$$

Example 16

Rationalize the denominator: $\dfrac{2}{3+\sqrt{7}}$

Solution: The denominator is a binomial containing a radical expression. Simply multiplying it by $\sqrt{7}$ will not remove the irrational part of the denominator. That is, $\sqrt{7}\left(3+\sqrt{7}\right) = 3\sqrt{7}+7$. Now is time to recall that multiplication of conjugate expressions result in no middle term. The conjugate of $3+\sqrt{7}$ is $3-\sqrt{7}$. So the process here is to multiply both the numerator and denominator by the conjugate expression.
$$\frac{2}{3+\sqrt{7}} = \frac{2}{\left(3+\sqrt{7}\right)} \cdot \frac{\left(3-\sqrt{7}\right)}{\left(3-\sqrt{7}\right)} = \frac{6-2\sqrt{7}}{9-7} = \frac{6-2\sqrt{7}}{2} = \frac{2(3-\sqrt{7})}{2} = 3-\sqrt{7}$$

Example 17

Rationalize the denominator: $\dfrac{3-\sqrt{2}}{4+5\sqrt{2}}$

Solution: $$\frac{3-\sqrt{2}}{4+5\sqrt{2}} = \frac{\left(3-\sqrt{2}\right)}{\left(4+5\sqrt{2}\right)} \cdot \frac{\left(4-5\sqrt{2}\right)}{\left(4-5\sqrt{2}\right)} = \frac{12-15\sqrt{2}-4\sqrt{2}-5 \cdot 2}{16-25 \cdot 2}$$
$$= \frac{12-19\sqrt{2}-10}{16-50} = \frac{2-19\sqrt{2}}{-34}$$

Check Your Understanding of Section 1.5

1. Simplify: $\sqrt[4]{96}$

(1) $4\sqrt{6}$ (2) $4\sqrt[4]{6}$ (3) $2\sqrt[4]{6}$ (4) $2\sqrt{6}$

2. The base of a triangle is $4\sqrt{3}$ cm long, and its height is $3\sqrt{6}$ cm long. Determine the number of square centimeters in the area of the triangle.

(1) $18\sqrt{2}$ (2) $36\sqrt{2}$ (3) $15\sqrt{2}$ (4) $54\sqrt{2}$

3. When expressed as a single term, $\sqrt{75}-\frac{4}{2\sqrt{3}}$ has what number as the numerical coefficient of $\sqrt{3}$?

(1) $\frac{1}{3}$ (2) $\frac{4}{3}$ (3) 5 (4) $\frac{13}{3}$

4. Expand: $\left(4-3\sqrt{2}\right)^2$

(1) $34-24\sqrt{2}$ (2) $-2-24\sqrt{2}$ (3) 34 (4) 48

5. Multiply: $2\sqrt{3}\left(4\sqrt{6}-3\sqrt{15}\right)$

(1) $-6\sqrt{3}$ (2) $24\sqrt{2}-18\sqrt{5}$ (3) $24-18\sqrt{5}$ (4) $24-18\sqrt{2}$

6. Evaluate: $64^{-\frac{2}{3}}$

(1) $\frac{1}{4}$ (2) $\frac{1}{8}$ (3) $\frac{1}{16}$ (4) 16

7. Simplify: $\frac{24\sqrt[3]{1296}}{4\sqrt[3]{54}}$

(1) $12\sqrt[3]{24}$ (2) $6\sqrt[3]{24}$ (3) $12\sqrt[3]{3}$ (4) $6\sqrt[3]{6}$

8. $\left(3+\sqrt{3}\right)\left(2-3\sqrt{3}\right)=$

(1) -3 (2) $-7-3\sqrt{3}$ (3) $-3-7\sqrt{3}$ (4) $-3-10\sqrt{3}$

9. If $z = 32$, determine the value of $z^{\frac{1}{5}}+2z^0+(8z)^{-\frac{1}{2}}$

(1) $\frac{41}{16}$ (2) 20 (3) $\frac{49}{16}$ (4) $\frac{65}{16}$

10. $6\sqrt{45}-3\sqrt{80}=$

(1) 6 (2) $6\sqrt{5}$ (3) $3\sqrt{5}$ (4) $15\sqrt{5}$

11. Multiply $\left(4a^{-2}b^{\frac{3}{5}}\right)\left(2a^{-3}b^{\frac{5}{6}}\right)$ and express the answer with only positive exponents.

(1) $8a^6b^{\frac{1}{2}}$ (2) $\dfrac{8b^{\frac{43}{30}}}{a^5}$ (3) $6a^6b^{\frac{1}{2}}$ (4) $\dfrac{6b^{\frac{43}{30}}}{a^5}$

12. Write $\dfrac{6}{p^{-\frac{4}{5}}}$ as an equivalent expression in radical form.

(1) $6\sqrt[4]{p^5}$ (2) $6\sqrt[5]{p^4}$ (3) $\dfrac{24}{5}p$ (4) $\dfrac{\sqrt[5]{p}^{-4}}{6}$

13. Write $\dfrac{6}{5\sqrt{3}}$ as an equivalent fraction with a rational denominator.

(1) $\dfrac{3\sqrt{5}}{5}$ (2) $\dfrac{2\sqrt{5}}{3}$ (3) $\dfrac{2\sqrt{3}}{5}$ (4) $\dfrac{2\sqrt{5}}{15}$

14. Rationalize the denominator: $\dfrac{2+\sqrt{6}}{1+3\sqrt{6}}$

15. Rationalize the denominator: $\dfrac{\sqrt{3}-\sqrt{5}}{\sqrt{3}+4\sqrt{5}}$

16. Perform the indicated operations: $\left(3-4\sqrt{2}\right)\left(3+4\sqrt{2}\right)-6\sqrt{2}\left(1+\sqrt{2}\right)$

17. Multiply: $4\sqrt{3}\cdot 5\sqrt[3]{3}$

18. Simplify: $\dfrac{s^{\frac{3}{4}}}{s^{-\frac{1}{3}}}$

19. Rationalize the denominator: $\dfrac{6}{\sqrt{2}-3}$

1.6 SOLVING EQUATIONS AND INEQUALITIES

Special note: The first part of this section is a quick review from Integrated Algebra and is not part of the Algebra 2/Trigonometry curriculum. However, the information should be reviewed before studying the part of this chapter that covers the Algebra 2/Trigonometry curriculum. The new content in this section involves solving special types of equations and inequalities.

KEY IDEAS

Do you remember solving equations and inequalities in Integrated Algebra? If so, recall that you need to use the addition, subtraction, multiplication, and division properties of equality to solve them.

Linear Equations and Inequalities

MATH FACTS

For the real numbers *a, b* and *c*:

Addition property of equality: If $a = b$, then $a + c = b + c$.

Subtraction property of equality: If $a = b$, then $a - c = b - c$.

Multiplication property of equality: If $a = b$, then $a \cdot c = b \cdot c$.

Division property of equality: If $a = b$, then $\frac{a}{c} = \frac{b}{c}$ if $c \neq 0$.

The following table shows the basic applications of the equality properties.

Addition Property	Subtraction Property	Multiplication Property	Division Property
$x - 3 = 5$ $+3 \ +3$ $x = 8$	$y + 8 = 12$ $-8 \ -8$ $y \ = 4$	$\frac{3z}{3} = \frac{6}{3}$ $z = 2$	$\frac{x}{5} = 4$ $5 \cdot \frac{x}{5} = 5 \cdot 4$ $x = 20$

Multistep equations use these same properties in combination with methods of simplifying the polynomial expressions on each side of the equation. Some review examples follow.

Example 1

Solve for x: $3x + 59 = 75 - 5x$

Solution:	$3x + 59 = 75 - 5x$
Add $5x$ to both sides of the equation	$8x + 59 = 75$
Subtract 59 from both sides of the equation	$8x = 16$
Divide both sides of the equation by 8	$x = 2$

Check: $3 \cdot 2 + 59 \stackrel{?}{=} 75 - 5 \cdot 2$
$6 + 59 \stackrel{?}{=} 75 - 10$
$65 = 65$ ✓

Example 2

Solve for x: $3(2x - 3) + 17 = (5x + 1) + 2x$

Solution:	$3(2x - 3) + 17 = (5x + 1) + 2x$
Distribute the 3	$6x - 9 + 17 = (5x + 1) + 2x$
Combine like terms on both sides of the equation	$6x + 8 = 7x + 1$
Subtract $6x$ from both sides of the equation	$8 = x + 1$
Subtract 1 from both sides of the equation	$7 = x$

Check: $3(2x - 3) + 17 = (5x + 1) + 2x$
$3(2 \cdot 7 - 3) + 17 \stackrel{?}{=} (5 \cdot 7 + 1) + 2 \cdot 7$
$3(14 - 3) + 17 \stackrel{?}{=} (35 + 1) + 14$
$3(11) + 17 \stackrel{?}{=} (35 + 1) + 14$
$33 + 17 \stackrel{?}{=} 36 + 14$
$50 = 50$ ✓

MATH FACTS

For the real numbers a, b and c:

Addition Property of Inequality: If $a < b$, then $a + c < b + c$.

Subtraction Property of Inequality: If $a < b$, then $a - c < b - c$.

Multiplication Property of Inequality: If $a < b$, then $a \cdot c < b \cdot c$ if $c > 0$.
If $a < b$, then $a \cdot c > b \cdot c$ if $c < 0$.

Division Property of Inequality: If $a < b$, then $\frac{a}{c} < \frac{b}{c}$ if $c > 0$.
If $a < b$, then $\frac{a}{c} > \frac{b}{c}$ if $c > 0$.

Each of these rules can be modified, replacing all less than signs (<) with greater than signs (>) and vice versa. The same applies to less than or equal to (≤) and greater than or equal to (≥) signs. Generally, when multiplying or dividing both sides of an inequality by a negative number, it is necessary to reverse the order of the inequality.

Example 3

Solve for x: $2x + 15 \leq 6 - x$

Solution:	$2x + 15 \leq 6 - x$
Add x to both sides of the inequality	$3x + 15 \leq 6$
Subtract 15 from both sides of the inequality	$3x \leq -9$
Divide both sides of the inequality by 3	$x \leq -3$

However, if you use a different set of steps to solve this inequality, the process may look different.

	$2x + 15 \leq 6 - x$
Subtract 6 from both sides of the inequality	$2x + 9 \leq -x$
Subtract $2x$ from both sides of the inequality	$9 \leq -3x$
Now when you divide both sides of the inequality by −3, you need to be careful to reverse the order of the inequality from a ≤ inequality to a ≥ inequality	$\frac{9}{-3} \geq \frac{-3x}{-3}$ $-3 \geq x$ or $x \leq -3$

MATH FACTS

To solve second-degree equations—quadratic equations, the Zero Product Property was introduced in Integrated Algebra. The Zero Product Property states that if $a \cdot b = 0$, then either $a = 0$ or $b = 0$ or both.

If $(x + 3)(x - 2) = 0$, then either $x + 3 = 0$ or $x - 2 = 0$. Solving each equation results in $x = -3$ and $x = 2$.

Example 4

Solve for x: $x^2 + 8x + 12 = 0$

Solution: $x^2 + 8x + 12 = 0$

Since the equation is already set equal to zero, factor the trinomial:

$$(x + 6)(x + 2) = 0$$

$$x + 6 = 0 \text{ or } x + 2 = 0$$

$$x = -6 \text{ or } x = -2$$

Example 5

Solve for y: $y^2 - 21 = 4y$

Solution: The equation $y^2 - 21 = 4y$ is not set equal to 0. So first subtract $4y$ from both sides of the equation.

Now $y^2 - 4y - 21 = 0$

and factoring $(y + 3)(y - 7) = 0$

so that $y + 3 = 0$ or $y - 7 = 0$

and $y = -3$ or $y = 7$

Fractional Equations

A fractional equation is an equation that contains fractions with variables in the denominator. There are two ways to work with many fractional equations. The first method is to perform all the arithmetic operations on the equation. The second method is to multiply the entire equation by the lowest common denominator and then do the arithmetic. Following are a few examples worked out in both ways in order to compare and decide which approach is best to use. Note that the first example is not actually a fractional equation but, rather, an equation that contains some fractions. It will still assist in the investigation of fractional equations, though.

Perform the Work with the Fractions	Multiply Through by the LCD
$3x+\frac{1}{5}=5x-\frac{1}{3}$ $3x-5x=-\frac{1}{3}-\frac{1}{5}=-\frac{5}{15}-\frac{3}{15}$ $-2x=\frac{-8}{15}$ $x=\frac{-\frac{8}{15}}{-2}=\frac{-8}{15}\cdot\frac{1}{-2}=\frac{4}{15}$	$3x+\frac{1}{5}=5x-\frac{1}{3}$ $15\cdot 3x+\cancel{15}^{3}\cdot\frac{1}{\cancel{5}}=15\cdot 5x-\cancel{15}^{5}\cdot\frac{1}{\cancel{3}}$ $45x+3=75x-5$ $3=30x-5$ $8=30x$ $\frac{8}{30}=\frac{30x}{30}$ $\frac{4}{15}=x$
$\frac{30}{x}+2=9-\frac{5}{x}$ (Note: x cannot equal 0) $\frac{30}{x}+\frac{5}{x}=9-2$ $\frac{35}{x}=7$ $\cancel{x}\cdot\frac{35}{\cancel{x}}=x\cdot 7$ $35=7x$ $5=x$	$\frac{30}{x}+2=9-\frac{5}{x}$ (Note: x cannot equal 0) $\cancel{x}\cdot\frac{30}{\cancel{x}}+x\cdot 2=x\cdot 9-\cancel{x}\cdot\frac{5}{\cancel{x}}$ $30+2x=9x-5$ $30+5=9x-2x$ $35=7x$ $5=x$
$\frac{35}{x-3}+\frac{20}{x}=\frac{70}{x}$ $\frac{35}{(x-3)}\cdot\frac{x}{x}+\frac{20}{x}\cdot\frac{(x-3)}{(x-3)}=\frac{70}{x}\cdot\frac{(x-3)}{(x-3)}$ $\frac{35x}{x(x-3)}+\frac{20(x-3)}{x(x-3)}=\frac{70(x-3)}{x(x-3)}$ $\frac{35x}{x(x-3)}+\frac{20x-60}{x(x-3)}=\frac{70x-210}{x(x-3)}$ $\frac{35x+20x-60}{x(x-3)}=\frac{70x-210}{x(x-3)}$ $\frac{55x-60}{x(x-3)}=\frac{70x-210}{x(x-3)}$ $55x-60=70x-210$ $210-60=70x-55x$ $150=15x$ $10=x$	$\frac{35}{x-3}+\frac{20}{x}=17-x$ $x\cancel{(x-3)}\frac{35}{\cancel{(x-3)}}+\cancel{x}(x-3)\frac{20}{\cancel{x}}$ $=\cancel{x}(x-3)\frac{70}{\cancel{x}}$ $35x+20(x-3)=70(x-3)$ $35x+20x-60=70x-210$ $55x-60=70x-210$ $210-60=70x-55x$ $150=15x$ $10=x$

A special case of fractional equations is a proportion. The process of solving proportions is not new. This process follows the fact that in a proportion, the product of the means is equal to the product of the extremes. In the proportion $\frac{x}{3}=\frac{9}{27}$, the means are the 3 and the 9 and the extremes are the x and the 27. It is easier to see why they are named as means and extremes if you rewrite the proportion in the form $x : 3 = 9 : 27$. The means are the middle two terms, and the extremes are the first and last term in the proportion.

Example 6

Solve for x: $\frac{x+4}{18}=\frac{x-1}{8}$

Solution: $\frac{x+4}{18}=\frac{x-1}{8}$

$$8(x + 4) = 18(x - 1)$$

$$8x + 32 = 18x - 18$$

$$18 + 32 = 18x - 8x$$

$$50 = 10x$$

$$5 = x$$

Check: $\frac{x+4}{18}\stackrel{?}{=}\frac{x-1}{8}$

$$\frac{5+4}{18}\stackrel{?}{=}\frac{5-1}{8}$$

$$\frac{9}{18}\stackrel{?}{=}\frac{4}{8}$$

$$\frac{1}{2}=\frac{1}{2}\ \checkmark$$

Some fractional equations also require recalling how to solve quadratic equations. That process, using the zero product property, was reviewed earlier in this section.

Example 7

Solve for x: $\frac{10}{x+2}+\frac{2}{x-2}=4$

Solution: $\frac{10}{x+2}+\frac{2}{x-2}=4$

$$\cancel{(x+2)}(x-2)\frac{10}{\cancel{(x+2)}}+(x+2)\cancel{(x-2)}\frac{2}{\cancel{(x-2)}}=4(x+2)(x-2)$$

$10(x-2)+2(x+2)=4(x^2-4)$

$10x-20+2x+4=4x^2-16$

$12x-16=4x^2-16$

$0=4x^2-12x$

$0=4x(x-3)$

Now use the zero product property:

$4x=0$ or $x-3=0$

$x=0$ or $x=3$

Check 3: $\frac{10}{3+2}+\frac{2}{3-2}\stackrel{?}{=}4$

$\frac{10}{5}+\frac{2}{1}\stackrel{?}{=}4$

$2+2\stackrel{?}{=}4$

$4=4$ ✓

Check 0: $\frac{10}{0+2}+\frac{2}{0-2}\stackrel{?}{=}4$

$\frac{10}{2}+\frac{2}{-2}\stackrel{?}{=}4$

$5-1\stackrel{?}{=}4$

$4=4$ ✓

This whole process of solving a quadratic equation will be reviewed again in Chapter 3, Section 3.3.

Example 8

Solve for x: $\frac{x}{x+3}+\frac{2}{x-3}=\frac{12}{x^2-9}$

Solution: $\frac{x}{x+3}+\frac{2}{x-3}=\frac{12}{x^2-9}$

The LCD is $(x + 3)(x - 3)$. So multiply by that factor on both sides of the equation:

$$(\cancel{x+3})(x-3)\frac{x}{\cancel{x+3}}+(x+3)(\cancel{x-3})\frac{2}{\cancel{x-3}}=(\cancel{x+3})(\cancel{x-3})\frac{12}{(\cancel{x+3})(\cancel{x-3})}$$

$(x - 3)x + (x + 3)2 = 12$

$x^2 - 3x + 2x + 6 = 12$

$x^2 - x + 6 = 12$

$x^2 - x - 6 = 0$

$(x - 3)(x + 2) = 0$

$x - 3 = 0$ or $x + 2 = 0$

$x = 3$ or $x = -2$

Now check these answers:

If $x = 3$, $\frac{x}{x+3}+\frac{2}{x-3}=\frac{12}{x^2-9}$

$$\frac{3}{3+3}+\frac{2}{3-3}\stackrel{?}{=}\frac{12}{3^2-9}$$

Two of these fractions are undefined because their denominators equal 0. So x cannot equal 3, and $x = 3$ is called an extraneous root for the problem. In other words, $x = 3$ is not a suitable answer for the question.

If $x = -2$, $\frac{x}{x+3}+\frac{2}{x-3}=\frac{12}{x^2-9}$

$$\frac{-2}{-2+3}+\frac{2}{-2-3}\stackrel{?}{=}\frac{12}{(-2)^2-9}$$

$$\frac{-2}{1}+\frac{2}{-5}\stackrel{?}{=}\frac{12}{4-9}$$

$$\frac{-2}{1}\cdot\frac{(-5)}{(-5)}+\frac{2}{-5}\stackrel{?}{=}\frac{12}{-5}$$

$$\frac{10}{-5}+\frac{2}{-5}\stackrel{?}{=}\frac{12}{-5}$$

$$\frac{12}{-5}=\frac{12}{-5}\ ✓$$

So $x = -2$ is a solution to the problem.

This example is important because it shows that checking answers to rational equations is always necessary as they may have extraneous roots.

Fractional Inequalities

The same rules for inequalities apply to fractional or rational inequalities as they do for other inequalities.

Example 9

Solve for x: $x < \dfrac{12}{x} - 1$

Solution: The LCD is x. However, consider whether or not x is positive or negative. Two sets of inequalities must be worked. Notice that the inequality sign has been reversed when $x < 0$.

If $x > 0$

$$x\cdot x < \frac{12}{x}\cdot x - 1\cdot x$$

$$x^2 < \frac{12}{\not x}\cdot \not x - 1\cdot x$$

$x^2 < 12 - x$

$x^2 + x - 12 < 0$

$(x + 4)(x - 3) < 0$

If $(x + 4)(x - 3) < 0$, then
either $x + 4 > 0$ and $x - 3 < 0$
$x > -4$ and $x < 3$
so $-4 < x < 3$
but since $x > 0$, this is
limited to $0 < x < 3$

If $x < 0$

$$x\cdot x > \frac{12}{x}\cdot x - 1\cdot x$$

$$x^2 > \frac{12}{\not x}\cdot \not x - 1\cdot x$$

$x^2 < 12 - x$

$x^2 + x - 12 > 0$

$(x + 4)(x - 3) < 0$

If $(x + 4)(x - 3) > 0$, then
either $x + 4 < 0$ and $x - 3 < 0$
$x < -4$ and $x < 3$
so $x < -4$
but since $x < 0$, $x < -4$

or $x + 4 < 0$ and $x - 3 > 0$
$x < -4$ and $x > 3$
so there is no solution in this case.

or $x + 4 > 0$ and $x - 3 > 0$
$x > -4$ and $x > 3$
so $x > 3$
but since $x < 0$, there is no solution in this case.

Therefore, the final solution is $\{x: 0 < x < 3 \text{ or } x < -4\}$.

Example 10

Solve for x: $\frac{x+3}{x-3} \geq 0$

Solution: The LCD for this example is $(x - 3)$.

If $x - 3 > 0$, $\frac{x+3}{x-3} \geq 0$

$$(\cancel{x-3})\frac{(x+3)}{(\cancel{x-3})} \geq (x-3)0$$

$x + 3 \geq 0$
$x \geq -3$(but x cannot equal 3)
$x > -3$
but $x - 3 > 0$
or $x > 3$
so $x > 3$

If $x - 3 > 0$, $\frac{x+3}{x-3} \leq 0$

$$(\cancel{x-3})\frac{(x+3)}{(\cancel{x-3})} \leq (x-3)0$$

$x + 3 \leq 0$
$x \leq -3$(but x cannot equal 3)
$x < -3$
but $x - 3 < 0$
or $x < 3$
so $x < -3$

Therefore the solution set is $\{x: x < -3 \text{ or } x > 3\}$.

Example 11

Solve for y: $\frac{y+3}{8} < \frac{y-1}{4}$

Solution: $\frac{y+3}{8} < \frac{y-1}{4}$

$$8 \cdot \frac{y+3}{8} < 8 \cdot \frac{y-1}{4}$$

$y + 3 < 2(y - 1)$

$y + 3 < 2y - 2$

$2 + 3 < 2y - y$

$5 < y$

Direct and Indirect Variation

Examine a scenario such as one in which a person earns an hourly wage and the task is to find out how much money the person can earn after a certain number of hours. For instance, Kim earns $8.50 an hour at her job. The chart below shows her total earnings over an interval of time where the number of hours worked is rounded to the nearest hour.

Number of Hours	1	2	3	4	5	6	7	8
Earnings	$8.50	$17.00	$25.50	$34.00	$42.50	$51.00	$59.50	$68.00

From the above chart, it can be observed that as Kim works more hours, her earnings increase. The relationship can be represented in the equation $e = 8.5h$. In such a relationship, where one variable increases as the other also increases (for where one variable decreases as the other also decreases), the variables are said to be in direct variation with each other.

There are other situations in which as one variable increases, the other decreases and vice versa. For instance, Arturo is planning a trip on his bicycle. The trip is 100 miles long. He calculated the number of hours it will take him for the trip depending on how fast he travels (in multiples of 5 mph.). The chart below shows a summary of his calculations.

Speed	5 mph	10 mph	15 mph	20 mph
Number of Hours	20 hr.	10 hr.	6 hr. 40 min.	5 hr.

From the above chart, it can be observed that when Arturo increases his speed, his trip time decreases. The relationship can be represented in the equation $t = \frac{100}{r}$. In such a relationship, where one variable increases as the other decreases, the variables are said to be in inverse or indirect variation with each other.

If y varies directly as x, the general equation $y = kx$ can be written, where k is the constant of variation. In the example above about Kim's earnings, the constant k is 8.5. Often the equation is not provided. There is a need to work with two different pieces of data to establish the equation in order to work with a specific example.

Imagine a problem in which there is a direct variation between x and y for which when $x = 3$, $y = 5$. The task is to determine the value of x corresponding to a y-value of 12. Two approaches can be used. In the first approach, since y varies directly as x, the general equation, $y = kx$, can be used. Solve for k by substituting in the first set of values for x and y. This results in $5 = 3k$. From that it can be determined that $k = \frac{5}{3}$. Now replace this constant of variation into the general equation to get $y = \frac{5}{3}x$. The task now reduces to placing in the y-value of 12 to determine what the corresponding x-value is, $12 = \frac{5}{3}x \rightarrow 36 = 5x \rightarrow x = \frac{36}{5}$ or $7\frac{1}{5}$.

A second approach is to recognize that $y = kx$ for two different sets of values of x and y yields a proportion since k is the ratio between y and x. The proportion is therefore $\frac{y_1}{x_1} = \frac{y_2}{x_2}$. In this problem, $\frac{5}{3} = \frac{12}{x}$. Using the product of the means is equal to the product of the extremes results in $5x = 36$ or $x = \frac{36}{5}$.

MATH FACTS

If y varies directly as x, the formula, $y = kx$ is used to solve for the constant of variation, k. The variables x and y are also proportional, and the formula $\frac{y_1}{x_1} = \frac{y_2}{x_2}$ can be used to solve problems involving those two variables.

Example 12

Sasha owns a business that needs to send out a mailing to a large number of possible clients. She noticed that when she has 2 employees doing a mailing for her, they are able to get 130 mailings ready to be posted in one afternoon of work. She is pretty confident that all of her workers can do the same task at approximately the same speed. Obviously, the more employees working independently on the task, the more mailings that can be ready to be posted in the same time frame. So she assumes there is a direct variation between the number of employees and number of mailings ready for posting. If she were to put 5 of her employees on this task, how many mailings should she expect to have ready for posting in one afternoon?

Solution: Here are the two possible approaches to solving this.

Approach 1	**Approach 2**
$y = kx$	$\frac{y_1}{x_1} = \frac{y_2}{x_2}$
$130 = 2k$ $k = \frac{130}{2} = 65$ $y = 65x$ $y = 65(5) = 325$	$\frac{130}{2} = \frac{y}{5}$ $2y = 650$ $y = 325$

Imagine a problem in which there is an inverse variation between x and y for which when $x = 7$, $y = 4$. The task is to determine the value of x corresponding to a y-value of 6. Two approaches can be used. In the first approach, since y varies inversely with x, the general equation $y = \frac{k}{x}$ can be used. Solve for k by substituting in the first set of values for x and y. This yields in $4 = \frac{k}{7}$. From that it can be determined that $k = 28$. Now replace this constant of variation into the general equation to get $y = \frac{28}{x}$. The task now reduces to placing in the y-value of 6 to determine what the corresponding x-value is,

$$6 = \frac{28}{x} \rightarrow 6x = 28 \rightarrow x = \frac{28}{6} = \frac{14}{3} \text{ or } 4\frac{2}{3}.$$

A second approach is to recognize that $y = \frac{k}{x}$ for two different sets of values of x and y yields a product equal to another product since $k = xy$. The resulting equation is $x_1 \cdot y_1 = x_2 \cdot y_2$. For this problem, the result is $7 \cdot 4 = 6x$. Solving this yields $6x = 28$ or $x = \frac{28}{6} = \frac{14}{3}$.

MATH FACTS

If y varies inversely as x, use the formula $y = \frac{k}{x}$ to solve for the constant of variation, k. The formula $x_1 \cdot y_1 = x_2 \cdot y_2$ can also be used to solve problems involving the variables x and y.

Example 13

Josh noticed that if he wanted to create a rectangle with a constant area and a base of 2 cm, the height would be 8 cm. He recognized that if he were to increase the length of the base, then the height would have to be smaller to maintain the area. He was working with an inverse variation. If the base were 6 cm long, how long would be the height of the rectangle?

Solution: In this inverse variation, you can use $b_1 \cdot h_1 = b_2 \cdot h_2$. Therefore, $2 \cdot 8 = 6 \cdot h$ and $h = \frac{16}{6} = \frac{8}{3}$ cm.

Radical Equations

Just as it sounds, a radical equation is an equation containing radical expressions. Examine the equation $\sqrt{x+1} = 3$. Think about it! To answer this, isn't it obvious that $x + 1$ has to be 9 in order for the radical to equal 3? So intuitively you see that the answer has to be $x = 8$.

The method used to solve this equation algebraically rather than intuitively is to square both sides of the equation which results in a rational expression. This is problematic, though. When squaring both sides of the equation, the number of possible solutions is doubled. Therefore, it is possible that there will be an extraneous root when you are finished with the problem. An extraneous root (or solution) is a solution that satisfies the revised equation but not the original equation.

So $\sqrt{x+1} = 3$ becomes $x + 1 = 9$ or $x = 8$. To check, substitute the $x = 8$ into the equation to see that $\sqrt{x+1} = 3 \rightarrow \sqrt{8+1} = 3 \rightarrow \sqrt{9} = 3 \rightarrow 3 = 3$.

Example 14

Solve for z: $\sqrt{32 - z} = 6$

Solution: $\sqrt{32 - z} = 6$

Square both sides: $32 - z = 36$

$-z = 4$

$z = -4$

Check: $\sqrt{32 - z} = 6$

$\sqrt{32 - (-4)} = 6$

$\sqrt{36} = 6$

$6 = 6$

Example 15

Solve for t: $\sqrt{t+5}+2=6$

Solution: Simply squaring each side of the equation to solve this problem is problematic when squaring produces a middle term. By subtracting 2 from each side first, then the problem becomes similar to the last two problems.

$$\sqrt{t+5}+2=6 \rightarrow \sqrt{t+5}=4$$

Square each side of the equation: $t + 5 = 16$
$t = 11$

Check: $\sqrt{t+5}+2=6 \rightarrow \sqrt{11+5}+2=6 \rightarrow \sqrt{16}+2=6 \rightarrow 4 + 2 + 6 \rightarrow 6 = 6$

Example 16

Solve for x: $\sqrt{x}+\sqrt{x-5}=5$

Solution: Just like the last example, simply squaring each side of the equation to solve this problem does not work because squaring produces a middle term. Now, there is no integer to subtract from both sides of the equation. However, just squaring the left hand side of the current form of this equation the middle term will yield a result with twice the product of the $\sqrt{x}$ and the $\sqrt{x+5}$. Subtracting $\sqrt{x+5}$ from both sides of the equation, even though it will still leave a radical middle term, is the less-complex approach.

	$\sqrt{x}+\sqrt{x-5}=5$
Subtract $\sqrt{x+5}$ from both sides	$\sqrt{x}=5-\sqrt{x-5}$
Square both sides of this equation	$x=25-10\sqrt{x-5}+(x-5)$
Combine like terms	$x=x+20-10\sqrt{x-5}$
Subtract x from both sides of the equation	$0=20-10\sqrt{x-5}$
Subtract 20 from both sides of the equation	$-20=-10\sqrt{x-5}$
Divide both sides of the equation by -10	$2=\sqrt{x-5}$
Now square both sides of the equation again	$4 = x - 5$
	$9 = x$

Check: $\sqrt{x}+\sqrt{x-5}=5 \rightarrow \sqrt{9}+\sqrt{9-5}=5 \rightarrow 3+\sqrt{4}=5 \rightarrow 3+2=5 \rightarrow$
$5 = 5$

Example 17

Solve for x: $\sqrt{x+9} = -5$

Solution: $\sqrt{x+9} = -5$
$x + 9 = 25$
$x = 16$

Checking this answer results in $\sqrt{x+9} = -5 \rightarrow \sqrt{16+9} = -5 \rightarrow \sqrt{25} = -5$, $x = 16$ does not satisfy the equation. It is an extraneous root. There is no solution for the equation.

Check Your Understanding of Section 1.6

1. Solve for x: $\dfrac{27}{x+8} - \dfrac{15}{x^2+18x+80} = \dfrac{40}{x+10}$

(1) $x = 5$ (2) $x = -5$ (3) $x = \dfrac{4}{13}$ (4) $x = -8$

2. Solve for y: $\dfrac{9}{2y} + 5 = \dfrac{12}{y}$

(1) $y = 15$ (2) $y = \dfrac{5}{2}$ (3) $y = \dfrac{3}{2}$ (4) $y = 1$

3. Solve for c: $\dfrac{5c+4}{7} = \dfrac{c+14}{8}$

(1) $c = \dfrac{10}{33}$ (2) $c = \dfrac{7}{8}$ (3) $c = \dfrac{130}{47}$ (4) $c = 2$

4. Solve for z: $\dfrac{12}{z-2} + \dfrac{1}{4} = \dfrac{z+7}{z-2}$

(1) $z = 6$ (2) $z = 13$ (3) $z = \dfrac{13}{4}$ (4) $z = \dfrac{1}{6}$

5. Solve for p: $\dfrac{5p-5}{p+1} < 10$

(1) $\{p: p < -3\}$
(2) $\{p: p < -3 \text{ or } p > -1\}$
(3) $\{p: p > 3 \text{ or } p < 1\}$
(4) $\{p: p > -1\}$

6. Pete is planning to take a motorboat ride for 80 miles and then take a rowboat for another 20 miles. He figures his motorboat can travel at a rate that is twice as fast as he can row the rowboat. If he wants to take 3 hours for the trip, how fast will he row the rowboat?
(1) 20 mph (2) 40 mph (3) 45 mph (4) 60 mph

7. Solve for x: $\frac{10}{x+1}+4>x+2$
(1) $\{x: x<-3\}$
(2) $\{x: -1<x<4\}$
(3) $\{x: x>4\}$
(4) $\{x: x<-3 \text{ or } -1<x<4\}$

8. If s varies directly as t when $s = 8$, $t = 3$, find s when $t = 12$.
(1) 2 (2) 6 (3) 24 (4) 32

9. If m varies inversely as n and $m = 5$ when $n = 36$, then find n when $m = 9$.
(1) 20 (2) 24 (3) $64\frac{4}{5}$ (4) 40

10. Carla baked a cake that requires 4 cups of flour. It came out so good that she decided to bake 3 more of the same type of cake. How many cups of flour will she need to bake the additional 3 cakes?
(1) 5 cups (2) 7 cups (3) 9 cups (4) 12 cups

11. According to Boyle's law, the volume of mass of a gas varies inversely as its absolute pressure (as long as the temperature is constant). The volume of a certain gas is 36 cubic inches when the absolute pressure is 45 pounds per square inch. When its absolute pressure is 50 pounds per square inch, what is the volume?
(1) $32\frac{4}{5}$ cubic inches
(2) 32.4 cubic inches
(3) 40 cubic inches
(4) 40.5 cubic inches

12. In a given rectangle, the length varies inversely as the width. If the length is tripled, the width will
(1) be divided by 3
(2) be multiplied by 3
(3) remain the same
(4) increase by 3

13. Solve for x: $\sqrt{x+3}-\sqrt{x-4}=1$
(1) 1 (2) 7 (3) 13 (4) 16

14. Solve for y: $\sqrt{8-y}+5=15$
(1) 92 (2) 100 (3) −92 (4) −100

15. Solve for x: $\frac{10}{x+8} - \frac{6}{x} = \frac{-60}{x^2+8x}$

16. Solve for z: $\frac{90}{z+2} - \frac{60}{z^2-4} \leq \frac{48}{z-2}$

17. Solve for x: $\frac{27}{2x} - \frac{9}{x} > \frac{81}{x^2}$

18. The speed of a stream is 4 mph. Fiona travels upstream for 12 miles in the same time that she travels downstream for 36 miles. What is the speed of Fiona's boat in still water?

19. The numerator of a fraction is 3. The sum of the fraction and a second fraction with the same numerator and whose denominator is double that of the first fraction is $\frac{9}{10}$. What is the denominator of the original fraction?

20. Thalia drove her car 180 miles north at a constant rate and returned 180 miles south traveling 15 mph slower than she did when she traveled north on the same route. If the return trip took her an hour longer, how fast was she traveling northbound?

21. The difference between two numbers is 64. If the difference between their square roots is 4, what are both numbers?

22. Solve for x: $\sqrt{x+8} + \sqrt{x-8} = 8$

23. For a rectangular garden of fixed area, the length of the garden varies inversely with the width. If the length of the garden is 9 feet, its width is 4 feet. Find the length if the width is 3 feet.

24. Mrs. Koster is constructing a test for her class. Originally, she had 20 questions at 5 points each. The number of points per question varies inversely as the number of questions on the test. How many questions can she create if she wants each question to count for 4 points?

25. Mrs. Lewis can bake 5 batches of cookies in 3 hours. How many batches of cookies can she bake in 12 hours?

26. Solve for c: $\frac{c}{c-7} - \frac{7}{c+4} = \frac{77}{c^2-3c-28}$

Chapter Two

FUNCTIONS AND RELATIONS

2.1 RELATIONS

KEY IDEAS

A relation is a matching or pairing of two sets. An example of a relation would be a matching of a street with people who live on that street

What Is a Relation?

Eight students from the same school are on the same bus route. The driver noted that Jared lives on Sycamore Street, Ben lives on Oak Street, Jill lives on Sycamore Street, Harlyn lives on Maple Avenue, Caren lives on Maple Avenue, Tom lives on Oak Street, Nadine lives on Maple Avenue, and Taryn lives on Oak Street. A relation that maps a street name on this bus route with students that live on that street can be defined in the set of ordered pairs {(Sycamore, Jared), (Sycamore, Jill), (Oak, Ben), (Oak, Tom), (Oak, Taryn), (Maple, Harlyn), (Maple, Caren), (Maple, Nadine)}.

Some relations pair elements from two different sets, and some relations pair elements of a set with elements in that same set. A relation is a mapping of each element of one set, called its domain, that matches it to an element in a second set, called its range. Such a mapping can be defined in words, in ordered pairs, in a table of values (as used for creating graphs), or as a mapping from one set to another.

For instance, the relation defined by the set of ordered pairs {(1, 4), (3, 7), (–2, 5), (3, 5), (8, 2)} can also be repeated as the table of values or the mapping shown below:

x	y
1	4
3	7
–2	5
3	5
8	2

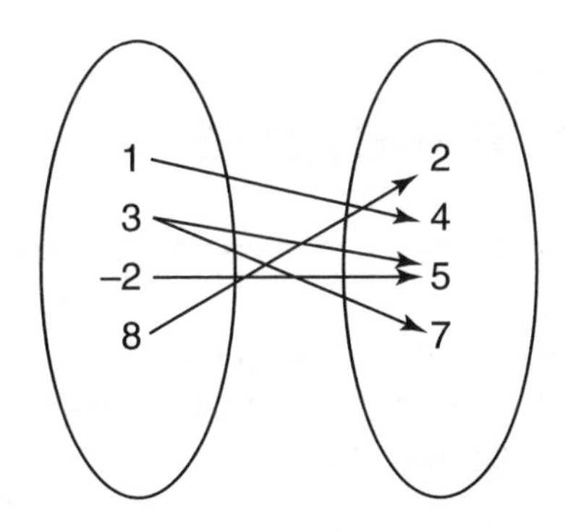

Example 1

A relation is defined by {(2, –1), (3, 4), (–1, 2), (4, 5), (1, 3), (5, 8), (4, –3), (6, 3)}

a. What is the domain of this relation?
b. What is the range of this relation?

Solution: The domain is the set defined by the numbers that are the first elements in each ordered pair. So the domain is equal to {2, 3, –1, 4, 1, 5, 6} or {–1, 1, 2, 3, 4, 5, 6}. Notice that 4 is a first element in two of the ordered pairs, but there is no need to list it twice. Identifying the set is easier after placing the elements in a recognizable order.

The range is the set defined by the numbers that are the second elements in each ordered pair. So the range is equal to {–1, 4, 2, 5, 3, 8, –3, 3} or {–3, –1, 2, 3, 4, 5, 8}.

Check Your Understanding of Section 2.1

1. Which of the following sets is a relation?
(1) {3, 5, 9, 13, 17}
(2) {chairs, desks, books}
(3) {(–3, 2), (4, 8), (3, –1), (5, 4)}
(4) $\{x: x + 8 = 6\}$

2. Which of the following sets is not a relation?
(1) {–12, –5, –1}
(2) $\{(x, y): y^2 = x^2 - 5\}$
(3) $\{(x, y): y = 2x + 3\}$
(4) {(1, 3), (8, –4), (–9, 7), (1, 6)}

3. In the relation defined by {(1, 4), (2, 5), (1, 6), (2, 4)}, what are both the domain and range of this relation?

2.2 FUNCTIONS

Some relations map one element from the domain to more than one element in the range. Relations that do not allow that to happen are called functions.

What Is a Function?

A function is a relation in which each element of the domain is mapped to one and only one member of the range. Since a relation is a set of ordered pairs, it is often written in the form (x, y), where x is an element of the domain and y is an element of the range. Without even knowing it, some work accomplished in this and previous courses have involved many functions. The relation $\{(1, 2), (3, 4), (-4, 8), (-2, 1)\}$ is an example of a relation that is a function, while $\{(2, 7), (3, 5), (2, 1)\}$ is not a function since 2 maps to both 7 and 1. Look at these same relations in the diagrams below and see how the first mapping takes each member of the first set to exactly one member of the second set. See also that the second mapping takes a member of the first set, 2, to two different members of the second set, 7 and 1.

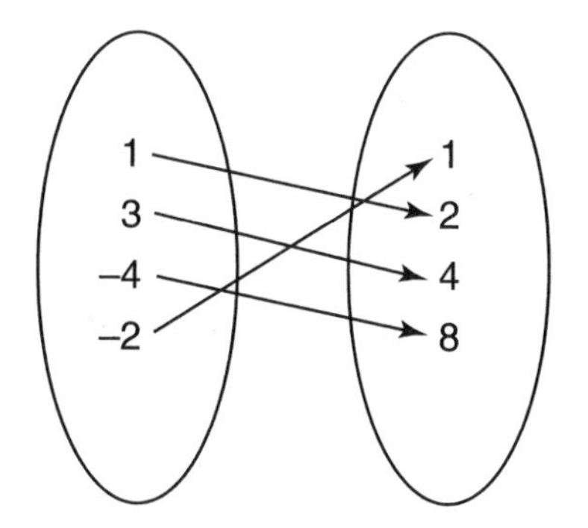

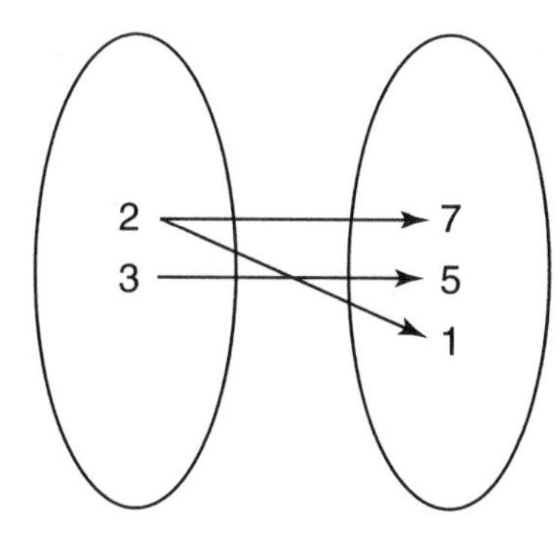

For instance, $y = x + 1$ is a function that describes an infinite number of ordered pairs. Most importantly, for each value of x there exists one and only one value of y. The domain of this function is the set of suitable values of x for the equation $y = x + 1$. The fact is that any type of real number can be used for x in this equation, so the domain is the set of all real numbers. The range of this function is the set of suitable values of y for the equation $y = x + 1$. Once again, it is possible to reach any value of y by replacing x with a real number that is one less than that y-value. So the range of this function is also the set of all real numbers, $\mathcal{R}$.

MATH FACTS

If f is a function mapping set A to set B, then the elements of set A are called the domain of function f and the elements of set B are called the range of function f. In other words, if $y = f(x)$, then the domain is the set of suitable values of x and the range is the set of suitable values of y.

Examine a partial table of values for the function $y = x + 1$ and also the function $y = \frac{1}{x}$:

x	$x + 1 = y$
2	3
4.7	5.7
$8\frac{2}{3}$	$9\frac{2}{3}$
–100	–99

x	$\frac{1}{x} = y$
–2	$\frac{1}{-2} = -\frac{1}{2}$
3	$\frac{1}{3}$
1	$\frac{1}{1} = 1$
0	$\frac{1}{0}$ is undefined

By examining these tables, it is easy to understand why both the domain and range of $y = x + 1$ are indeed all real numbers. What is the impact of having an undefined y-value for an x-value of 0 in the second function? That means that $x = 0$ cannot be used in the function $y = \frac{1}{x}$ because there is no y-value in the range of this function that can be paired with an x-value of 0. So the domain becomes restricted. It can be said that the domain of $y = \frac{1}{x}$ is the set of all real numbers except 0. This can be written in several ways: $\{x: x \neq 0\}$ or $\mathcal{R}/\{0\}$. Also notice that there is no x-value in $y = \frac{1}{x}$ that will map to $y = 0$. So the range of this function must also exclude 0.

Graphical Representation of a Function

Graphing was studied in the Integrated Algebra course. In order to graph a line, two or three possible methods were explored. In this section, the only method that will be used is a table of values. For instance, for the functions $y = 2x + 3$ and $y = x^2 - 1$, you can create table(s) such as those on the next page.

x	$2x + 3 = y$
0	$2(0) + 3 = 0 + 3 = 3$
1	$2(1) + 3 = 2 + 3 = 5$
–2	$2(-2) + 3 = -4 + 3 = -1$

x	$x^2 - 1 = y$
0	$(0)^2 - 1 = 0 - 1 = -1$
1	$(1)^2 - 1 = 1 - 1 = 0$
–2	$(-2)^2 - 1 = 4 - 1 = 3$
–1	$(-1)^2 - 1 = 1 - 1 = 0$
2	$(2)^2 - 1 = 4 - 1 = 3$

It is important to determine as many ordered pairs as necessary to determine how they are connected on the graph. Based on the ordered pairs listed in the tables, here are the graphical representations for these two functions:

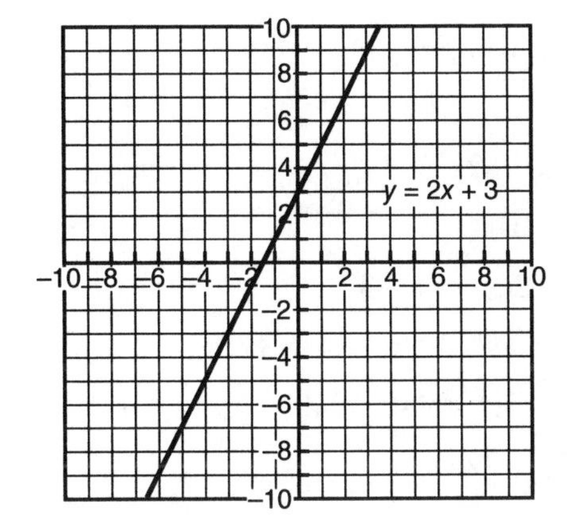

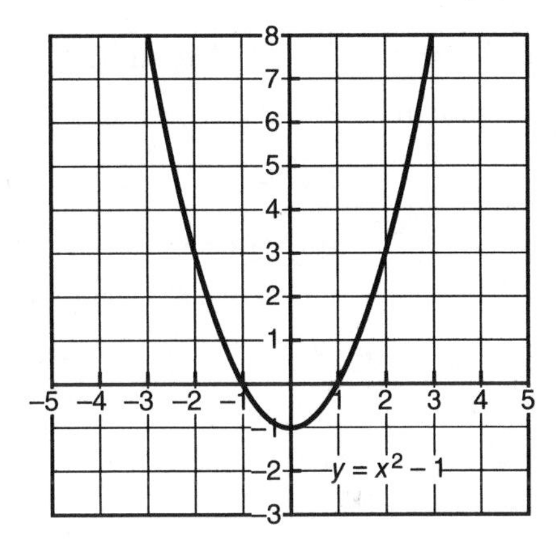

Functional Notation

There is a special notation or way to write a function. For the function $y = 2x + 3$ that was just graphed, the function can be written to show that y is a function of x by writing $f(x) = 2x + 3$. Usually an f, g, or h is used for the name of a function. However, any letter may be chosen for this notation. In this notation, it can be seen that the x-values or domain values map to $f(x)$, which are the y-values or range values. Look at the previous table of values for $y = 2x + 3$. From this table, it is possible to see that $f(0) = 3$, $f(1) = 5$, and $f(-2) = -1$. In addition, the function can be evaluated for any x-value in the domain of the function, which is all real numbers. For instance, $f(5) = 2(5) + 3 = 10 + 3 = 13$.

Example 1

If $f(x) = x^3 - 2x + 3$, determine $f(1)$, $f(3)$, and $f(-2)$.

Solution: $f(1) = 1^3 - 2(1) + 3 = 1 - 2 + 3 = 2$

$$f(3) = 3^3 - 2(3) + 3 = 27 - 6 + 3 = 24$$

$$f(-2) = (-2)^3 - 2(-2) + 3 = -8 + 4 + 3 = -1$$

A function can also be evaluated for different variables or expressions. For instance, the same function, $f(x) = 2x + 3$, can be evaluated for z or for $x + 2$. To do this, simply substitute the new variable or expression for the x-value in the function rule. So $f(z) = 2z + 3$ and $f(x + 2) = 2(x + 2) + 3 = 2x + 4 + 3 = 2x + 7$.

Example 2

If $f(x) = x^3 - 2x + 3$, determine $f(t)$, $f(x^2)$, and $f(z - 7)$.

Solution: $f(t) = t^3 - 2t + 3$

$$f(x^2) = (x^2)^3 - 2(x^2) + 3 = x^6 - 2x^2 + 3$$

$$\begin{aligned} f(z-7) &= (z-7)^3 - 2(z-7) + 3 \\ &= (z-7)^2(z-7) - 2(z-7) + 3 \\ &= (z^2 - 14z + 49)(z-7) - 2z + 14 + 3 \\ &= z^3 - 14z^2 + 49z - 7z^2 + 98z - 343 - 2z + 14 + 3 \\ &= z^3 - 21z^2 + 145z - 326 \end{aligned}$$

Notice that to raise the $z - 7$ to the third power, it was first squared and then multiplied by the third factor by the distributive law.

Domain and Range

The domain of a function is the set of pre-images or inputs (usually x-values) for the function. In other words, the domain is the set of suitable values for x. The range is the set of images or outputs (usually y-values) from the function. So the range is the set of suitable values of y. Examine the graph that follows.

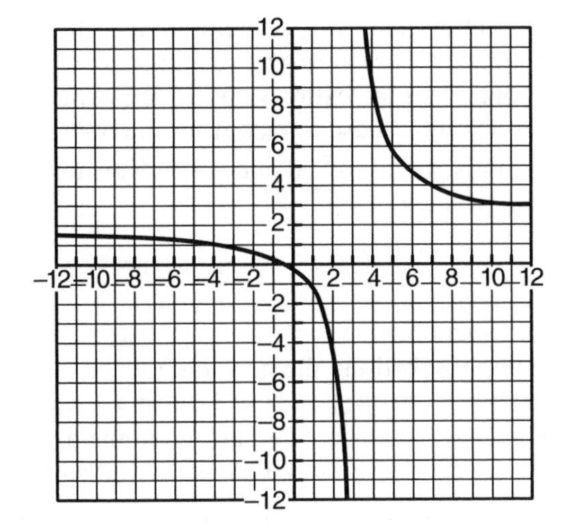

Observe that there is no x-value of 3 or y-value of 2. Therefore, the domain of this function is $\{x: x \neq 3\}$ and the range is $\{y: y \neq 2\}$.

Example 3

What are the domain and range of the function pictured below?

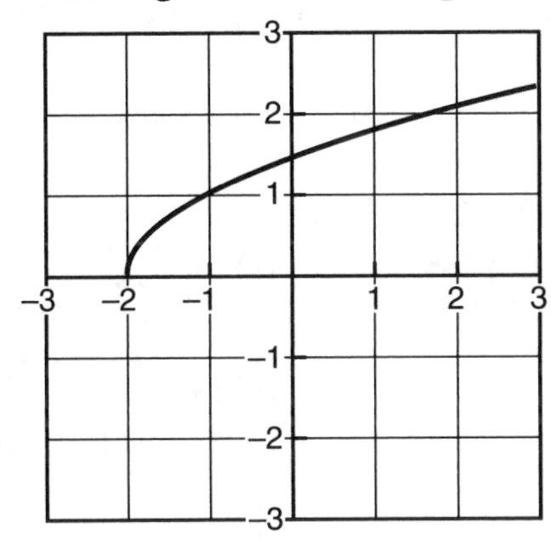

Solution: Domain = $\{x: x \geq -2\}$
Range = $\{y: y \geq 0\}$

Algebraically, determining the domain of a function really comes down to understanding what x-values cannot be used. For instance, division by 0 makes a fraction undefined. Therefore, if the function rule is in the form of a rational expression, set the denominator equal to 0 and exclude that answer from the domain of the function. So if $f(x) = \frac{3}{x+2}$, set $x + 2 = 0$ and solve. Now state that the domain is $\{x: x \neq -2\}$.

A second operation that makes an expression undefined is an even index root of a negative number. Therefore, if a function rule has a radical expression, solve to find out when the expression inside the radical (the radicand) is less than 0. Then exclude those values from the domain of the function. If $g(x) = \sqrt[4]{x-4}$, simply set $x - 4 < 0$ and solve. This results in $x < 4$. Now state that the domain is $\{x: x \geq 4\}$.

Example 4

What are the domain and range of $h(x) = \dfrac{1}{\sqrt{x-1}}$?

Solution: There are two concerns about the domain. The function contains a fraction as well as an even index radical. First solve $x - 1 < 0$ to get $x < 1$, which has to be excluded from the domain. Then, solve $\sqrt{x-1} = 0$. This yields $x - 1 = 0$ or $x = 1$. This also needs to be excluded from the domain. So the domain is $\{x: x > 1\}$.

Examine the range of this function to see what the elements in the domain map to. Perhaps the best way to do this would be to create a table of values on the graphing calculator for values of $x > 1$. Use the keystrokes shown below to enter the function: **Y =** **1** **÷** **2ND** **X^2** **X** **–** **1** **)**.

Then to set up the table, use these keystrokes: **2ND** **WINDOW** **1**

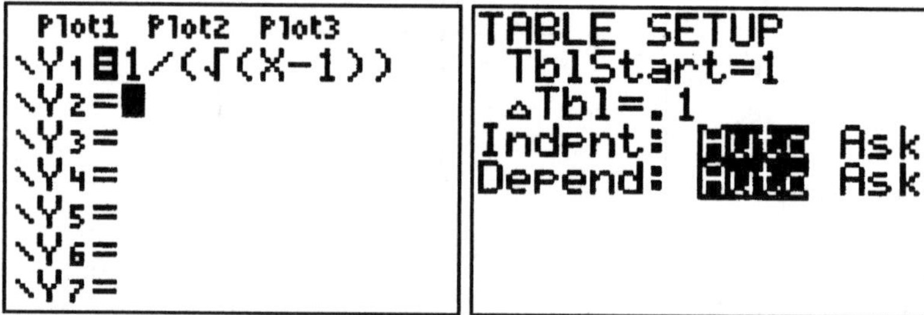

X	Y1
1	ERROR
1.1	3.1623
1.2	2.2361
1.3	1.8257
1.4	1.5811
1.5	1.4142
1.6	1.291

X=1

The error shown at $x = 1$ in the table supports the fact that $x = 1$ is not in the domain of the function. Let's use the TABLE ASK function of the calculator to get an even better idea of what is happening. Use these keystrokes: 2ND WINDOW ▼ ▼ ► ENTER 2ND GRAPH.

Now enter whatever numbers you want desirable to use to examine the behavior of this function. See the table entries below.

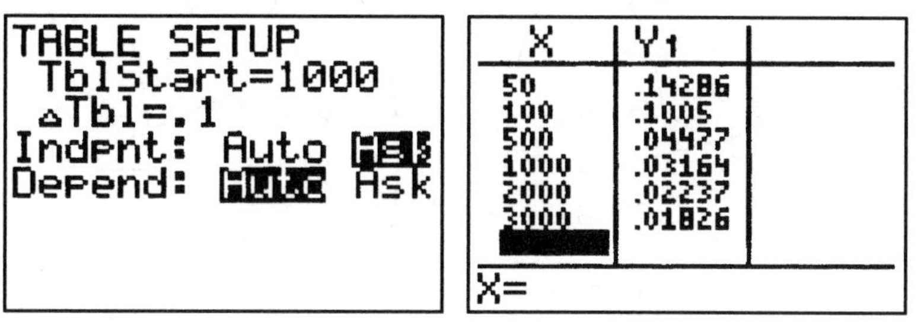

Notice that as x gets very large, y comes closer and closer to 0. Now it can be stated that the range is $\{y: y > 0\}$.

The Vertical Line Test

Perhaps the easiest way to determine that a relation is a function is by viewing the graph of the function. In order for a relation to be a function, each x-value has to map to a unique y-value. When viewing a graph of a function, a vertical line can intersect its graph at only one point. Below are graphs of two different relations. The first is for the relation $x = y^2$, and the second is for $y = x^2$. Notice that a vertical line intersecting the graph of $x = y^2$ intersects it in two points, while the graph of $y = x^2$ intersects it in only one point. Therefore, $x = y^2$ is not a function, and $y = x^2$ is a function.

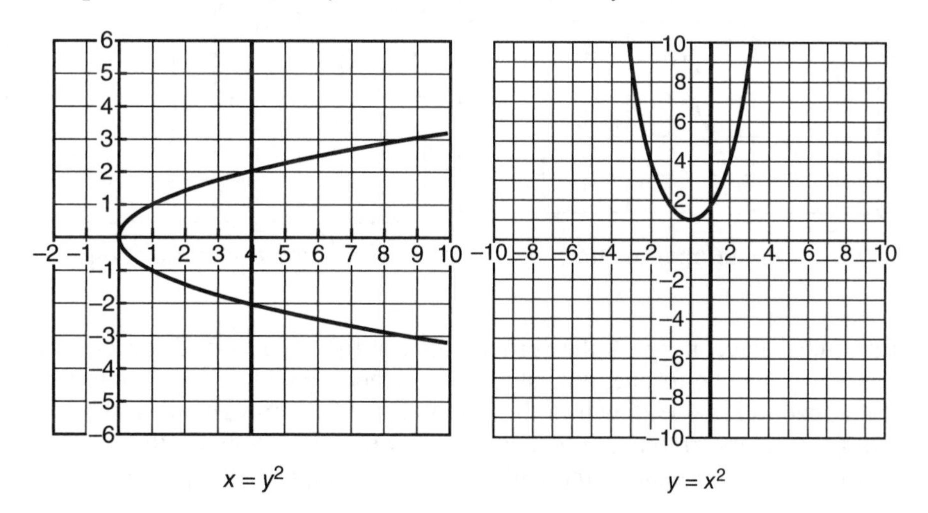

$x = y^2$ $y = x^2$

One-to-One and onto Functions

Some functions have special properties. A one-to-one function is a function such that each y-value comes from one and only one x-value. That combined with the fact that in any function each x-value has one and only one y-value means that whenever $f(a) = f(b)$, then $a = b$.

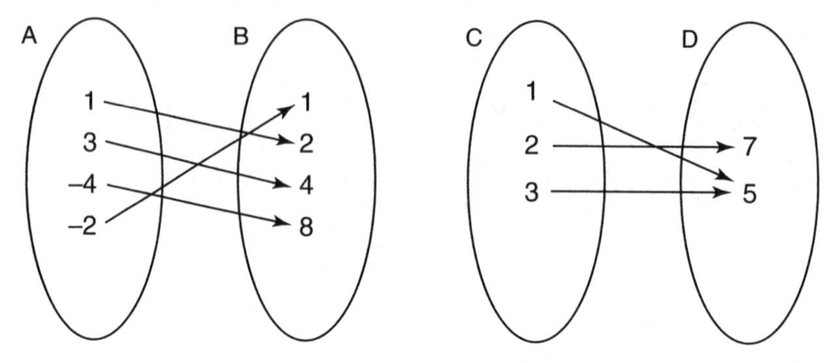

In the diagrams above, the function mapping set A to set B is a one-to-one function. The mapping from set C to set D is not one-to-one. It can even be said that the mapping from C to D is a two-to-one function.

Let's examine the graphs of two functions, one that is one-to-one and the other that is not.

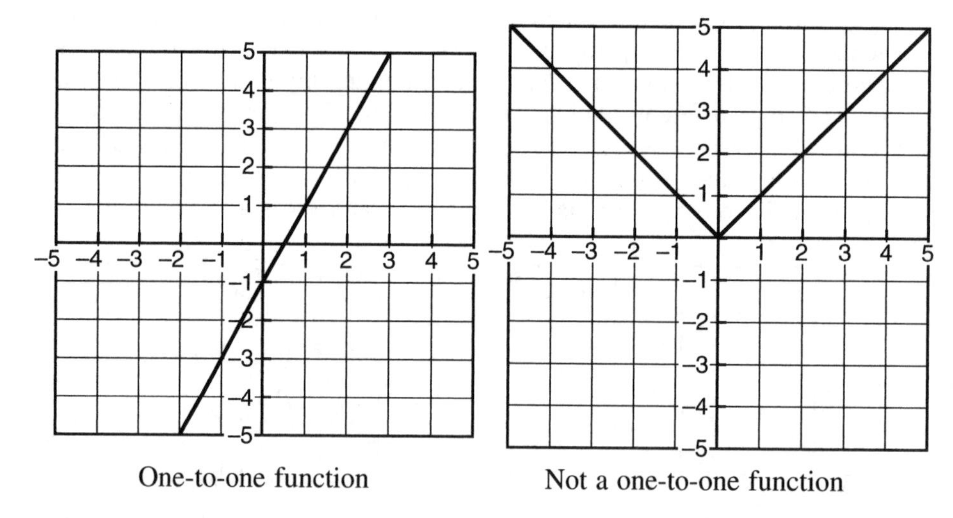

One-to-one function

Not a one-to-one function

Remember the concept of the vertical line test for a function? In order for a function to be one-to-one, it must also pass a horizontal line test. Notice that if a horizontal line is drawn in the graph on the left through any y-value, it will intersect the graph in only one point. Try the same thing for the diagram on the right. A horizontal line will intersect the graph in two places.

Another special classification of functions is called onto. A function is onto if each member of the outcome set or range has a pre-image or is actually mapped to a member of the domain or input set. In other words, every element in the outcome set is used. In the next set of diagrams, the mapping from *A* to *B* is onto while the mapping from *C* to *D* is not onto. The 9 does not correspond to any element in set *C*.

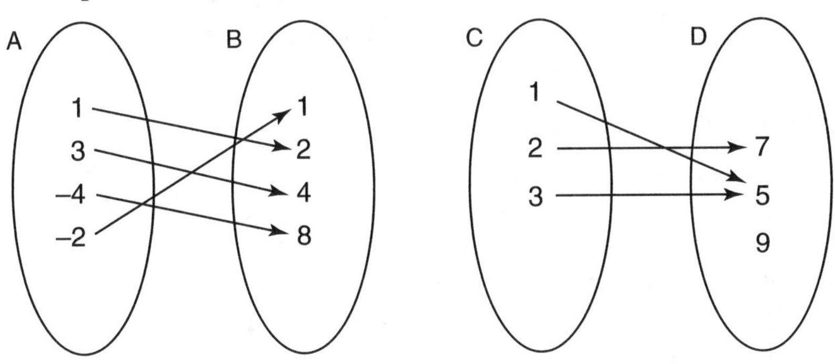

In fact, the mapping from set *A* to *B* is also one-to-one. It can be called a one-to-one onto function.

Examine the following graph. Assume that the function is a mapping from the set of all real numbers, $\mathcal{R}$, to all real numbers (f: $\mathcal{R} \rightarrow \mathcal{R}$).

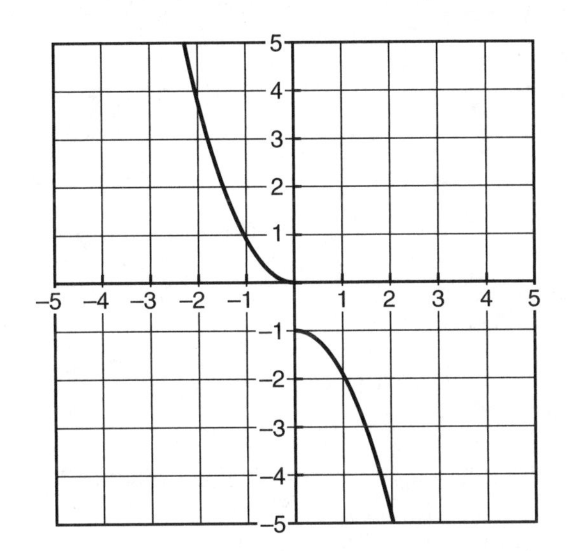

Notice that there are no *x*-values that map to *y*-values greater than –1 and less than 0. These *y*-values do not correspond to any value in the domain $\mathcal{R}$. Therefore, f is not onto because all the elements of the outcome set are not used.

Check Your Understanding of Section 2.2

1. Which of the following functions (f: $\mathcal{R} \to \mathcal{R}$) is one-to-one and onto?

(1) $f(x) = x^2$ (3) $f(x) = 5x - 3$
(2) $f(x) = \sqrt{x}$ (4) $f(x) = |x + 2|$

2. What is the domain of $g(x) = \dfrac{3}{x-5}$?

(1) $\{x: x \neq -5\}$ (3) $\{x: x > 5\}$
(2) $\{x: x \neq 5\}$ (4) $\{x: x \neq 3\}$

3. What is the range of $h(x) = |x + 4|$?

(1) $\{y: y > 0\}$ (3) $\{y: y \leq 4\}$
(2) $\{y: y > 4\}$ (4) $\{y: y \geq 0\}$

4. Which diagram below shows a function that is one-to-one but not onto?

(1)
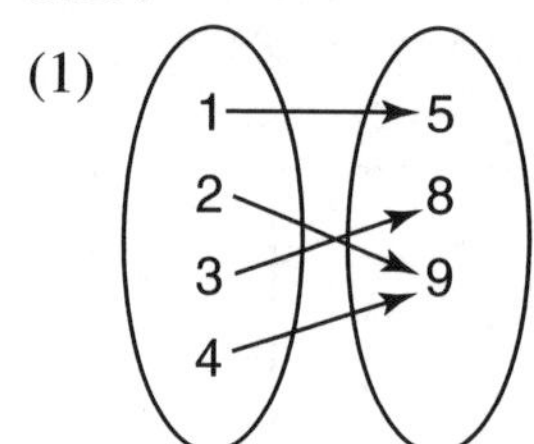

(3)
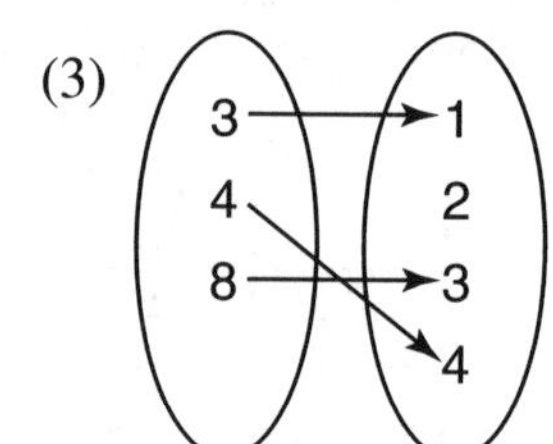

(2)
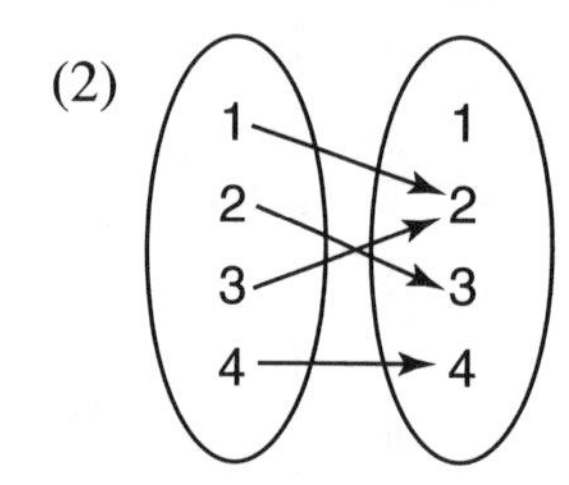

(4)
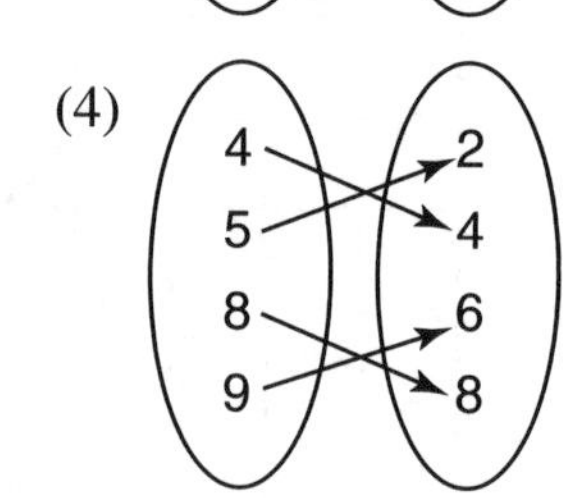

5. Which of the following mappings is not a function?

(1) $\{(3, -1), (4, -1), (2, 5), (3, 5)\}$
(2) $\{(3, -4), (4, -1), (2, 5), (-3, -5)\}$
(3) $\{(3, -1), (4, -2), (2, 5), (-1, 5)\}$
(4) $\{(3, -1), (4, -1), (-2, 5), (-3, 5)\}$

6. If $f(x) = \dfrac{x^2}{2x-5}$, determine $f(5)$.

(1) 5 (2) undefined (3) -5 (4) $\dfrac{1}{5}$

7. If $g(x)=3\sqrt{x-2}$, determine $g(t+3)$.

(1) $g(t+3)=6\sqrt{t-2}$
(2) $g(t+3)=3\sqrt{t+3}$
(3) $g(t+3)=3\sqrt{t+1}$
(4) $g(t+3)=6\sqrt{t+1}$

8. For the function $f(x)=\dfrac{2}{\sqrt{x+5}}$, what are the domain and range?

(1) Domain = $\{x: x > -5\}$ and Range = $\{y: y \geq 0\}$
(2) Domain = $\{x: x < -5\}$ and Range = $\{y: y < 0\}$
(3) Domain = $\{x: x > -5\}$ and Range = $\{y: y > 0\}$
(4) Domain = $\{x: x \leq -5\}$ and Range = $\{y: y < 0\}$

9. If $f(x) = (x - 5)^2$, evaluate $f(-2)$.

(1) -7 (2) -49 (3) 9 (4) 49

10. If $g(x) = x^2 - 3x + 4$, evaluate $g(-2)$.

(1) 2 (2) 14 (3) 11 (4) 6

11. If $h(x) = |3x| + 2$, which of the following is true?

(1) $h(x)$ is not a function.
(2) $h(x)$ is one-to-one and onto.
(3) $h(x)$ is one-to-one but not onto.
(4) $h(x)$ is neither one-to-one or onto.

12. $h(x)=\dfrac{(x-1)^3}{x^2}$, evaluate each of the following:

a. $h(1) =$
b. $h(-1) =$
c. $h(s) =$
d. $h(3) =$
e. $h\left(\dfrac{1}{3}\right)=$
f. $h(y + 1) =$

13. Graph the function $f(x) = 2x - 4$ and explain why this function is one-to-one and onto.

14. Given the function, $g(x) = |2x - 3|$, determine each of the following:

a. $g(1) =$ b. $g(2) =$ c. $g(0) + 2 =$
d. What is the domain of function g?
e. What is the range of g?
f. Is $g(x)$ one-to-one?
g. Is $g(x)$ onto?

15. Use the graph below to determine each of the following:

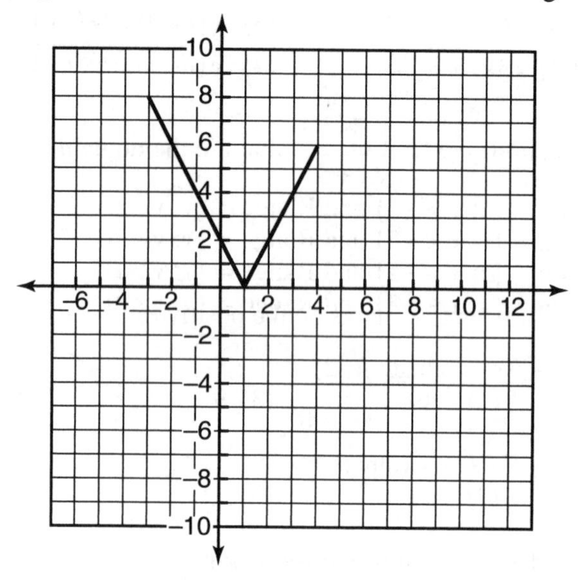

a. Is this a graph of a function or a relation? Explain your answer.
b. What is its domain?
c. What is its range?
d. Is it one-to-one? Explain your answer.

2.3 FUNCTIONS AND THE GRAPHING CALCULATOR

One example of how to view a function on a graphing calculator has already been investigated. Now let's examine viewing a function on the calculator in different ways that may give better views of what is happening in specific functions.

What Is the Standard Graphing Window?

The standard graphing window has x-values from -10 to $+10$ and y-values from -10 to $+10$. Notice, though, that the calculator screen is not a square. That means that your picture will be slightly distorted. The following pictures show the graph grid screen in a standard window as it is displayed on the graphing calculator and the WINDOW screen.

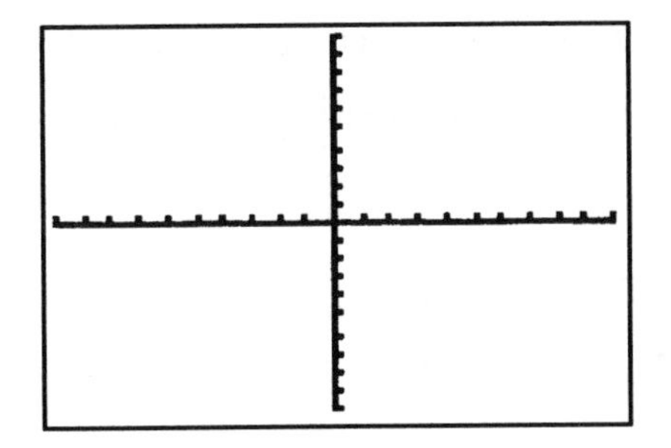

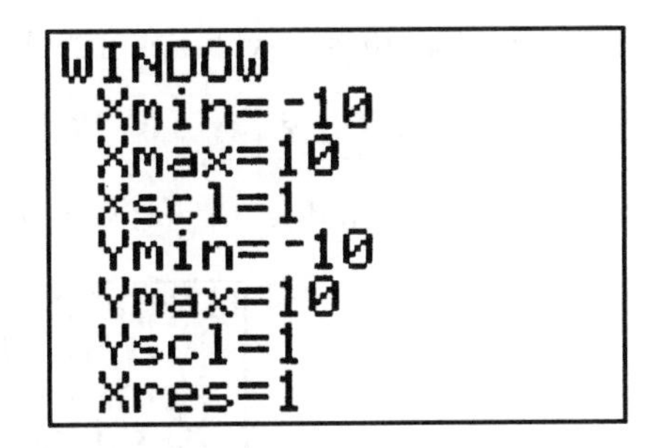

Notice that there are ten tick marks on the positive and negative axes of both the *x*-axis and the *y*-axis. These numbers are consistent with what the WINDOW indicates as the maximum and minimum *x*- and *y*-values.

Sometimes the standard viewing window is not the best window to use to view the graph of a particular function. For instance, in the function $f(x) = x^2 + 12$, the smallest *y*-value for this function is $y = 12$. This value is not even in the range of *y*-values that can be viewed in the standard window.

Enter **Y =** **x** **x^2** **+** **1** **2** **ENTER** **GRAPH**.

Now the graph displayed is a blank standard window. To see the graph, one option to try is called a Zoom Fit by entering **ZOOM** **0**. The resulting graph and the viewing window are displayed below.

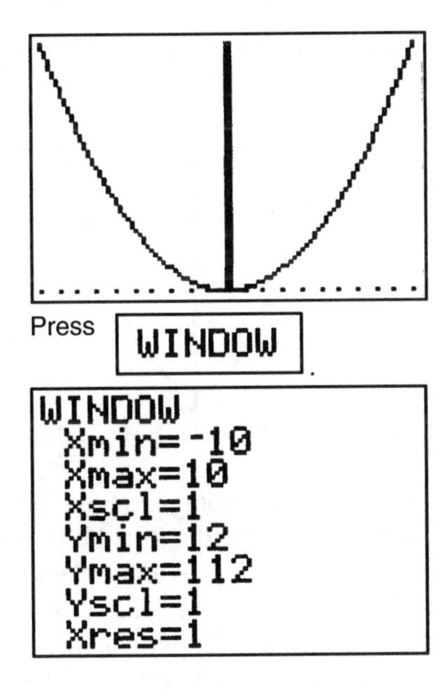

Since the smallest *y*-value is 12, the viewing window can be changed by pressing ▼ ▼ ▼ 1 0 ▼ 2 5 ▼ **GRAPH** to change the *y*-values in that window to go from 10 to 25.

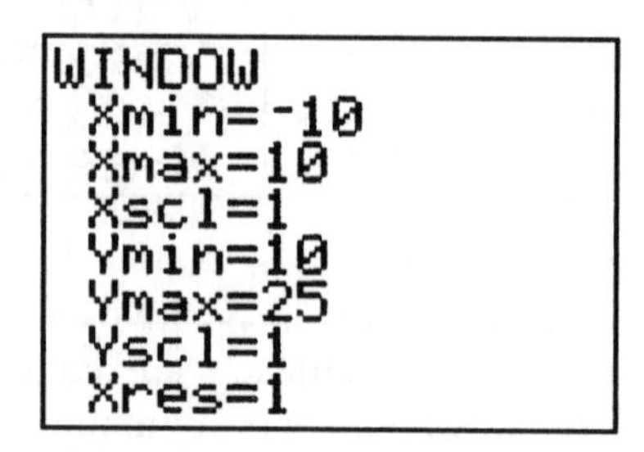

The graph now looks like:

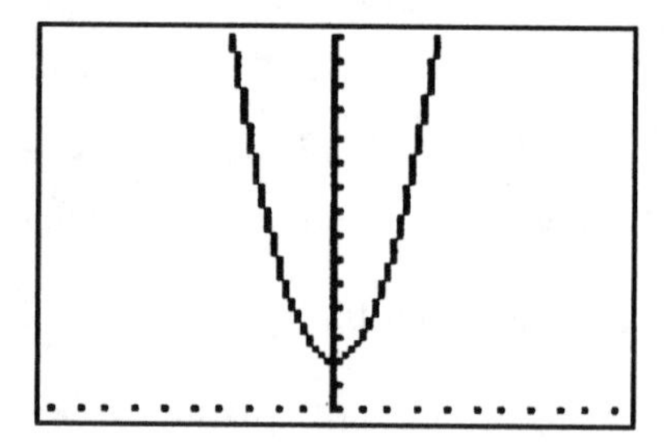

Change the window again, modifying the *x*-values to go from –6 to +6. The graph now looks like:

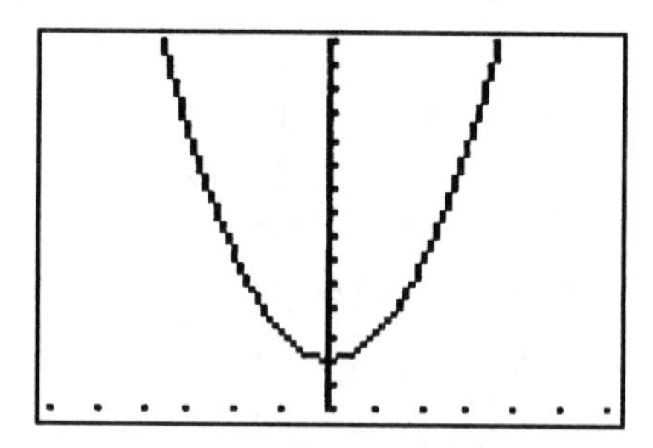

Notice that all three of the graphs displayed above for the same function are reasonable. When using these techniques to help graph a function on an examination, it will be necessary to provide a sketch of the function along with a description of the maximum and minimum *x*- and *y*-values used in that sketch. Usually that can be done by writing these values underneath the sketch of the graph showing [X Min, X Max] × [Y Min, Y Max]. For example, for the last calculator graph of $y = x^2 + 12$ above, it is appropriate to write $[-6, 6] \times [10, 25]$ or use the inequalities $-6 \leq x \leq 6$ and $10 \leq y \leq 25$.

The standard window displays a graph using the minimum x-value of −10, a maximum x-value of +10, a minimum y-value of −10, and a maximum y-value of +10. These values can be changed to get a better view of the graph by pressing **WINDOW** and scrolling to the values that need to be changed, entering the new values, and pressing **ENTER**. The number of tick marks on the axes can also be changed by entering new values in the Xscl and Yscl in **WINDOW**.

Another helpful thing that the graphing calculator can do is to create a table of values using the TABLE function on the calculator as examined in the previous section of this chapter. First determine the specific domain values that need to be used when creating a table. For instance, to see the table of values from $x = -3$ to $x = +3$, enter **2ND** **WINDOW** **(−)** **3** **▼** **1** **▼** **ENTER** and **2ND** **GRAPH**.
Now the table is displayed:

X	Y1	
-3	21	
-2	16	
-1	13	
0	12	
1	13	
2	16	
3	21	

X= -3

Notice that in this table, the apparent minimum y-value is 12. That assumes that the minimum y-value occurs at an integral value of x. This assumption is true for this example but may not be true for other examples. So either change the ΔTbl to a decimal value to check or do an Indpnt: Ask. An Indpnt: Ask can be used to choose whatever values of the function are being looking for. For instance, for the function $f(x) = x^2 + 12$, this process can be used to evaluate the function at specific x-values. The table that follows was generated to evaluate $f(x)$ at $x = 3$, $x = 5$, and $x = -2$. Start by setting up the table as follows on the next page.

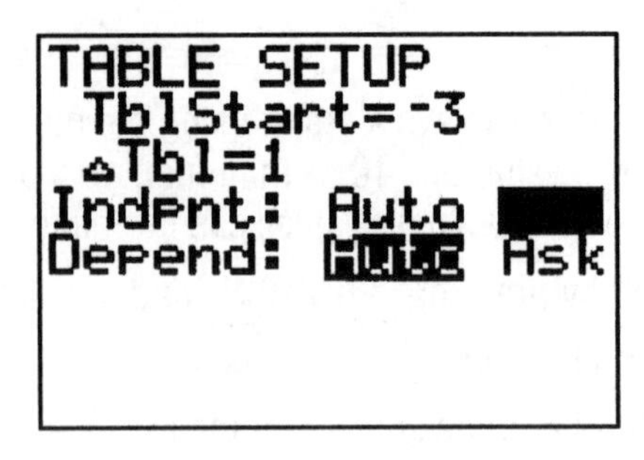

The resulting table is

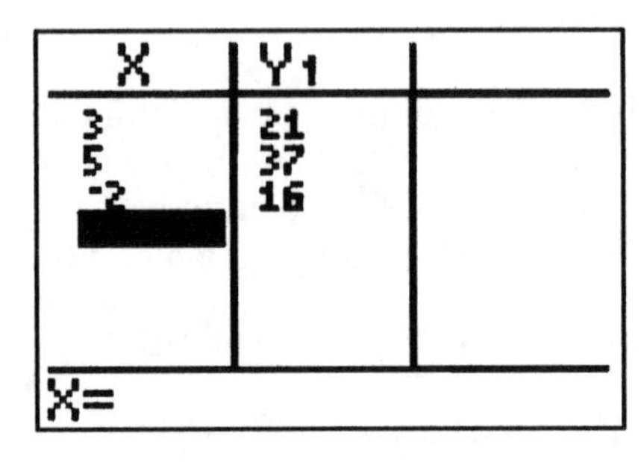

To get these x-values in the table, enter: **2ND** **GRAPH** **3** **ENTER** **5** **ENTER** **(–)** **2** **ENTER**.

Now let's examine the calculator-generated graph of $f(x)=\frac{2}{3}x+1$. First, enter the function as Y_1 in the calculator by pressing **Y =** **(** **2** **÷** **3** **)** **×** **+** **1**. Now go to the **GRAPH** button (making sure that zoom standard is selected) to see the graph displayed as follows:

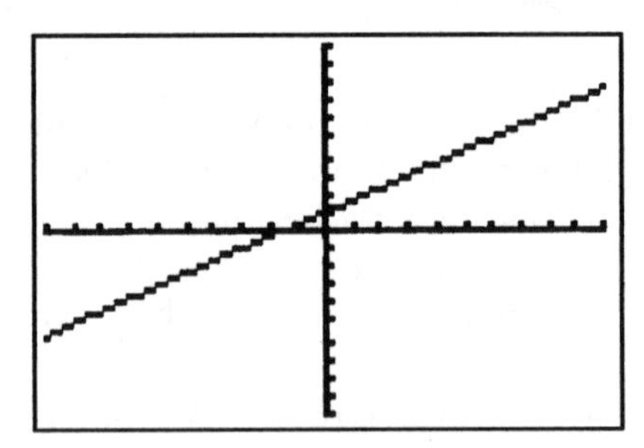

Now use the **ZOOM** button and choose choice 5: ZSquare to get the graph displayed below:

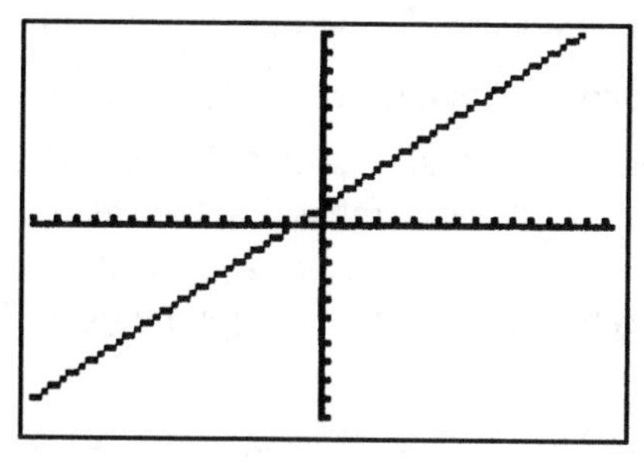

The window is

```
WINDOW
 Xmin=-15.16129...
 Xmax=15.161290...
 Xscl=1
 Ymin=-10
 Ymax=10
 Yscl=1
 Xres=1█
```

Change the window as follows:

```
WINDOW
 Xmin=-15.16129...
 Xmax=15.161290...
 Xscl=3
 Ymin=-10
 Ymax=10
 Yscl=3
 Xres=1
```

Now the graph display changes so that there are less tick marks on the axes. Specifically, there are only at multiples of 3.

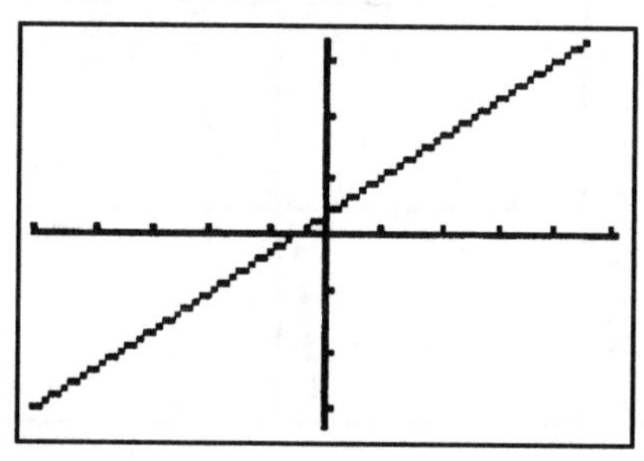

TIP

There are several different ways to change the viewing window in order to get the best view of a particular function. The default values can be changed in the WINDOW screen as previously described. It is also possible to choose one of ten different ways to zoom into the graph of a function. Remember that the calculator remembers the last viewing screen that was used, even if the calculator has been turned off. So to be sure, start any graph by using Zoom Standard or ZOOM 6.

There are ten choices in the ZOOM screen of the calculator. Seven of these choices come up as soon as ZOOM is pressed on the calculator:

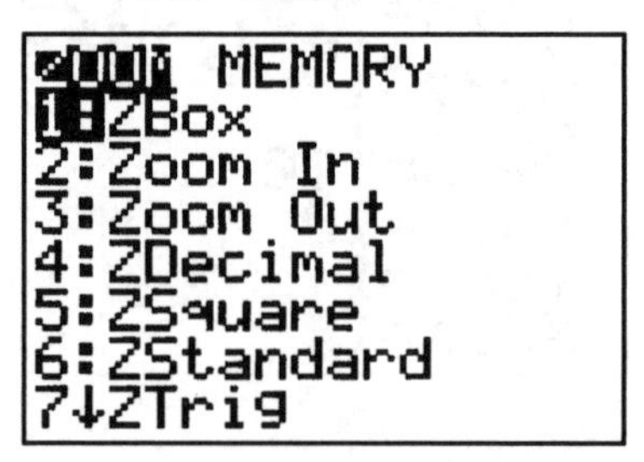

Let's explore some of the other Zoom choices with the function $f(x) = (x - 1)^3 + 2$. First enter the function as Y_1 on the calculator by pressing

Now press GRAPH. The graph is displayed in a standard window on the calculator as long as the standard window was used the last time you displayed a graph. If the standard window is not displayed, press ZOOM 6, which is ZStandard.

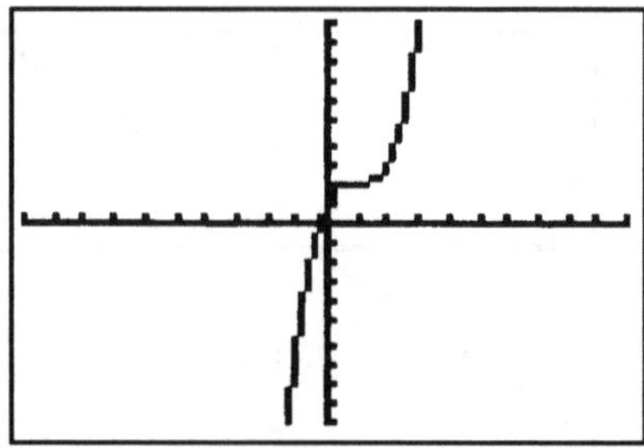

The Zoom choices of ZFit and Zsquare have already been shown to work. To do a Zoom box (ZBox), press **ZOOM** **1**. Now use the arrow keys to get to a position on the graph display to start drawing the box, and press **ENTER**. Then use arrow up or down from that starting point. Go in one direction and press **ENTER** when it is as far in the desired direction. Then move the arrow up or down and go as far as you want, and press **ENTER**. This is an example of what will be seen:

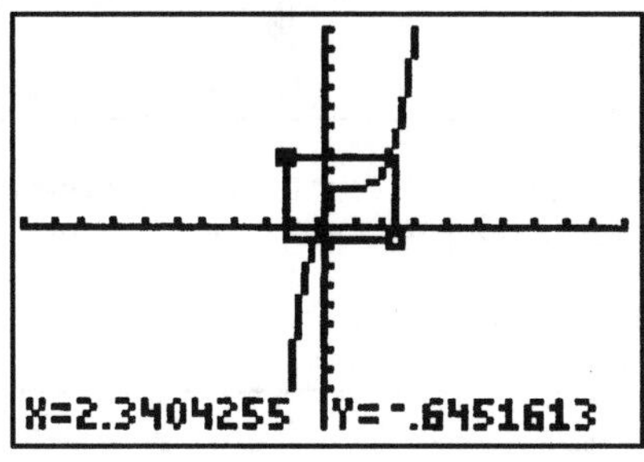

Pressing **ENTER** again causes the display to look like:

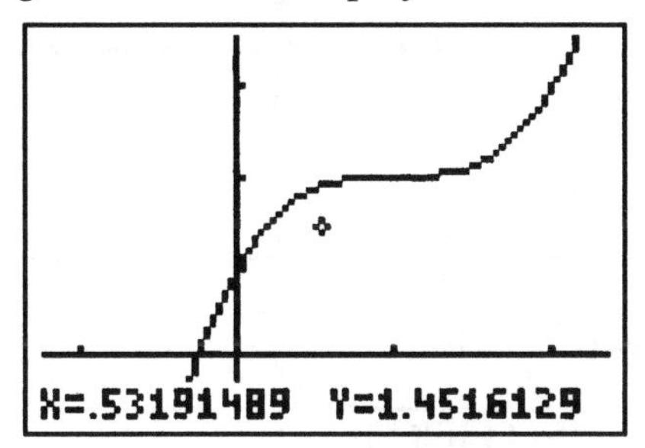

As can be seen, the inside of the box created is displayed on the calculator screen.

Zoom In focuses in closer on the graph, like a microscope. From the current display, press **ZOOM** **6** **ZOOM** **2** **ENTER** to get:

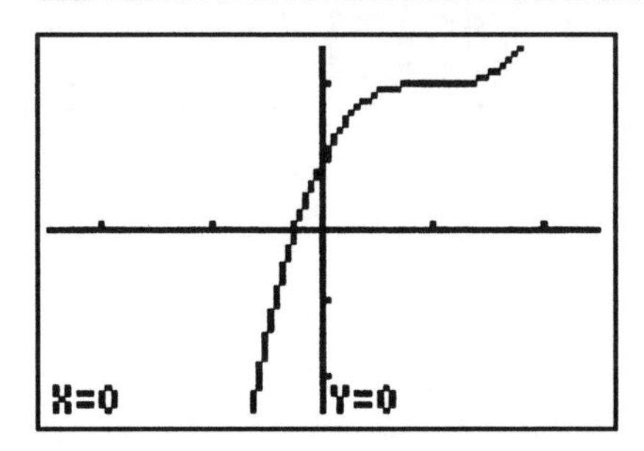

Zoom Out magnifies the view of the graph.

Press **ZOOM** **6** **ZOOM** **3** **ENTER**. Now the calculator screen looks like:

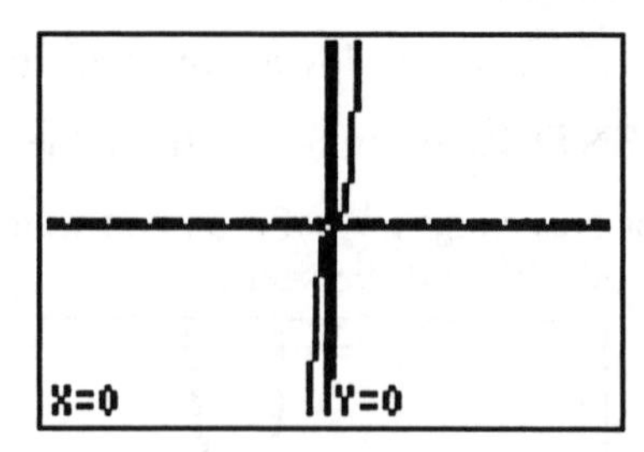

Zoom decimal (ZDecimal) adjusts for the dimensions of the individual pixels on the screen. It squares each pixel rather than using a rectangular pixel. The view from **ZOOM** **4** is

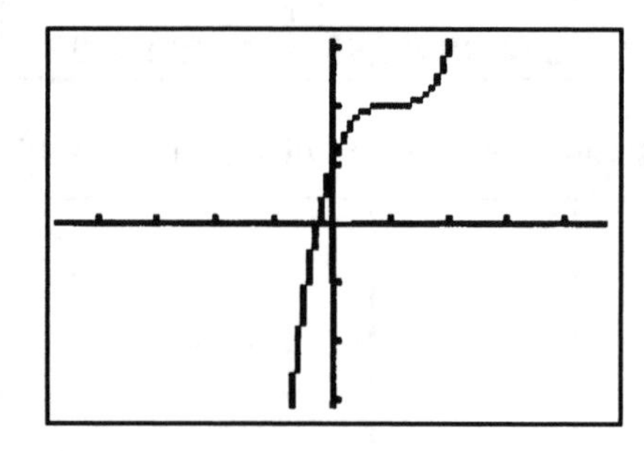

Its viewing window is

```
WINDOW
 Xmin=-4.7
 Xmax=4.7
 Xscl=1
 Ymin=-3.1
 Ymax=3.1
 Yscl=1
 Xres=1
```

Zoom integer (ZInteger) has a viewing window of

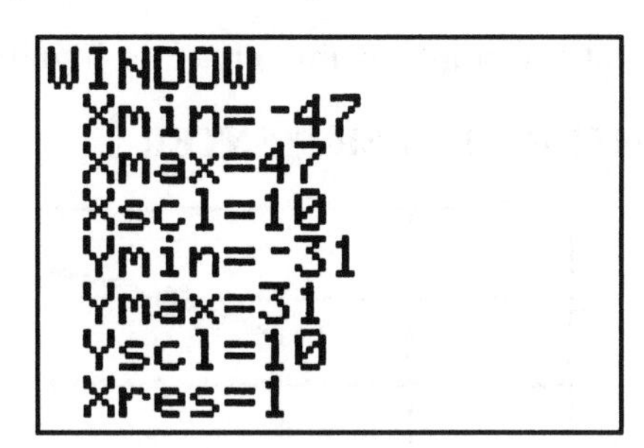

To get ZInteger, press **ZOOM** **8** **ENTER** to get the graph:

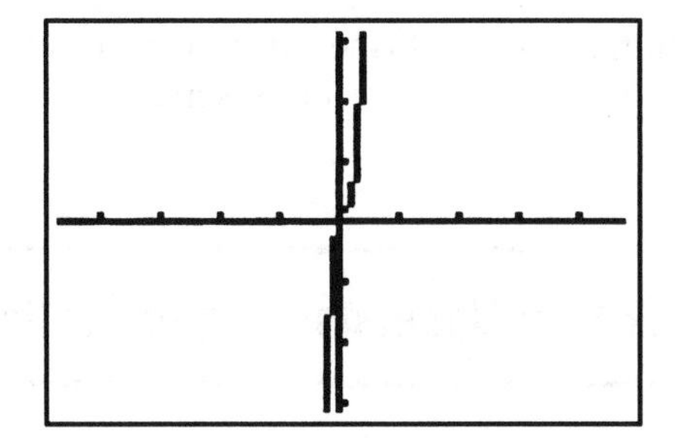

Each Zoom choice has its own benefit. There is no reason in this chapter to examine ZTrig or ZStat as these options will be examined when they are needed in later chapters.

One other thing that can be done with the graphing calculator is to solve equations. In Section 1-6, the solution of $\sqrt{x} + \sqrt{x-5} = 5$ was shown. To use the calculator to solve this, enter the left-hand side of the equation, $\sqrt{x} + \sqrt{x-5}$ as Y_1 and enter 5 as Y_2. Press **Y =** **2ND** **x^2** **x** **)** **+** **2ND** **x^2** **x** **–** **5** **)** **ENTER** **5** **ENTER** and then **GRAPH**. Since the graphs of the two functions intersect at nearly the end of the screen, modify the window to graph the functions from $x = -6$ to $+14$ for a better view. This graph is shown:

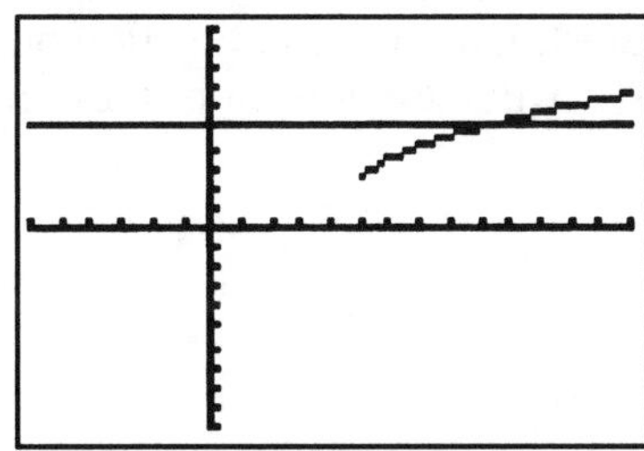

To determine the solution to this equation, just trace the graph of $y_1 = \sqrt{x} + \sqrt{x-5}$ until a point on this curve is highlighted. Press **2ND** **TRACE** **5** **ENTER** **ENTER** **ENTER**.

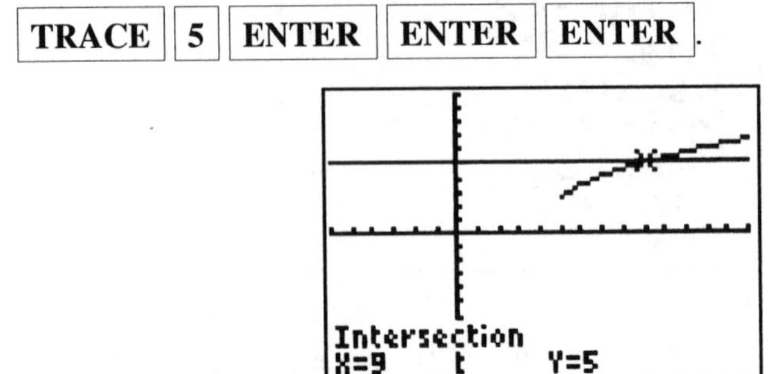

Notice that the solution of $x = 9$ (the same answer obtained in Example 16 on page 67) is displayed on the calculator screen.

Check Your Understanding of Section 2.3

1. Use your calculator to graph $f(x) = -|x + 3|$. Then, based on your graph or a combination of a graph and a table view, determine the domain and range of $f(x)$.

2. Use your calculator to solve for x if $2x + 5 = 14 - x$.

3. Use your calculator to solve for x if $\frac{x+1}{x-3} = 2$.

4. Anastacia wants to invest \$20,000 for three years. She knows that the formula for interest is $I = prt$, where I = the interest, p = the amount invested, in this case \$20,000, r = the rate of interest in decimal form, and t = time in years, in this case 3. She sees that she can invest her money in different types of accounts, one at 2% interest, one at 2.4% interest, and one at 3.1% interest. Use the table function on your calculator to see how much interest she will earn at each of these rates.

Chapter Three TYPES OF FUNCTIONS

3.1 LINEAR FUNCTIONS

KEY IDEAS

A linear function is a function whose graph is a line. There are several forms that define a linear function. The most general form is $ax + by = c$.

What Is a Linear Function?

The standard form of a linear equation is $ax + by = c$ where a, b, and c are constants and the variables x and y are both first degree. Solving a linear equation for y will show how to express the equation as a function. Subtracting ax from both sides of the equation results in $by = -ax + c$. Then dividing by b gets $y = \frac{-a}{b}x + \frac{c}{b}$. This transformation is probably not a form of the equation of a line that looks extremely familiar. However, it should look vaguely familiar as it is similar to the slope-intercept form of the equation of a line, $y = mx + b$. Compare the slope-intercept form of the equation of a line to the equation above. Notice that the slope is $m = \frac{-a}{b}$ and the y-intercept, b, is $\frac{c}{b}$. In this slope-intercept form, $y = mx + b$, the equation can stated as a function $f(x) = mx + b$.

From the standard form of the equation of a line, $ax + by = c$, the intercepts of the graph of the function (where the graph crosses the axes) can be determined. For instance, if $3x + 4y = 24$, then the x-intercept can be determined by setting $y = 0$. This yields $3x = 24$ or $x = 8$. So the graph of the line crosses the x-axis when $x = 8$. Setting $x = 0$ yields the y-intercept. This means that $4y = 24$ can be solved to give $y = 6$, so $y = 6$ is the y-intercept.

From the slope-intercept form of the equation of a line, the slope and the x-intercept can be determined. For example, if $y = \frac{-3}{4}x + 6$, the slope is $\frac{-3}{4}$ and the y-intercept is 6. The equation $3x + 4y = 24$ is the equation of the same line as $y = \frac{-3}{4}x + 6$. To check, solve for y when $3x + 4y = 24$.

The linear equation $3x + 4y = 24$ can be examined in functional notation as $f(x) = \frac{-3}{4}x + 6$. This function can be graphed by several approaches that were learned in Integrated Algebra. Evaluate the function at $x = 0$, 4, and -4:

$$f(0) = \frac{-3}{4} \cdot 0 + 6 = 6$$

$$f(4) = \frac{-3}{4} \cdot 4 + 6 = -3 + 6 = 3$$

$$f(-4) = \frac{-3}{4} \cdot (-4) + 6 = 3 + 6 = 9$$

Furthermore, using the TABLE function of the graphing calculator can show more ordered pairs that fit this function. For instance,

results in

X	Y1
-4	9
-2	7.5
0	6
2	4.5
4	3
6	1.5
8	0

X= -4

Here is a graph of this function on a standard viewing screen:

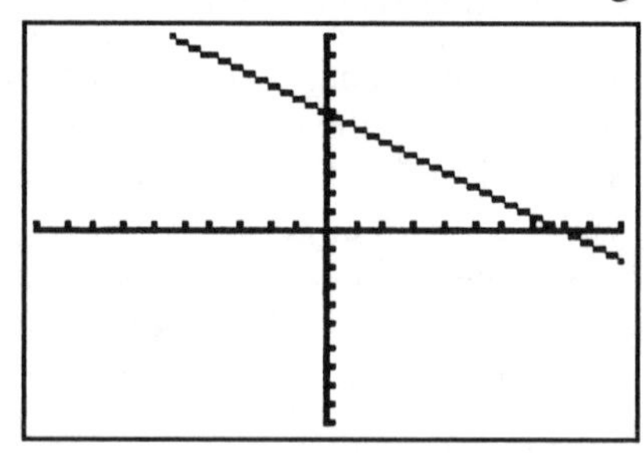

Notice that since a linear function has a graph that is a line, it passes the vertical line test and is a function. It also passes the horizontal line test, so it is one-to-one. Notice that the domain of this function is $\mathcal{R}$ and the range is $\Re$. Therefore, this function is also onto. In fact, every linear function, except $y = c$ where c is a constant, is one-to-one and onto with domain $\Re$ and range $\Re$. A function of the form $f(x) = c$ has a graph that is a horizontal line. It does not pass the horizontal line test, so it is not one-to-one and it certainly is not onto.

MATH FACTS

A linear function can be stated by using the slope-intercept form of a linear equation, $f(x) = mx + b$. Both the domain and range of a linear function whose slope is unequal to 0 is the set of real numbers, $\Re$. A linear function whose slope is unequal to 0 is one-to-one and onto.

Check Your Understanding of Section 3.1

In questions 1–3, rewrite each of the following linear equations by solving for y and expressing them in functional notation:

1. $5x - 5y = 60$ **2.** $3x + 2y = 12$ **3.** $x - 4y = 16$

4. Evaluate $f(x) = 2x - 5$ at
a. $x = 2$ b. $x = 5$ c. $x = -3$ d. $x = 10$

5. Which of the following functions is a linear function?

(1) $f(x) = 3x^3 - 5$

(2) $f(x) = \sqrt{2x+1}$

(3) $f(x) = \dfrac{2}{3x-4}$

(4) $f(x) = \dfrac{2}{3}x - 4$

3.2 ABSOLUTE VALUE FUNCTIONS

KEY IDEAS

An absolute value function is a function whose functional rule contains an absolute value. An example of an absolute value function is $f(x) = |2x - 5|$.

Functions Containing an Absolute Value

Recall that the absolute value of a number measures how far that number is from 0. Consider the absolute value function, $f(x) = |2x - 5|$. Examine the table of values:

x	$f(x) = |2x - 5|$	$g(x) = 2x - 5$
−1	$|2 \cdot -1 - 5| = |-2 - 5| = |-7| = 7$	$2 \cdot -1 - 5 = -2 - 5 = -7$
0	$|2 \cdot 0 - 5| = |0 - 5| = |-5| = 5$	$2 \cdot 0 - 5 = 0 - 5 = -5$
1	$|2 \cdot 1 - 5| = |2 - 5| = |-3| = 3$	$2 \cdot 1 - 5 = 2 - 5 = -3$
2	$|2 \cdot 2 - 5| = |4 - 5| = |-1| = 1$	$2 \cdot 2 - 5 = 4 - 5 = -1$
3	$|2 \cdot 3 - 5| = |6 - 5| = |1| = 1$	$2 \cdot 3 - 5 = 6 - 5 = 1$
4	$|2 \cdot 4 - 5| = |8 - 5| = |3| = 3$	$2 \cdot 4 - 5 = 8 - 5 = 3$
5	$|2 \cdot 5 - 5| = |10 - 5| = |5| = 5$	$2 \cdot 5 - 5 = 10 - 5 = 5$

As expected, the numerical values found at each x-value for $f(x)$ is the absolute value of the numerical value for the same x-value in $g(x)$. Below is a graph of $f(x)$.

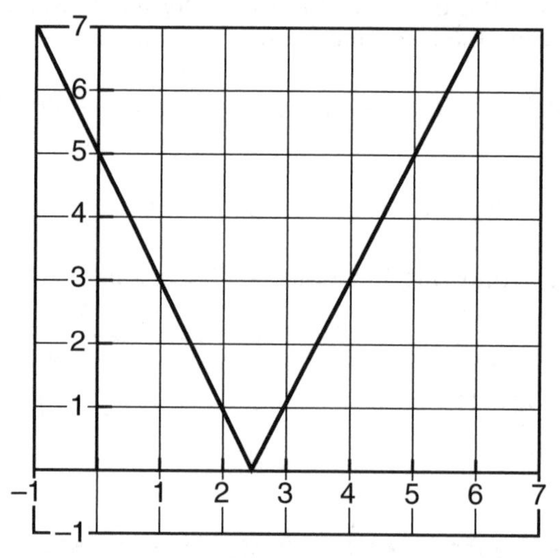

Something that should be clear from this graph is that the range of $f(x)$ is $\{y: y \geq 0\}$. Clearly, the domain of $f(x)$ is $\mathcal{R}$.

Solving Absolute Value Equations

Think about it. What needs to be done to solve an equation such as $|2x - 5| = 3$? Graphically, the solution is easy to see when $x = 1$ or $x = 4$ because a horizontal line intersects the graph of this function at $y = 3$. In order to solve this equation for other values, it is important to expand the definition of absolute value.

MATH FACTS

$$\text{Definition: } f(x) = |x| = \begin{cases} x \text{ if } x \geq 0 \\ -x \text{ if } x < 0. \end{cases}$$

With this definition, the algebraic method to solve this same absolute value equation should be clear. $|2x - 5| = 3$ can be solved by recognizing that $|2x - 5|$ means $2x - 5$ when $2x - 5 \geq 0$ or $x \geq 5/2$. Also see that $-(2x - 5) = -2x + 5$ when $2x - 5 < 0$ or $x < 5/2$.

Therefore, either $2x - 5 = 3$ when $x \geq 5/2$ or $-2x + 5 = 3$ when $x < 5/2$.

$2x = 8$ | $-2x = -2$

$x = 4$ and $4 \geq 5/2$ | $x = 1$ and $1 < 5/2$

So $x = 4$ or $x = 1$, which are the answers found from the graph.

Alternately, it may be easier to set up the two conditions based on the fact that when $2x - 5 \geq 0$, then $2x - 5 = 3$ and when $2x - 5 < 0$, then $2x - 5 = -3$. The same two solutions will then be determined.

Example 1

Solve for x: $|4x + 3| = 7$

Solution:

Either $4x + 3 = 7$ when $4x + 3 \geq 0$ or $4x + 3 = -7$ when $4x + 3 < 0$.

$4x = 4$ when $4x \geq -3$ or $4x = -10$ when $4x < -3$

$x = 1$ when $x \geq -3/4$ or $x = -10/2 = -5/2$ when $x < -3/4$

and $1 \geq -3/4$ and $-5/2 < -3/4$

So $x = 1$ or $x = -5/2$ both solve this equation.

Example 2

Solve for z: $|4 - z| = 3$

Solution: If $4 - z \geq 0$ or $4 \geq z$

$4 - z = 3$

$-z = -1$

$z = 1$

If $4 - z < 0$ or $4 < z$

$4 - z = -3$

$-z = -7$

$z = 7$

The solution set is $\{z: z = 1 \text{ or } z = 7\}$.

Example 3

Solve for y: $|5y + 3| = -2$

Solution: One way of solving this is to recognize that the absolute value of any number must be positive, therefore there is no solution. Alternatively, set up the two conditions:

If $5y + 3 \geq 0$
or $y \geq -3/5$
then
$5y + 3 = -2$
$5y = -5$
$y = -1$
-1 does not satisfy the condition that $y \geq -3/5$. Therefore, there is no solution.

If $5y + 3 < 0$
or $y < -3/5$
then
$-5y - 3 = -2$
$-5y = 1$
$y = -1/5$
$-1/5$ does not satisfy the condition that $y < -3/5$. Therefore, there is no solution.

Example 4

Solve for x: $|x - 3| = |2x + 1|$

Solution: To solve this equation, it necessary to recognize that both absolute value expressions have two possibilities and each of these needs to be matched with the two conditions for the other side of the equation. So four conditions actually need to be solved.

When $x - 3 \geq 0$ or $x \geq 3$ and $2x + 1 \geq 0$ or $x \geq -1/2$, which means that $x \geq 3$	$x \geq 3$ and $x < -1/2$, which invalidate this case since there are no common x-values.	$x < 3$ and $x \geq -1/2$, which means that $-1/2 \leq x < 3$	$x < 3$ and $x < -1/2$, which means that $x < -1/2$
$x - 3 = 2x + 1$ $-4 = x$, which is not a number in the above condition. No solution.	No solution.	$-x + 3 = 2x + 1$ $-3x = -2$ $x = 2/3$, which is a number in the above-stated condition.	$-x + 3 = -2x - 1$ $x = -4$, which is a number in the above-stated condition.

The solution is $\{x: x = 2/3 \text{ or } x = -4\}$.

Solving Absolute Value Inequalities

In order to understand how to solve absolute value inequalities, it is easiest to use the following definitions.

MATH FACTS

Definition: Assume that a is a positive number.
If $|x| < a$, then $-a < x < a$, and similarly, if $|x| \leq a$, then $-a \leq x \leq a$.
If $|x| > a$, then $x > a$ or $x < -a$, and similarly, if $|x| \geq a$, then $x \geq a$ or $x \leq -a$.

Example 5

Solve for x: $|x - 7| < 5$

Solution: $|x - 7| < 5$ implies that $-5 < x - 7 < 5$.

$$2 < x < 12$$

The solution set is $\{x: 2 < x < 12\}$.

Example 6

Solve for x: $|2x + 1| \geq 9$

Solution: $|2x + 1| \geq 9$ implies that $2x + 1 \geq 9$ or $2x + 1 \leq -9$.

$$2x \geq 8 \quad \text{or} \quad 2x \leq -10$$

$$x \geq 4 \quad \text{or} \quad x \leq -5$$

The solution set is $\{x: x \geq 4 \text{ or } x \leq -5\}$.

Check Your Understanding of Section 3.2

1. What is the solution set for $|3x| = 12$?
(1) $\{x: x = 4\}$
(2) $\{x: x = -4\}$
(3) $\{x: x = 4 \text{ or } x = -4\}$
(4) $\{x: x = 3 \text{ or } x = -3\}$

2. If $|x + 5| = 8$, then x may be
(1) -13 (2) 4 (3) 8 (4) -8

3. If $|y - 7| < 4$, then y may be
(1) 3 (2) 5 (3) 11 (4) 13

4. If $|12 + z| = -2$, then
(1) z is negative (2) z is positive
(3) there is no solution for z (4) $z = 0$

5. What is the solution set for $|3x - 5| \geq 8$?

(1) $\{x: x \leq -1\}$ (3) $\left\{x: x \geq \frac{-13}{3}\right\}$

(2) $\left\{x: x \leq -1 \text{ or } x \geq \frac{13}{3}\right\}$ (4) $\left\{x: x \geq -1 \text{ or } x \leq \frac{13}{3}\right\}$

6. Which graph represents the solution set for $|2x + 1| \leq 5$?

(1)
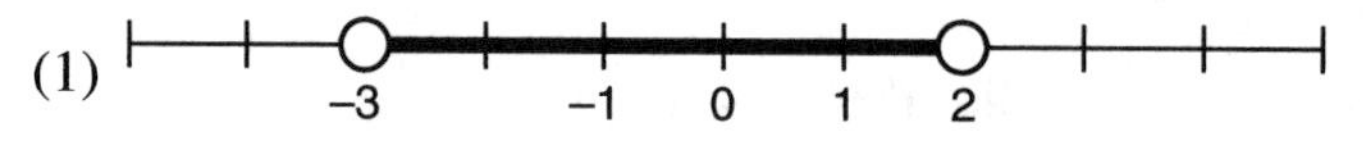

(2)
–3 –1 0 1 2

(3)
–3 –1 0 1 2

(4)
–3 –1 0 1 2

7. Solve for x: $|4x - 3| + 2 = 7$

8. Solve for y: $|18 - 3y| = 9$

9. Solve for x: $|3x + 2| = |x - 4|$

10. Solve for z: $|6z - 3| + 5 \leq 8$ and sketch the solution on a number line.

11. Solve for c: $|15 - 4c| \geq 19$

12. The average cost for a hamburger platter at Joe's Hamburger Palace is \$9.25. The manager wants to keep the price for the hamburger platter within \$1.05 of the current price as he plans a new menu. Write an absolute value inequality that he can use to model the price for his new menu. Then solve this inequality for the range of acceptable charges he can list for the hamburger platter.

13. The manager at Prime Grocery Store set the price for 10 cans of soup at \$28.90. She knows that her cost for 10 cans varies by \$8.00 each time she restocks. Write an absolute value inequality that she can use to model the price for a single can of soup in the future. Use that inequality to find a range of prices she would charge for a single can of soup.

3.3 QUADRATIC FUNCTIONS

A quadratic function is a function that can be written in the form, $f(x) = ax^2 + bx + c$, where $a \neq 0$. In Integrated Algebra, quadratic equations were solved by factoring and using the Zero Product Property.

Quadratic Functions

If $f(x) = x^2 - 5x + 2$, it can be evaluated as can any other function, by substituting different x-values into the function rule. For instance,

$$\begin{aligned} f(2) &= 2^2 - 5 \cdot 2 + 2 \\ &= 4 - 10 + 2 \\ &= -6 + 2 \\ &= -4 \end{aligned}$$

However,

$$\begin{aligned} f(-2) &= (-2)^2 - 5 \cdot (-2) + 2 \\ &= 4 - (-10) + 2 \\ &= 14 + 2 \\ &= 16 \end{aligned}$$

Notice that it is important to recall the difference between squaring -2, which yields $(-2)^2 = 4$, and taking the additive inverse of 2 squared or -2^2, which squares only the 2 and yields -4.

Example 1

Evaluate $f(x) = x^2 + 2x - 3$ for each integer from $x = -5$ to $x = 3$. Use that information to sketch the graph of $f(x)$ on the interval $-5 < x < 3$.

Solution:

x	$f(x) = x^2 + 2x - 3$
−5	$f(-5) = (-5)^2 + 2(-5) - 3 = 25 - 10 - 3 = 12$
−4	$f(-4) = (-4)^2 + 2(-4) - 3 = 16 - 8 - 3 = 5$
−3	$f(-3) = (-3)^2 + 2(-3) - 3 = 9 - 6 - 3 = 0$
−2	$f(-2) = (-2)^2 + 2(-2) - 3 = 4 - 4 - 3 = -3$
−1	$f(-1) = (-1)^2 + 2(-1) - 3 = 1 - 2 - 3 = -4$
0	$f(0) = 0^2 + 2(0) - 3 = -3$
1	$f(1) = 1^2 + 2(1) - 3 = 1 + 2 - 3 = 0$
2	$f(2) = 2^2 + 2(2) - 3 = 4 + 4 - 3 = 5$
3	$f(3) = 3^2 + 2(3) - 3 = 9 + 6 - 3 = 12$

Notice that the y-values have a symmetric pattern. The graph of a quadratic function is a parabola. (This was established in Integrated Algebra.) Below is a graph of this function:

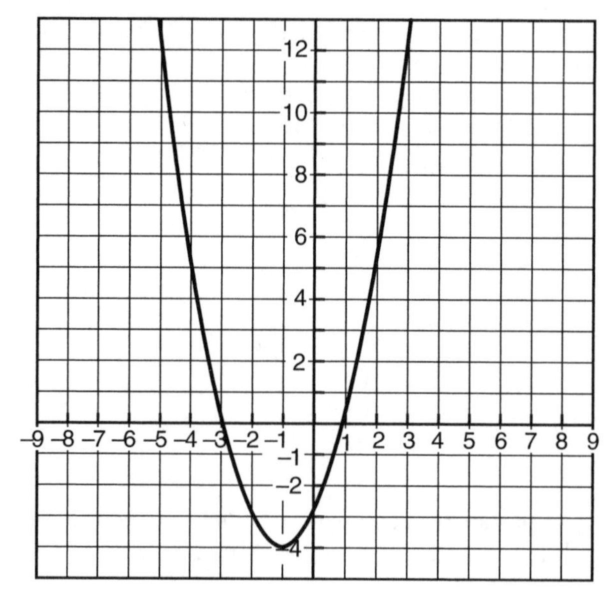

Notice that this graph passes the vertical line test, so it is a function. However, it does not pass the horizontal line test, so it is not one-to-one. In fact, all quadratic functions are not one-to-one. Similar to the absolute value function, not all the y-values in $\mathcal{R}$ are mapped, so a quadratic function is also not onto.

Solving Quadratic Equations Using the Zero Product Property

Back in Section 1.5, the Zero Product Property was reviewed to solve quadratic equations. One more look at this follows in the next example.

Example 2

Solve for x: $x^2 + 3x - 28 = 0$

Solution: $x^2 + 3x - 28 = 0$

Factoring $(x + 7)(x - 4) = 0$

Apply the Zero Product Property $x + 7 = 0$ or $x - 4 = 0$

$x = -7$ or $x = 4$

The Zero Product Property is not the only approach that can be used to solve a quadratic equation. There are some specific situations that do not need this property. For example, such an example includes a quadratic equation that is an incomplete quadratic equation where the quadratic expression in the equation has either b or c equal to 0. When $b = 0$, there is another approach that can be applied.

Example 3

Solve for x: $3x^2 = 75$

Solution: $3x^2 = 75$

Divide by 3 $x^2 = 25$

Take the square root of both sides of the equation $x = \pm 5$

Solving Quadratic Equations by Completing the Square

To understand how to complete the square of a trinomial, first recall what it means to have a perfect square trinomial. To do that, square the expression $x + a$: $(x + a)^2 = x^2 + 2ax + a^2$.

To square a binomial, it is necessary to square the first term, double the product of the first and last terms, and then square the last term. Notice that taking half of the middle term's coefficient and then squaring it is equal to the last term. Imagine an expression such as $x^2 + 12x$. To find the constant that would make this a perfect square trinomial, recognize that half of 12 is 6 and $6^2 = 36$. Therefore the expression $x^2 + 12x + 36$ is a perfect square trinomial that factors as $(x + 6)^2$.

Now let's revisit Example 2, $x^2 + 3x - 28 = 0$. Since the trinomial on the left side of this equation is not a perfect square trinomial, first add the 28 to both sides of the equation, yielding $x^2 + 3x = 28$. Now complete the square on the left side of the equation. Half of 3 is $\frac{3}{2}$ and $\left(\frac{3}{2}\right)^2 = \frac{9}{4}$. So add $\frac{9}{4}$ to both sides of this equation to get $x^2 + 3x + \frac{9}{4} = 28 + \frac{9}{4}$. Now factor the left side to get $\left(x + \frac{3}{2}\right)^2 = \frac{121}{4}$. Take the square root of both sides of this equation to get $x + \frac{3}{2} = \pm\frac{11}{2}$. So $x = \frac{-3}{2} \pm \frac{11}{2} = \begin{cases} \frac{8}{2} = 4 \\ \frac{-14}{2} = -7 \end{cases}$. These are the same two answers found in Example 2 by applying the Zero Product Property.

Example 4

Solve for x: $x^2 - 8x + 9 = 0$

Solution: $x^2 - 8x + 9 = 0$

$$x^2 - 8x = -9$$

Half of -8 is -4 and $(-4)^2 = 16$

$$x^2 - 8x + 16 = -9 + 16$$

$$x^2 - 8x + 16 = 7$$

$$(x - 4)^2 = 7$$

$$x - 4 = \pm\sqrt{7}$$

$$x = 4 \pm \sqrt{7} = 4 + \sqrt{7} \text{ or } 4 - \sqrt{7}$$

Example 5

Solve for y: $6y^2 - y - 15 = 0$

Solution: $6y^2 - y - 15 = 0$

Divide by 6 $y^2 - \frac{1}{6}y - \frac{15}{6} = 0$

Add $\frac{15}{6}$ to both sides $y^2 - \frac{1}{6}y - \frac{15}{6}$

$\frac{1}{2}$ coefficient of middle term squared added to both sides of the equation

$$y^2 - \frac{1}{6}y + \frac{1}{144} = \frac{15}{6} + \frac{1}{144}$$

$$y^2 - \frac{1}{6}y + \frac{1}{144} = \frac{360}{144} + \frac{1}{144}$$

$$y^2 - \frac{1}{6}y + \frac{1}{144} = \frac{361}{144}$$

$$\left(y - \frac{1}{12}\right)^2 = \frac{361}{144}$$

$$y - \frac{1}{12} = \pm\frac{19}{12}$$

$$y = \frac{1}{12} \pm \frac{19}{12}$$

$$y = \frac{20}{12} \text{ or } -\frac{18}{12}$$

$$y = \frac{5}{3} \text{ or } -\frac{3}{2}$$

Now let's examine the general quadratic equation, $ax^2 + bx + c = 0$, and apply the completing the square method to that.

$$ax^2 + bx = -c$$

$$\frac{ax^2}{a} + \frac{bx}{a} = -\frac{c}{a}$$

$$ax^2 + \frac{bx}{a} = -\frac{c}{a}$$

$$x^2 + \frac{bx}{a} + \frac{b^2}{4a^2} = -\frac{c}{a} + \frac{b^2}{4a^2}$$

$$x^2 + \frac{bx}{a} + \frac{b^2}{4a^2} = -\frac{4ac}{4a^2} + \frac{b^2}{4a^2}$$

$$x^2 + \frac{bx}{a} + \frac{b^2}{4a^2} = \frac{b^2 - 4ac}{4a^2}$$

$$\left(x + \frac{b}{2a}\right)^2 = \frac{b^2 - 4ac}{4a^2}$$

$$x + \frac{b}{2a} = \pm\sqrt{\frac{b^2 - 4ac}{4a^2}}$$

$$x+\frac{b}{2a}=\pm\frac{\sqrt{b^2-4ac}}{2a}$$

$$x=\frac{-b}{2a}\pm\frac{\sqrt{b^2-4ac}}{2a}$$

$$x=\frac{-b\pm\sqrt{b^2-4ac}}{2a}$$

This final result is called the quadratic formula.

MATH FACTS

To solve a quadratic function in x set equal to 0, $ax^2 + bx + c = 0$, first identify the values of a, b, and c and substitute these into the quadratic formula, $x=\frac{-b\pm\sqrt{b^2-4ac}}{2a}$.

Example 6

Solve for x: $6x^2 - x - 15 = 0$

Solution: Notice that this is essentially the same equation that was solved in Example 5. In this problem, $a = 6$, $b = -1$, and $c = -15$. Now substitute these values into the quadratic formula:

$$x=\frac{-b\pm\sqrt{b^2-4ac}}{2a}$$

$$x=\frac{-(-1)\pm\sqrt{(-1)^2-4(6)(-15)}}{2(6)}$$

$$x=\frac{1\pm\sqrt{1-24(-15)}}{12}$$

$$x=\frac{1\pm\sqrt{1+360}}{12}$$

$$x=\frac{1\pm\sqrt{361}}{12}$$

$$x=\frac{1\pm 19}{12}$$

$$x = \frac{20}{12} \text{ or } -\frac{18}{12}$$

$$x == \frac{5}{3} \text{ or } -\frac{3}{2}$$

Remember that a quadratic equation can also result from an equation that starts out as a fractional equation.

Example 7

Solve for x: $\frac{5}{x} + \frac{1}{x+2} = 2$

Solution: $\frac{5}{x} + \frac{1}{x+2} = 2$

$$\frac{5}{\cancel{x}} \cdot \cancel{x}(x+2) + \frac{1}{\cancel{x+2}} \cdot x(\cancel{x+2}) = 2 \cdot x(x+2)$$

$5(x + 2) + x = 2x(x + 2)$

$5x + 10 + x = 2x^2 + 4x$

$6x + 10 = 2x^2 + 4x$

$0 = 2x^2 - 2x - 10$

$0 = x^2 - x - 5$ (dividing both sides of the equation by 2)

$a = 1$, $b = -1$, and $c = -5$

$$x = \frac{-1 \pm \sqrt{(-1)^2 - 4(1)(-5)}}{2(1)} = \frac{-1 \pm \sqrt{1+20}}{2} = \frac{-1 \pm \sqrt{21}}{2}$$

Check Your Understanding of Section 3.3

1. Solve for x by the completing the square method: $6x^2 + x - 5 = 0$

(1) $\left\{\frac{-5}{6}, -1\right\}$ (2) $\left\{\frac{5}{6}, -1\right\}$ (3) $\left\{\frac{5}{3}, \frac{-1}{2}\right\}$ (4) $\left\{\frac{-5}{3}, \frac{1}{2}\right\}$

2. Solve for z using the quadratic formula: $16z^2 - 15 = 8z$

(1) $\left\{\frac{-3}{4}, \frac{-5}{4}\right\}$ (2) $\left\{\frac{-5}{4}, \frac{3}{2}\right\}$ (3) $\left\{\frac{-5}{4}, \frac{3}{4}\right\}$ (4) $\left\{\frac{5}{4}, \frac{-3}{4}\right\}$

3. Solve for y by any method: $y^2 + 6y + 3 = 0$

(1) $y = -3 \pm \sqrt{6}$

(2) $y = \dfrac{-6 \pm \sqrt{60}}{2}$

(3) $y = -3 \pm 2\sqrt{6}$

(4) $y = -6 \pm \sqrt{6}$

4. Solve the quadratic equation $v^2 + 7v = 2$

(1) $v = \dfrac{7 \pm \sqrt{57}}{6}$

(2) $v = \dfrac{-1}{3}$ or $v = -2$

(3) $v = \dfrac{7 \pm \sqrt{57}}{2}$

(4) $v = \dfrac{-7 \pm \sqrt{57}}{2}$

5. Solve for p: $3p^2 + 12 = 14p$

6. Solve for x: $\dfrac{x}{x+1} - \dfrac{2}{3x+5} = 4$

7. Solve for y: $\dfrac{y}{y-2} + \dfrac{2}{y+2} = \dfrac{8}{y^2-4}$

8. A ball is propelled upward from the ground according to the function $h(t) = -16t^2 + 80t$ where t represents the time in seconds after the ball leaves the ground. After how many second(s) will the ball be 36 feet above the ground?

3.4 SOLVING A SYSTEM OF LINEAR AND QUADRATIC EQUATIONS

The graph of a quadratic function is a parabola, and the graph of a linear function is a line. A line intersects a parabola in 0, 1, or 2 points. Since the number of common points of intersection is the same as the number of solutions of the system of equations, there are 0, 1, or 2 solutions to be found when solving a linear-quadratic system of equations.

Solving a System of Equations Algebraically

Do you remember how to solve a system of two linear equations? One method is commonly referred to as the substitution technique. For instance, when $y - 3x = 4$ and $2x + y = 19$, the first equation can be solved to show that $y = 3x + 4$. By replacing the y in the second equation with $3x + 4$, the result is $2x + 3x + 4 = 19$ or $5x + 4 = 19$. Subtract 4 from both sides of the equation to get $5x = 15$ or $x = 3$. Now replace x by 3 in the first equation to get $y - 9 = 4$ or $y = 13$. So the intersection of these two lines is at (3, 13), which is the solution for this system of equations.

Solving a Linear-Quadratic System of Equations Algebraically

What if one of the equations represents a quadratic function? The same basic approach can be applied.

Example 1

Determine the solution for the system of equations: $y = x^2 + 4$ and $y = x + 6$

Solution: Replace $x + 6$ for y in $y = x^2 + 4$.

$$x + 6 = x^2 + 4$$

$$0 = x^2 - x - 2$$

$$0 = (x - 2)(x + 1)$$

$$x - 2 = 0 \text{ or } x + 1 = 0$$

$$x = 2 \text{ or } x = -1$$

If $x = 2$ and $y = x + 6$, then $y = 2 + 6 = 8$ by substitution. If $x = -1$ and $y = x + 6$, then $y = -1 + 6$ or $y = 5$. There are two ordered pair solutions, (2, 8) and (–1, 5). Let's examine a check for these solutions. Each ordered pair must be replaced into both equations.

Check (2, 8)

$y = x^2 + 4$ and $y = x + 6$

$8 \stackrel{?}{=} 2^2 + 4$ $\quad$ $8 \stackrel{?}{=} 2 + 6$

$8 \stackrel{?}{=} 4 + 4$ $\quad$ $8 = 8$ ✓

$8 = 8$ ✓

Check (–1, 5)

$y = x^2 + 4$ and $y = x + 6$

$5 \stackrel{?}{=} (-1)^2 + 4$ $\quad$ $5 \stackrel{?}{=} -1 + 6$

$5 \stackrel{?}{=} 1 + 4$ $\quad$ $5 = 5$ ✓

$5 = 5$ ✓

This same problem can be solved using a graphing calculator. First enter the functions into the calculator. Press [Y=] [X] [^] [2] [+] [4] [ENTER] [X] [+] [6] [2ND] [TRACE] [5]. The screen now shows a blinking X on the parabola and asks, "First curve?" Scroll with the arrow key until the blinking X is near the intersection of the parabola and the line and then press **ENTER** . Now the display shows the blinking X on the line and asks, "Second curve?" The blinking X should be near the intersection (if not scroll near the intersection). Once there, press **ENTER** again. The calculator says, "Guess?" Press **ENTER** again. Now the display shows the intersection is at $x = -1$ and $y = 5$ (one of the two algebraic solutions). This is what is displayed on the calculator:

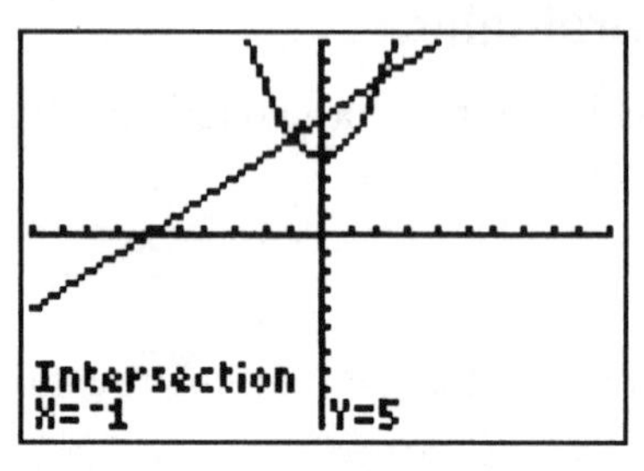

To get the other solution, repeat the process above but scroll close to the other intersection between the parabola and the line.

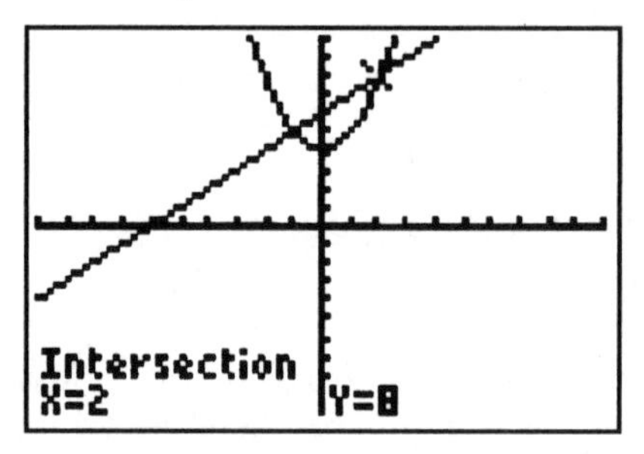

TIP

Caution! When solving a linear-quadratic system of equations, be careful not to solve for the squared variable and then substitute. If this is done, an extraneous root may be found.

Look again at the system $y = x^2 + 4$ and $y = x + 6$ to see what can happen if this tip is not followed. Since $y = x^2 + 4$, $x^2 = y - 4$ and $x = \pm\sqrt{y-4}$. Replacing $\pm\sqrt{y-4}$ for x in $y = x + 6$ results in $y = \pm\sqrt{y-4} + 6$ or $y - 6 = \pm\sqrt{y-4}$. Squaring both sides of this equation yields $y^2 - 12y + 36 = y - 4$. So $y^2 - 13y + 40 = 0$ and $(y - 8)(y - 5) = 0$ to give $y = 8$ or $y = 5$. This result does not cause a problem in and of itself. However, should these solutions for y be substituted into the first equation, $y = x^2 + 4$, the result will provide two answers for x for each value of y. If $y = 8$, then $8 = x^2 + 4$ so that $x^2 = 4$ and $x = \pm 2$. Similarly if $y = 5$, $5 = x^2 + 4$ and $x^2 = 1$, yielding $x = \pm 1$. The solution now appears to be (1, 5), (−1, 5), (2, 8) and (−2, 8). However, (1, 5) and (−2, 8) do not check in the second equation, $y = x + 6$, and are extraneous roots.

MATH FACTS

To solve a linear-quadratic system of equations:

1. Solve the linear equation for one of the variables in terms of the other.
2. Substitute the expression found in Step 1 for that variable in the quadratic equation to eliminate one variable.
3. Solve the resulting quadratic equation for the variable. That provides two solutions for the variable that is left in this equation.
4. Separately, replace both solutions from the quadratic equation in Step 2 into the linear equation and solve for the other variable. That gives two ordered pair solutions to the system of equations.
5. Always remember to check both ordered pair solutions using both equations!

Example 2

Determine the solution for the system of equations: $y = x^2 + 3x - 5$ and $y = 4x + 1$

Solution: Replace $4x + 1$ for y in the equation $y = x^2 + 3x - 5$.

$$4x + 1 = x^2 + 3x - 5$$

$$0 = x^2 - x - 6$$

$$0 = (x - 3)(x + 2)$$

$$x - 3 = 0 \text{ or } x + 2 = 0$$

$$x = 3 \text{ or } x = -2$$

If $x = 3$ and $y = 4x + 1$

$$y = 4 \cdot 3 + 1$$

$$y = 12 + 1$$

$$y = 13$$

If $x = -2$ and $y = 4x + 1$

$$y = 4(-2) + 1$$

$$y = -8 + 1$$

$$y = -7$$

Solutions: (3, 13) and (−2, −7)

Check (3, 13)

$y = 4x + 1$

Check (−2, −7)

$y = 4x - 1$

$13 \stackrel{?}{=} 4 \cdot 3 + 1$

$13 \stackrel{?}{=} 12 + 1$

$13 = 13$ ✓

$-9 \stackrel{?}{=} 4(-2) - 1$

$-9 \stackrel{?}{=} -8 - 1$

$-9 = -9$ ✓

Check (3, 13)

$y = x^2 + 3x - 5$

$13 \stackrel{?}{=} (3)^2 + 3 \cdot 3 - 5$

$13 \stackrel{?}{=} 9 + 9 - 5$

$13 = 13$ ✓

Check (–2, –7)

$y = x^2 + 3x - 5$

$-7 \stackrel{?}{=} (-2)^2 + 3(-2) - 5$

$-7 \stackrel{?}{=} 4 - 6 - 5$

$-7 = -7$ ✓

Example 3

Determine the solution for the system of equations: $y = 3x^2 - 2x - 7$ and $6y = 3x - 10$

Solution: First solve $6y = 3x - 10$ for y by dividing through by 6. That yields $y = \frac{1}{2}x - \frac{5}{3}$. Now substitute the expression $\frac{1}{2}x - \frac{5}{3}$ for y in $y = 3x^2 - 2x - 7$.

$$\frac{1}{2}x - \frac{5}{3} = 3x^2 - 2x - 7$$

$$0 = 3x^2 - \frac{5}{2}x - \frac{16}{3}$$

$$0 = 18x^2 - 15x - 32$$

The resulting quadratic expression is not factorable. Use the quadratic formula to solve for x.

$$x = \frac{-(-15) \pm \sqrt{(-15)^2 - 4(18)(-32)}}{2(18)}$$

$$x = \frac{15 \pm \sqrt{225 + 2304}}{36}$$

$$x = \frac{15 \pm \sqrt{2529}}{36}$$

$$x = \frac{15 \pm 3\sqrt{281}}{36} = \frac{\cancel{3}\left(5 \pm \sqrt{281}\right)}{\cancel{3} \cdot 12} = \frac{5 \pm \sqrt{281}}{12}$$

Now it is these two x-values can be substituted into the equation $6y = 3x - 10$ to solve for the corresponding y-values.

$$6y = 3x - 10 \text{ and } x = \frac{5+\sqrt{281}}{12}$$

$$6y = \cancel{3}\left(\frac{5+\sqrt{281}}{\cancel{12}_4}\right) - 10$$

$$6y = \frac{5+\sqrt{281}}{4} - 10$$

$$6y = \frac{5+\sqrt{281}}{4} - \frac{40}{4}$$

$$6y = \frac{-35+\sqrt{281}}{4}$$

$$y = \frac{-35+\sqrt{281}}{24}$$

$$6y = 3x - 10 \text{ and } x = \frac{5-\sqrt{281}}{12}$$

$$6y = \cancel{3}\left(\frac{5-\sqrt{281}}{\cancel{12}_4}\right) - 10$$

$$6y = \frac{5-\sqrt{281}}{4} - 10$$

$$6y = \frac{5-\sqrt{281}}{4} - \frac{40}{4}$$

$$6y = \frac{-35-\sqrt{281}}{4}$$

$$y = \frac{-35-\sqrt{281}}{24}$$

The solution set is $\left\{\left(\frac{5+\sqrt{281}}{12}, \frac{-35+\sqrt{281}}{24}\right), \left(\frac{5-\sqrt{281}}{12}, \frac{-35-\sqrt{281}}{24}\right)\right\}$.

Example 4

Determine the solution for the system of equations: $\frac{x}{y^2 - y - 12} = 1$ and $x = y - 4$

Solution: Replace $y - 4$ for x in $\frac{x}{y^2 - y - 12} = 1$

$$x = y^2 - y - 12$$

$$y - 4 = y^2 - y - 12$$

$$0 = y^2 - 2y - 8$$

$$0 = (y + 2)(y - 4)$$

$$y + 2 = 0 \text{ or } y - 4 = 0$$

$$y = -2 \text{ or } y = 4$$

However, $y = 4$ causes both the numerator and denominator of $\frac{x}{y^2 - 7y - 12}$ to be equal to 0, which is undefined. Therefore, $y = 4$ does not fit the equation and it is not a solution for this system of equations. If $y = -2$, then $x = -6$. So the solution set is $\{(-6, -2)\}$.

Check Your Understanding of Section 3.4

1. What is the solution set for the system of equations: $y = 2 - 4x - 4x^2$ and $y = -8x - 6$?

(1) $\{(-1, 2), (-2, 10)\}$
(2) $\{(1, -14), (2, -22)\}$
(3) $\{(-1, 2), (2, -22)\}$
(4) $\{(1, -14), (-2, 10)\}$

2. What is the solution set for the system of equations: $x = y^2 + 12y + 38$ and $x - 11y = 38$?

(1) $\{(27, -1), (38, 0)\}$
(2) $\{(49, 1), (38, 0)\}$
(3) $\{(49, 1), (27, -1)\}$
(4) $\{(51, 1), (-38, 0)\}$

3. What is the solution set for the system of equations: $y = x^2 - 4x + 9$ and $3x + y = 15$?

(1) $\{(2, -1), (-3, 6)\}$
(2) $\{(-2, 21), (3, 6)\}$
(3) $\{(-2, 21), (-3, 6)\}$
(4) $\{(2, -1), (3, 6)\}$

4. The price of a product, $P(x)$, decreased and then increased over a six-month period of time according to the function $P(x) = 0.25x^2 - 1.5x + 4.25$, where x equals the number of months. A competitor produced the same product, and the product's price steadily increased according to the function $R(x) = 0.5x + 0.5$. Use a graphing calculator to sketch these functions. Determine when the prices of the two competitors' products were equal and how much the products cost at those times.

5. Solve the system of equations: $y = 10x^2 - 31x + 15$ and $19x - y = 45$

6. Solve the system of equations: $y = 7x^2 + 3x - 14$ and $17x - y = 14$

7. Solve the system of equations: $r = -3s^2 + 2s + 5$ and $r + 13s = 17$

8. Solve the system of equations: $b = 4a^2 - 4a - 15$ and $100b = -40a - 1556$

9. Solve the system of equations: $s = 25 - t^2$ and $s = 31 - 5t$

10. Solve the system of equations: $27x = y + 21$ and $5x + y = 8x^2 + 3$

11. Solve the system of equations: $\dfrac{x^2+5x+6}{y+1}=2$ and $y = x + 2$

12. Solve the system of equations: $\dfrac{2a+1}{a}+\dfrac{3}{a+2}-\dfrac{b}{a^2+2a}=\dfrac{1}{a}$ and $b = 4a + 2$

13. Solve the system of equations: $y = 2x - 1$ and $y = x^2 - 5x$

14. Solve the system of equations: $y = 4x - 2$ and $y + 3 = 2x^2$

15. A relief food package is dropped from a hovering helicopter from a height of 64 feet. Its height above the ground is modeled by the function $h(t) = -16t^2 + 32$. A rebel on a nearby hill shoots a rifle to destroy this package. The bullet follows a trajectory modeled by the function $s(t) = -32t - 16$. If the variable t represents time in seconds, will the bullet hit the package before it lands safely on the ground? Justify your answer.

3.5 SOLVING A QUADRATIC INEQUALITY

To solve a quadratic inequality, first solve the related quadratic equation. To solve $x^2 < 4$, first solve $x^2 = 4$. That separates the number line into three distinct regions. Now test each region to see where the values of x^2 are less than 4.

Solving a Quadratic Inequality Algebraically

Let's look at the inequality posed in the Key Ideas above. To solve $x^2 < 4$, first solve the equation $x^2 = 4$. The solution to this is $x = \pm 2$. This separates the number line into three distinct regions, $x < -2$, $-2 < x < 2$, and $x > 2$. Choose a number that is less than -2. The easiest would be -3. Check to see if -3 fits the inequality $x^2 < 4$ by replacing the x by -3. That yields $(-3)^2 = 9$, and 9 is not less than 4. Therefore the entire region $x < -2$ is not part of the solution set. Since 0 is in the region $-2 < x < 2$, substitute 0 for x into $x^2 < 4$. Since $0^2 = 0$, which is less than 4, $-2 < x < 2$ is part of the solution set for this inequality. Finally, it is easy to see that $3^2 > 4$, so the region $x > 2$ is not part of the solution set. The solution set is $\{x: -2 < x < 2\}$. Just to convince anyone who is doubtful about this approach, check out the following table of values for x and x^2.

x	x^2
–4	16
–3	9
–2	4
–1	1
0	0
1	1
2	4
3	9
4	16

The process was based on the fact that the only time the equation $x^2 = 4$ crosses the x-axis is at $x = -2$ and $x = 2$. Therefore, the graph of the function $f(x) = x^2$ must lie either entirely above the x-axis or entirely below the x-axis in each separate region bounded by the zeros of the function.

Example 1

Determine the solution for $x^2 - 8x + 15 \geq 0$.

Solution: To solve $x^2 - 8x + 15 \geq 0$, first solve $x^2 - 8x + 15 = 0$. Notice also that the inequality includes equal to 0. So the solutions for $x^2 - 8x + 15 = 0$ include these answers. Factoring and setting each factor equal to 0 by using the Zero Product Property yields the two equations:

$$x - 3 = 0 \text{ and } x - 5 = 0$$

$$x = 3 \text{ and } x = 5$$

This creates the regions of the x-axis at $x \leq 3$, $3 \leq x \leq 5$, and $x \geq 5$. Testing numbers in each region:

Region	x	$x^2 - 8x + 15$	**Conclusion**
$x \leq 3$	1	$1^2 - 8(1) + 15 = 8 > 0$	Part of the solution set
$3 \leq x \leq 5$	4	$4^2 - 8(4) + 15 = -1 < 0$	Not part of the solution set
$x \geq 5$	6	$6^2 - 8(6) + 15 = 3 > 0$	Part of the solution set

The solution set is $\{x: x \leq 3 \text{ or } x \geq 5\}$.

Example 2

Determine the solution for $2x^2 \le 9x + 5$.

Solution: Like a quadratic equality, first move all parts of the quadratic to one side of the inequality so as to have a 0 on the other side.

$$2x^2 \le 9x + 5$$
$$2x^2 - 9x - 5 \le 0$$

Now solve the quadratic equation

$$2x^2 - 9x - 5 = 0.$$
$$(2x + 1)(x - 5) = 0$$
$$2x + 1 = 0 \text{ or } x - 5 = 0$$
$$2x = -1 \text{ or } x = 5$$
$$x = -1/2$$

This creates the regions of the x-axis at $x \le \frac{-1}{2}$, $\frac{-1}{2} \le x \le 5$, and $x \ge 5$.

Region	x	$2x^2 - 9x - 5$	Conclusion
$x \le \frac{-1}{2}$	-1	$2(-1)^2 - 9(-1) - 5 =$ $2(1) + 9 - 5 = 2 + 9 - 5 =$ $11 - 5 = 6 > 0$	Not part of the solution set
$\frac{-1}{2} \le x \le 5$	0	$2(0)^2 - 9(0) - 5 =$ $0 - 0 - 5 = -5 < 0$	Part of the solution set
$x \ge 5$	6	$2(6)^2 - 9(6) - 5 =$ $72 - 54 - 5 = 13 > 0$	Not part of the solution set

The solution set is $\left\{x: \frac{-1}{2} \le x \le 5\right\}$.

Solving a Quadratic Inequality Graphically

Let's examine similar problems from a graphical perspective. For instance, let's look at the graph of $y = x^2 - 9$ to determine how to solve $x^2 - 9 < 0$. Look at the graph of $y = x^2 - 9$.

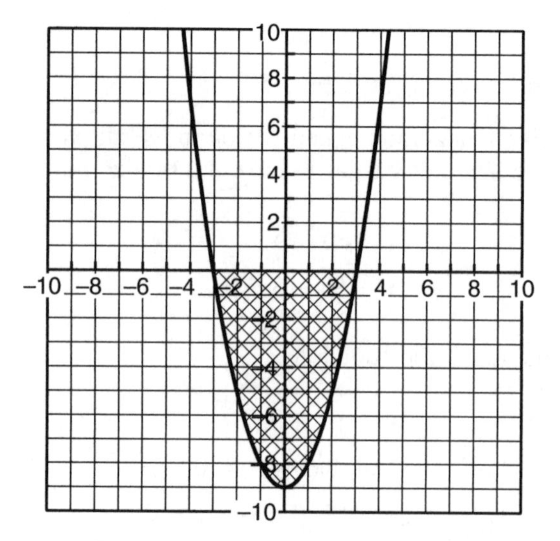

In the graph above, the region where the graph is displayed below the x-axis is shaded. Clearly the region between $x = -3$ and $x = 3$ is the region that is shaded. Based on this, the solution to this inequality is $\{x: -3 < x < 3\}$.

Example 3

Graphically determine solution for $2x^2 - 7x - 15 \geq 0$.

Solution: Examine the graph of $y = 2x^2 - 7x - 15$ below.

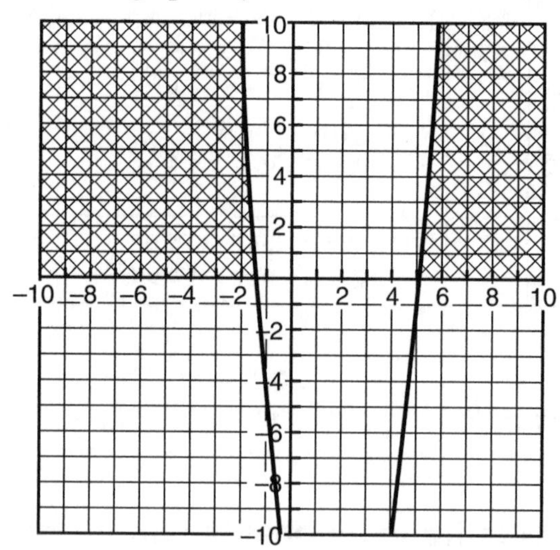

The shaded portion indicates where the solution is. However, one of the x-intercepts is not an integer value of x. A graphing calculator can be used to find the nonintegral x-intercept. Enter the function $2x^2 - 7x - 15$ in **Y=**. Then press **2ND** **TRACE** **2** and scroll to the left of the intercept and press **ENTER**. Now scroll to the right of the intercept and press **ENTER** again and then **ENTER** again. The display on the calculator shows the x-intercept at -1.5 as shown below.

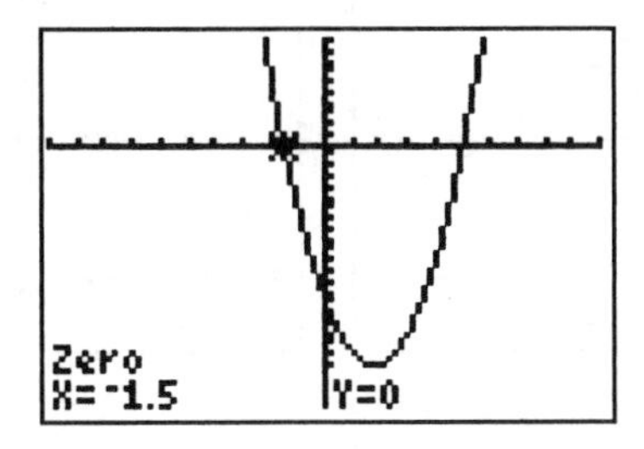

So the two intercepts are at $x = -1.5$ and $x = 5$. The intercept at $x = 5$ can also be verified with the graphing calculator. The solution set is $\{x: x \leq -1.5$ or $x \geq 5\}$.

Example 4

Graphically determine the solution for $4x^2 \leq 3 - x$.

Solution: Examine the graph of $y = 4x^2 + x - 3$ and find the x-intercepts for this function. Now compare that to the graphs of $y = 4x^2$ and $y = 3 - x$ and see where these two functions intersect each other.

For $y = 4x^2 + x - 3$:

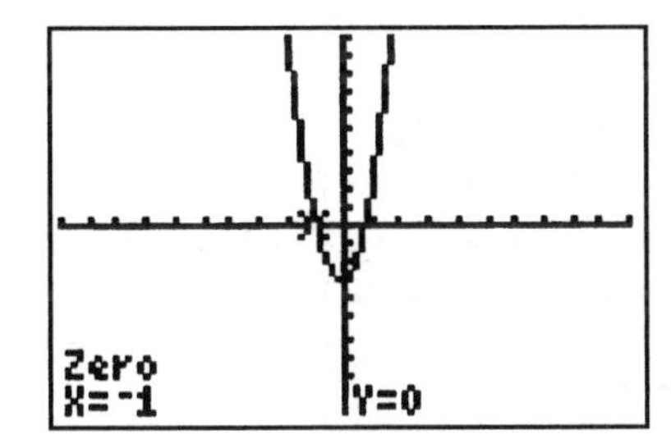

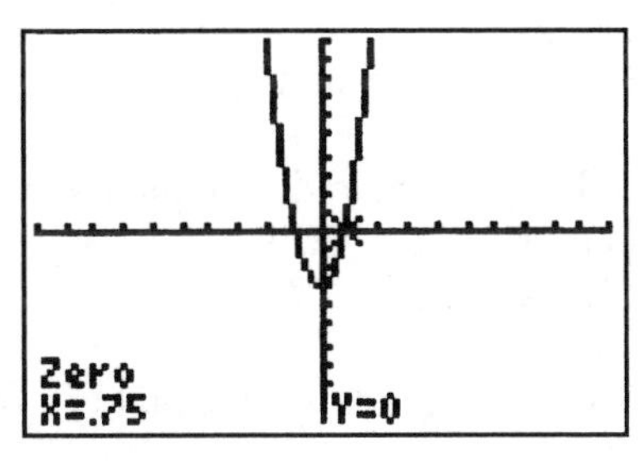

From this graph, solve $4x^2 \leq 3 - x$ by recognizing that it is the same as $4x^2 + x - 3 \leq 0$. The graph lies under the x-axis between $x = -1$ and $x = 0.75$.

So the solution set is

$\{x: -1 \le x \le 0.75\}$.

For $y = 4x^2$ and $y = 3 - x$:

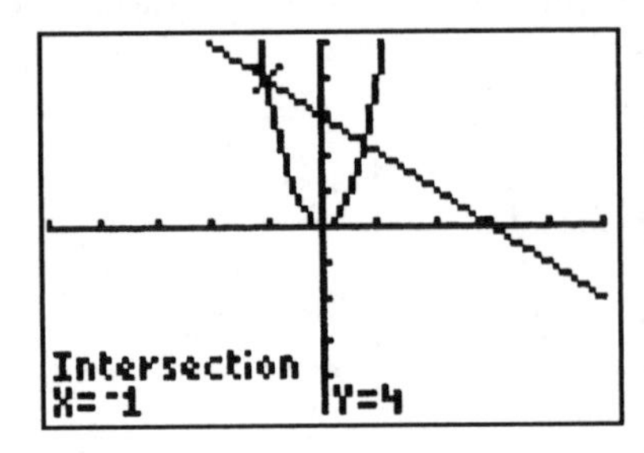

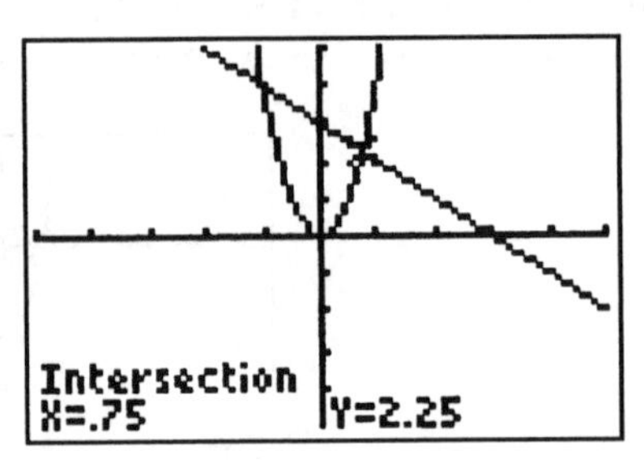

To solve $4x^2 \le 3 - x$, notice that the y-values of the parabola $y = 4x^2$ lie under the y-values for $y = 3 - x$ between $x = -1$ and $x = 0.75$. The solution set is once again revealed to be $\{x: -1 \le x \le 0.75\}$.

Obviously, this provides two entirely different approaches to solving an inequality such as $4x^2 \le 3 - x$. The choice of methods is up to the person solving the inequality.

Check Your Understanding of Section 3.5

1. Solve the quadratic inequality: $x^2 - 3x - 10 > 0$
(1) $\{x: x < -2 \text{ or } x > 5\}$
(2) $\{x: x < -5 \text{ or } x > 2\}$
(3) $\{x: -2 < x < 5\}$
(4) $\{x: -5 < x < 2\}$

2. Solve the quadratic inequality: $x^2 - 6x + 7 \le 0$
(1) $\{x: -7 \le x \le 1\}$
(2) $\{x: -1 \le x \le 7\}$
(3) $\{x: x \le -7 \text{ or } x \ge 1\}$
(4) $\{x: x \le -1 \text{ or } x \ge 7\}$

3. Solve the quadratic inequality: $6x^2 + 5x \ge 4$
(1) $\left\{x: -\frac{1}{2} \le x \le \frac{4}{3}\right\}$
(2) $\left\{x: -2 \le x \le \frac{3}{4}\right\}$
(3) $\left\{x: x \le \frac{-1}{2} \text{ or } x \ge \frac{4}{3}\right\}$
(4) $\left\{x: x \le \frac{-4}{3} \text{ or } x \ge \frac{1}{2}\right\}$

4. Solve the quadratic inequality: $3t^2 + 11t \leq 4$

(1) $\left\{t: -\dfrac{1}{3} \leq t \leq 4\right\}$

(2) $\left\{t: -4 \leq t \leq \dfrac{1}{3}\right\}$

(3) $\left\{t: -3 \leq t \leq \dfrac{1}{4}\right\}$

(4) $\left\{t: t \geq -4 \text{ or } t \leq \dfrac{1}{3}\right\}$

5. Solve the quadratic inequality: $6r^2 - 24r < 0$

(1) $\{r: -4 < r < 0\}$

(2) $\{r: r < 4\}$

(3) $\{r: 0 < r < 4\}$

(4) $\{r: r < 0 \text{ or } r > 4\}$

6. Solve the quadratic inequality $x^2 - 9x + 12 \leq 4$ graphically.

7. Solve the quadratic inequality $8x^2 > 14x + 15$ graphically.

8. Solve the quadratic inequality $5x^2 + 4x - 12 < 0$ graphically.

9. Solve the quadratic inequality $2c^2 + 15 \geq 13c$ algebraically.

10. Solve the quadratic inequality $6 - 7t \leq 20t^2$ algebraically.

11. Solve the quadratic inequality $8z^2 \geq 10z + 3$ algebraically.

12. A ball is thrown vertically upward with an initial velocity of 64 feet per second. (Velocity is the absolute value of speed.) After t seconds, the distance s (in feet) of the ball from the ground is modeled by the formula $s = 64t - 16t^2$. During what interval of time is the ball higher than 48 feet above the ground?

13. A farmer has 100 feet of fencing and wants to enclose a rectangular pasture. If l = length of the pasture, what values of l will ensure that the area is at least 225 square feet?

14. The producer of a musical knows what his fixed costs are for a production. He figures that he must take in at least $1,271 to cover these costs. He is pretty sure that he can sell at least 22 tickets if he charges $50. He offers the following deal: The cost of a ticket is $50 less a $1 discount for each and every ticket sold above the 22 tickets and an additional $1 per ticket for every ticket sold if less than 22 tickets are sold. What range of number of tickets sold will give him the $1,271 he needs to cover the costs of the production?

3.6 HIGHER-DEGREE POLYNOMIAL EQUATIONS

Some equations are not linear and are not quadratic. If the degree of a polynomial equation is higher than 2, the equation may be solved. In this section, some higher-degree polynomials will be solved by using already familiar approaches. Others will be solved only approximately by using a calculator.

Solving a Higher-Degree Polynomial Equation by Factoring

Some higher-degree polynomials can be solved because the polynomial expressions contained in them are easily factorable. For instance, when factoring $x^2 - 5x + 4$, the result is $(x - 4)(x - 1)$. Examine the polynomial $x^4 - 5x^2 + 4$. This is quite similar to $x^2 - 5x + 4$. In fact, the only difference is that the exponents in $x^2 - 5x + 4$ have each been squared to yield the expression $x^4 - 5x^2 + 4$. To factor $x^4 - 5x^2 + 4$, simply replace each x in $(x - 4)(x - 1)$ by x^2. The result is $(x^2 - 4)(x^2 - 1)$. Notice that by multiplying these two factors together, the polynomial $x^4 - 5x^2 + 4$ is obtained. Therefore, the factors of $x^4 - 5x^2 + 4$ are $(x^2 - 4)(x^2 - 1)$.

Example 1

Solve for x: $x^4 - 5x^2 + 4 = 0$

Solution: $x^4 - 5x^2 + 4 = 0$

Factoring results in $(x^2 - 4)(x^2 - 1) = 0$. By the Zero Product Property, either $x^2 - 4 = 0$ or $x^2 - 1 = 0$. Each of these quadratic equations are incomplete quadratic equations that can be solved by subtracting the constant from both sides of the equation(s) and finding both the positive and negative square roots. Therefore, $x = \pm 2$ or $x = \pm 1$.

Example 2

Solve for z: $4z^4 - 29z^2 + 45 = 0$

Solution: Remember that when the lead coefficient of a trinomial does not equal 1, two factoring methods can be attempted. Let's first take the product of the lead coefficient, 4, and the constant term, 45. That product is 180. Now find two numbers whose product is 180 that add up to the coefficient of the middle term, –29. Those two numbers are –20 and –9.

$$\text{So } 4z^4 - 29z^2 + 45 = 4z^4 - 20z^2 - 9z^2 + 45.$$

The first two terms of this expression have a common factor of $4z^2$, and the last two terms have a common factor of –9. Therefore, $4z^4 - 20z^2 - 9z^2 + 45 = 4z^2(z^2 - 5) - 9(z^2 - 5) = (z^2 - 5)(4z^2 - 9)$. Now set each of these factors equal to 0 and solve:

$$z^2 - 5 = 0 \text{ or } 4z^2 - 9 = 0$$
$$z^2 = 5 \text{ or } 4z^2 = 9$$
$$z^2 = \frac{9}{4}$$
$$z = \pm\sqrt{5} \text{ or } z = \pm\frac{3}{2}.$$

The solution set is $\left\{\sqrt{5}, -\sqrt{5}, \frac{3}{2}, -\frac{3}{2}\right\}$.

Notice that in both of the examples above, the second-degree factors were each a difference between the squared variable and a constant. If either of the factors are a sum of a squared variable and a constant, then there would be no real number solution associated with that factor.

The following example has a common monomial factor to start off the problem.

Example 3

Solve for r: $r^3 + r^2 - 20r = 0$

Solution: The common factor is r.

$$r(r^2 + r - 20) = 0$$
$$r(r + 5)(r - 4) = 0$$
$$r = 0 \text{ or } r + 5 = 0 \text{ or } r - 4 = 0$$
$$r = 0 \text{ or } r = -5 \text{ or } r = 4$$

The approach used in this last example can be mixed with the approach in the two previous examples to solve an example such as the one that follows.

Example 4

Solve for t: $4t^6 - 12t^4 + 8t^2 = 0$

Solution: A common monomial of $4t^2$ can be factored out to begin the problem.

$$4t^6 - 12t^4 + 8t^2 = 0$$
$$4t^2(t^4 - 3t + 2) = 0$$
$$4t^2(t^2 + 3)(t^2 - 1) = 0$$
$$4t^2 = 0 \text{ or } t^2 + 3 = 0 \text{ or } t^2 - 1 = 0$$
$$t = 0 \text{ or } t = \pm 1$$

Note that $t^2 + 3 = 0$ has no real number solution. The solution set has two zeros in it because when $t^2 = 0$, either $t = 0$ or $t = 0$. So the solution set is $\{0, 0, 1, -1\}$.

Another type of situation that can be solved through a factoring technique involves just common binomial factors.

Example 5

Solve for x: $3x^3 - 7x^2 - 6x + 14 = 0$

Solution: $3x^3 - 7x^2 - 6x + 14 = 0$

$$x^2(3x - 7) - 2(3x - 7) = 0$$
$$(3x - 7)(x^2 - 2) = 0$$
$$3x - 7 = 0 \text{ or } x^2 - 2 = 0$$
$$x = \frac{7}{3} \text{ or } x^2 = 2$$
$$x = \pm\sqrt{2}$$

The solution set is $\left\{\frac{7}{3}, \sqrt{2}, -\sqrt{2}\right\}$.

Example 6

Solve for c: $c^3 - 4c^2 + 21c = c^2 + 4c - 21$.

Solution: By observation, it is easy to see how the expression on the left side of this equation is equal to the right side of the equation multiplied by c. When the right side of the equation is subtracted from both sides of the equation, be careful not to combine like terms.

$$c^3 - 4c^2 + 21c - c^2 - 4c + 21 = 0$$

$$c(c^2 - 4c + 21) - 1(c^2 - 4c + 21) = 0$$

$$(c - 1)(c^2 - 4c + 21) = 0$$

$$(c - 1)(c - 7)(c + 3) = 0$$

$$c - 1 = 0 \text{ or } c - 7 = 0 \text{ or } c + 3 = 0$$

$$c = 1 \text{ or } c = 7 \text{ or } c = -3$$

The solution set is $\{1, 7, -3\}$.

Solving a Higher-Degree Polynomial Equation Using a Combination of Factoring and the Quadratic Formula

In some examples of higher-degree polynomial equations, the polynomial expression resembles a quadratic equation (or is a quadratic equation in x^2). For instance, let's reexamine Example 2, solving for z: $4z^4 - 29z^2 + 45 = 0$. An approach that can be used to solve this is to substitute another variable for each z^2 in this equation. For instance, by letting $x = z^2$, the equation now becomes $4x^2 - 29x + 45 = 0$. When this problem was investigated before, a factoring method was used. Since it can easily be seen as a quadratic equation, the quadratic formula can now be used where $a = 4$, $b = -29$, and $c = 45$. By replacing these values into the quadratic formula, the values of x^2 can be obtained.

$$x = \frac{-(-29) \pm \sqrt{(-29)^2 - 4(4)(45)}}{2(4)}$$

$$x = \frac{29 \pm \sqrt{841 - 720}}{8}$$

$$x = \frac{29 \pm \sqrt{121}}{8}$$

$$x = \frac{29 \pm 11}{8}$$

$$x = \frac{18}{8} \text{ or } \frac{40}{8}$$

$$z^2 = \frac{9}{4} \text{ or } z^2 = 5$$

$$z = \pm\frac{3}{2} \text{ or } z = \pm\sqrt{5}$$

Example 7

Solve for s: $2s^4 - 5s^2 + 3 = 0$

Solution: If $2s^4 - 5s^2 + 3 = 0$, then replace x for s^2. So $2x^2 - 5x + 3 = 0$, and $a = 2$, $b = -5$, and $c = 3$.

$$x = \frac{-(-5) \pm \sqrt{(-5)^2 - 4(2)(3)}}{2(1)}$$

$$x = \frac{5 \pm \sqrt{25 - 24}}{2}$$

$$x = \frac{5 \pm \sqrt{1}}{2}$$

$$x = \frac{5 \pm 1}{2}$$

$$x = \frac{6}{2} \text{ or } x = \frac{4}{2}$$

$$x = 3 \text{ or } x = 2$$

$$s^2 = 3 \text{ or } s^2 = 2$$

$$s = \pm\sqrt{3} \text{ or } s = \pm\sqrt{2}$$

The solution set is $\{\sqrt{2}, -\sqrt{2}, \sqrt{3}, -\sqrt{3}\}$.

Approximating the Solution from a Graph

A graphing calculator can be used to solve a higher-degree polynomial equation and can even be used to approximate its roots. To do this, enter the function rule into the calculator, and use the calculator to solve for the intersection. Let's examine how to do this for the problem from Example 1, $x^4 - 5x^2 + 4 = 0$. First enter the function by pressing **Y =** **X** **^** **4** **–** **5** **×** **^** **2** **+** **4** **ENTER**. Then press **GRAPH** and **2ND** **TRACE** **2**. Move the blinking X just to the left of the farthest left x-intercept and click **ENTER**. Then move the blinking X just to the right of that same x-intercept and press **ENTER** twice. The calculator now shows the value of this x-intercept, which is the root or solution to this equation. Repeat the process of pressing **ENTER** immediately to the left and then to the right of each of the other three x-intercepts. Below is the display that the calculator makes for each of these x-intercepts:

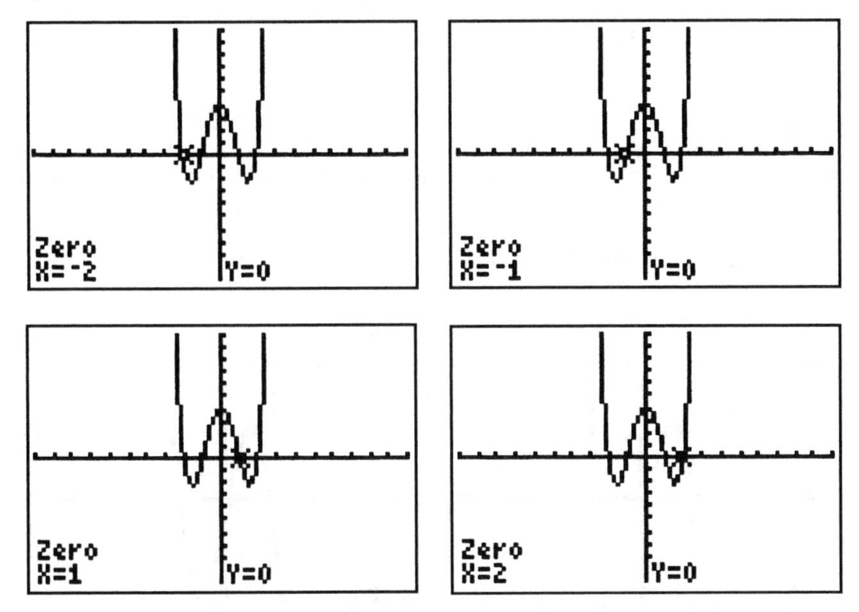

This problem involved a higher-degree polynomial where the roots were integers. Let's examine one where the roots are irrational. For instance, use this method for the problem from Example 5, $3x^3 - 7x^2 - 6x + 14 = 0$. Enter the function in the calculator, and repeat the same process for each of its three roots. The results are displayed below.

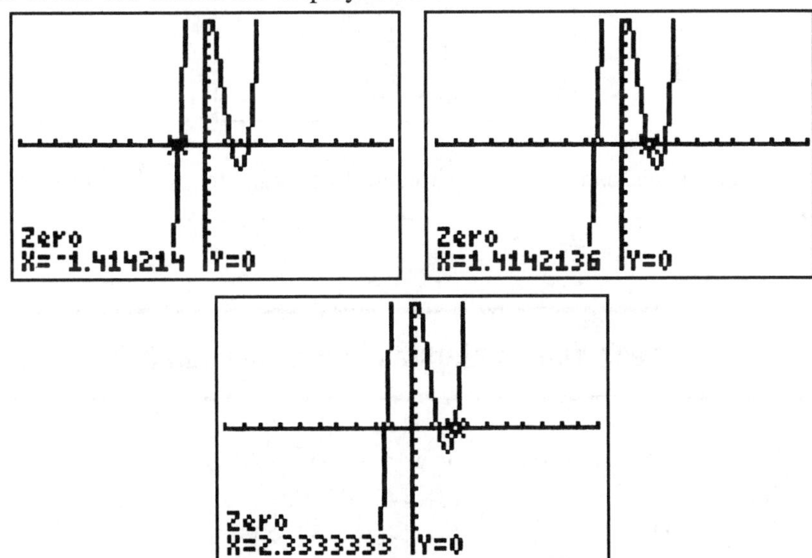

Notice that these results are consistent with the solution obtained when this example was investigated earlier, namely $\left\{\frac{7}{3}, \sqrt{2}, -\sqrt{2}\right\}$, because $\pm\sqrt{2} \approx \pm 1.41421356\overline{2}$ and $\frac{7}{3} = 2.3\overline{3}$.

Example 8

Solve for a: $5a^3 + a^2 - 30a - 6 = 0$ and approximate the solutions to the nearest thousandth.

Solution: Enter the function rule $5a^3 + a^2 - 30a - 6$ for Y_1. Use 2ND TRACE 2 to display the solutions.

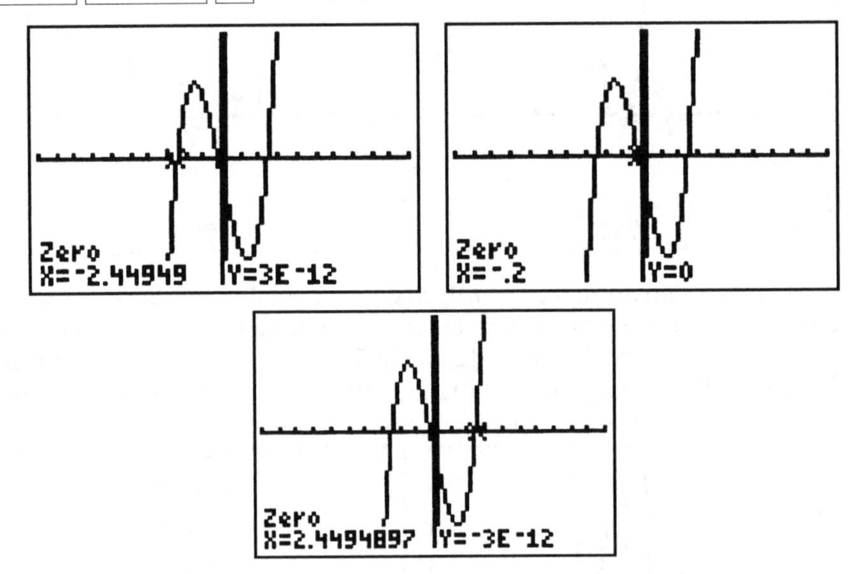

The solution set can be approximated from this to be {–2.449, 2.449, –0.2}.

Check Your Understanding of Section 3.6

1. Determine the solution set for y: $y^4 - 41y^2 + 400 = 0$

   (1) {–5, 5}
   (2) {4, –4}
   (3) {4, –4, 10, –10}
   (4) {4, –4, 5, –5}

2. Determine the solution set for z: $2z^6 - 4z^5 - 30z^4 = 0$

(1) $\{-3, 5\}$
(2) $\{0, 0, 0, 0, -3, 5\}$
(3) $\{3, -5\}$
(4) $\{0, 0, 0, 0, 3, -5\}$

3. Determine the solution set for r: $5r^3 - 4r^2 - 20r + 16 = 0$

(1) $\left\{\frac{4}{5}, -2, 2\right\}$
(2) $\{5, 4, -4\}$
(3) $\{4, 10, -10\}$
(4) $\left\{\frac{4}{5}, -\frac{4}{5}, 2\right\}$

4. Determine the solution set for b: $b^6 - 81b^4 - b^2 + 81 = 0$

(1) $\{1, 1, 9, -9\}$
(2) $\{1, 1, -1, -1, 9, -9\}$
(3) $\{1, -1, 9, -9\}$
(4) $\{3, 3, -3, -3\}$

5. Determine the solution set for p: $9p^4 + 5 = 46p^2$

(1) $\left\{5, \frac{1}{9}, -\frac{1}{9}\right\}$
(2) $\left\{5, -5, \frac{1}{3}, -\frac{1}{3}\right\}$
(3) $\left\{\sqrt{5}, -\sqrt{5}, \frac{1}{3}, -\frac{1}{3}\right\}$
(4) $\left\{\sqrt{5}, -\sqrt{5}, \frac{1}{9}, -\frac{1}{9}\right\}$

6. Determine the solution set for x: $3x^4 + 45x^2 = 24x^3$

(1) $\{0, 0, 3, -3, 5\}$
(2) $\{0, 0, 3, 5\}$
(3) $\{0, 0, -3, -5\}$
(4) $\{0, 0, 3, -5\}$

7. Solve $10ab - 20a + 6b - 12 = 0$ for a and b

8. Determine the solution set for x: $3x^3 - 7x^2 - 6x + 14 = 0$

9. Determine the solution set for k: $k^3 - 7k^2 = 9k - 63$

10. Determine the solution set for s: $s^3 - 6s^2 - 9s + 54 = 0$

11. One edge of a cube is increased by 10 cm, a second edge is increased by 12 cm, and the third edge is decreased by 5 cm. If the volume of the resulting rectangular solid is double that of the original cube, how long was the edge of the original cube if that edge has a length that is a whole number measured in centimeters?

Chapter Four

COMPOSITION AND INVERSES OF FUNCTIONS

4.1 COMPOSITION OF FUNCTIONS

KEY IDEAS

Composite materials are materials made from two or more constituent materials. So a composition of functions occurs when one or more functions are found within the rule of another function.

What Is a Composition of Functions?

Consider the function $f(x) = (x + 1)^8$. Notice that $f(x)$ is a function whose rule combines two different function rules. First of all, there is an overall function $g(x) = x^8$. Then there is a function $h(x) = x + 1$. The function $f(x)$ can be defined in terms of both of these functions, $f(x) = g(x + 1) = g(h(x))$. Many situations can occur where it is possible to think that one function is made up of two or more functions. However, a composition of functions is more than a situation in which two functions are combined by addition, subtraction, multiplication, or even division. A composition of functions occurs when one or more functions can be found inserted within the rule for another function.

Let's look at it from another standpoint. Let $f(x) = 2x - 7$ and $g(x)=\frac{1}{x}$. These two functions can be composed in two possible ways, function f following function g, which is $f(g(x))$, and function g following function f, which is $g(f(x))$. The function $f(g(x))=f\left(\frac{1}{x}\right)=2\left(\frac{1}{x}\right)-7=\frac{2}{x}-7$. However, the function $g(f(x))=g(2x-7)=\frac{1}{2x-7}$.

There is also a new symbol to learn that stands for a composition of two functions. The symbol for a composition is a small raised circle "∘" between two function names, $(f \circ g)(x) = f(g(x))$.

MATH FACTS

Function $h(x)$ is the composition of functions $f(x)$ and $g(x)$. It is written as $h(x) = (f \circ g)(x) = f(g(x))$. This composition indicates function f following function g. It can also be thought of as function g followed by function f.

Example 1

If $f(x) = x^2 - 5$ and $g(x) = x^2$, determine $(f \circ g)(x)$ and $(g \circ f)(x)$.

Solution: $(f \circ g)(x) = f(g(x)) = f(x^2) = (x^2)^2 - 5 = x^4 - 5$

$(g \circ f)(x) = g(f(x)) = g(x^2 - 5) = (x^2 - 5)^2 = x^4 - 10x^2 + 25$

Example 1 shows that the composition of functions is not commutative. The question arises, however, are there any functions for which $(f \circ g)(x) = (g \circ f)(x)$? The answer is yes. The next example is an example of one such situation.

Example 2

If $f(x) = x - 3$ and $g(x) = x + 2$, determine $(f \circ g)(x)$ and $(g \circ f)(x)$.

Solution: $(f \circ g)(x) = f(g(x)) = f(x + 2) = (x + 2) - 3 = x - 1$

$(g \circ f)(x) = g(f(x)) = g(x - 3) = (x - 3) + 2 = x - 1$

Example 3

If $f(x) = \dfrac{5}{x^2}$ and $g(x) = 3x - 4$, evaluate:

a. $(f \circ g)(2)$ b. $(g \circ f)(1)$

c. $(f \circ f)(-1)$ d. $(g \circ g)\left(\frac{2}{3}\right)$ e. $(f \circ g)\left(\frac{4}{3}\right)$

Solution:

a. $(f \circ g)(2) = f(g(2)) = f(3 \cdot 2 - 4) = f(2) = \dfrac{5}{2^2} = \dfrac{5}{4}$

b. $(g \circ f)(1) = g(f(1)) = g\left(\dfrac{5}{1^2}\right) = g\left(\dfrac{5}{1}\right) = g(5) = 3 \cdot 5 - 4 = 15 - 4 = 11$

c. $(f \circ f)(-1) = f(f(-1)) = f\left(\dfrac{5}{(-1)^2}\right) = f\left(\dfrac{5}{1}\right) = f(5) = \dfrac{5}{5^2} = \dfrac{5}{25} = \dfrac{1}{5}$

d. $(g \circ g)\left(\frac{2}{3}\right) = g\left(g\left(\frac{2}{3}\right)\right) = g\left(3\left(\frac{2}{3}\right) - 4\right) = g(2-4) = g(-2)$

$= 3(-2) - 4 = -6 - 4 = -10$

e. $(f \circ g)\left(\frac{4}{3}\right) = f\left(g\left(\frac{4}{3}\right)\right) = f\left(3\left(\frac{4}{3}\right) - 4\right) = f(4-4) = f(0) = \frac{5}{0^2}$

$(f \circ g)\left(\frac{4}{3}\right)$ is clearly undefined

Consider the function $h(x) = \frac{5}{x^2 - 1}$. The domain of $h(x)$ is $\{x\text{: } x \neq 1 \text{ and } x \neq -1\}$. However, $h(x)$ is a composite function of $f(x) = \frac{5}{x}$ following $g(x) = x^2 - 1$. The domain of $f(x)$ is $\{x\text{: } x \neq 0\}$ and the domain of $g(x)$ is $\mathcal{R}$. Since $h(x) = f(g(x))$, it follows that the domain of $h(x)$ includes any element in the domain of $g(x)$ that maps to an element in the domain of $f(x)$. In that way, the domain of $h(x)$ is any element of $\mathcal{R}$ that when inserted in the functional rule for $f(x)$ yields a number in the $\{x\text{: } x \neq 0\}$. So examining all real numbers to find when $x^2 - 1 \neq 0$ shows that $x \neq \pm 1$. That agrees with the domain of function $h(x)$ as indicated in the beginning of this paragraph.

Example 4

What is the domain of $h(x) = \frac{-1}{x+3}$?

Solution: $h(x) = f(g(x)) = \frac{-1}{x+3}$ where $f(x) = \frac{-1}{x}$ and $g(x) = x + 3$.

The domain of function $f(x)$ is $\{x\text{: } x \neq 0\}$ and $g(-3) = 0$. Therefore, the domain of $h(x)$ is $\{x\text{: } x \neq -3\}$. Note that there are other ways to express $h(x)$ as a composition of two functions.

Check Your Understanding of Section 4.1

1. If $h(x) = (f \circ g)(x) = \sqrt{x^2 + 2}$ then which of the following could be true?

(1) $f(x) = x^2 + 2$ and $g(x) = \sqrt{x}$
(2) $f(x) = \sqrt{x^2}$ and $g(x) = x + 2$
(3) $f(x) = \sqrt{x}$ and $g(x) = x^2 + 2$
(4) $f(x) = x^2$ and $g(x) = \sqrt{x+2}$

2. If $p(x) = x^2 - 3x + 5$ and $q(x) = x^2 - 3$, determine the function rule for $r(x)$ such that $r(x) = (p \circ q)(x)$.
(1) $r(x) = x^4 - 9x^2 + 5$
(2) $r(x) = x^4 - 3x^2 + 5$
(3) $r(x) = x^4 - 3x^2 + 23$
(4) $r(x) = x^4 - 9x^2 + 23$

3. If $a(x) = x^3 + 2x$ and $c(x) = x^{\frac{-3}{2}}(1 - 2x)$, determine $b(x)$ such that $c(x) = (a \circ b)(x)$.
(1) $b(x) = x^{\frac{-1}{2}}$ (2) $b(x) = x^{\frac{-3}{2}}$ (3) $b(x) = x^2$ (4) $b(x) = x^{-1}$

4. If $f(x) = 3x - 2$ and $g(x) = 2\sqrt{x-1}$, determine $(f \circ g)(4)$.
(1) 6 (2) $6\sqrt{3} - 2$ (3) 4 (4) $6\sqrt{3} - 12$

5. If $h(x) = (f \circ g)(x)$, $f(x) = \dfrac{12}{x-1}$ and $g(2) = 5$, determine $h(2)$.
(1) 12 (2) 4 (3) 3 (4) 1

6. If $f(x) = \dfrac{1}{x}$, determine $(f \circ f)(7)$.
(1) $\dfrac{1}{49}$ (2) $\dfrac{1}{7}$ (3) 49 (4) 7

7. If $f(x) = x^2$ and $g(x) = x^{\frac{2}{3}}$, which of the following has the value of 16?
(1) $(f \circ g)(2)$ (2) $(f \circ g)(8)$ (3) $(g \circ f)(4)$ (4) $(g \circ f)(2)$

8. If $f(x) = \sqrt{x+4}$ and $g(x) = x^2 - 4$, evaluate$(g \circ f)(-6)$.
(1) 6 (2) 36 (3) -6 (4) 32

9. If $r(x) = |x^2 - 2|$ and $q(x) = 3 - x$, evaluate $(r \circ q)(3)$.
(1) -4 (2) 4 (3) -2 (4) 2

10. If $g(x) = \sqrt[3]{x^2 - 1}$ and $f(x) = x + 4$, evaluate $(g \circ f)(-3)$.
(1) undefined (2) $\sqrt[3]{8} + 4$ (3) 0 (4) $2\sqrt[3]{8}$

11. Let $f(x) = x^4$, $g(x)=\dfrac{3}{x}$, $h(x)+2+\sqrt{x}$, determine each of the following:

a. $h(f(x)) =$ b. $(g \circ f)(x)=$ c. $((h \circ g) \circ f)(x) =$

d. $(g \circ f)(1) =$ e. $\left(h\left(g\left(\dfrac{1}{3}\right)\right)\right)=$ f. $(h \circ f)(2) =$

g. $(g \circ h)(25) =$ h. $(f \circ g)(-6) =$ i. $((g \circ f) \circ h)(1) =$

12. Marla examined the functions $f(x) = x^4$ and $g(x)=\sqrt{x}$. Then she used them to determine $(f \circ g)(4)$ and $(g \circ f)(4)$. She used her answers to decide whether or not composition of functions is commutative. What answers should she have found for $(f \circ g)(4)$ and $(g \circ f)(4)$? Based on this work, how should she answer the question about whether or not composition of functions is commutative?

13. If $f(x) = 2x + 1$, determine the function $g(x)$ such that $(f \circ g)(x) = x$. Then use this function to determine $(g \circ f)(x)$.

4.2 THE INVERSE OF A FUNCTION

A function maps x-values to y-values. An inverse of a function is another function that maps each y-value from the range of the original function back to its pre-image—in other words, to the x-value that mapped to it.

What Is the Inverse of a Function?

To answer this question, let's look at the function $f(x) = x - 5$. Notice that $f(3) = -2$ and $f(9) = 4$. The inverse of this function must map the -2 back to 3 and the 4 back to 9. Notice that in each of these two requirements, the preimages were 5 more than their images. Try $g(x) = x + 5$ and see that $g(-2) = 3$ and $g(4) = 9$. Does $g(x)$ map all the values in the range of $f(x)$ back to their preimages in the domain of $f(x)$? It is impossible to show all the ordered pairs involved. However, the following chart seems to lead to a confirmation that $g(x)$ is the inverse of $f(x)$. What needs to be true is that for all x, if $f(x) = y$, then $g(y)$ must be equal to x.

x	$f(x) = x - 5 = y$	y	$g(y) = y + 5$
1	$1 - 5 = -4$	-4	$-4 + 5 = 1$
2	$1 - 5 = -3$	-3	$-3 + 5 = 2$
6	$6 - 5 = 1$	1	$1 + 5 = 6$
10	$10 - 5 = 5$	5	$5 + 5 = 10$
22	$22 - 5 = 17$	17	$17 + 5 = 22$
3.4	$3.4 - 5 = -1.6$	-1.6	$-1.6 + 5 = 3.4$
8.3	$8.3 - 5 = 3.3$	3.3	$3.3 + 5 = 8.3$
$\frac{1}{2}$	$\frac{1}{2} - 5 = \frac{1}{2} - \frac{10}{2} = \frac{-9}{2}$	$\frac{-9}{2}$	$\frac{-9}{2} + 5 = \frac{-9}{2} + \frac{10}{2} = \frac{1}{2}$
$7\frac{2}{3}$	$7\frac{2}{3} - 5 = 2\frac{2}{3}$	$2\frac{2}{3}$	$2\frac{2}{3} + 5 = 7\frac{2}{3}$

Let's look at this in another way. Since $f(x) = x - 5$ and $g(x) = x + 5$, $f(g(x)) = f(x + 5) = (x + 5) - 5 = x$. Think about what that actually means. The $x + 5$ was found because $g(x) = x + 5$ is a y-value. When that y-value is placed into the function $f(x)$, the result yields the original value of x. That is exactly what was needed. Now examine $g(f(x))$. We see that, $g(f(x)) = g(x - 5) = (x - 5) + 5 = x$. So it works both ways. This is true because if function $g(y)$ is the inverse of function $f(x)$, then function $f(x)$ must also be the inverse of function $g(y)$.

In a sense, the inverse function must switch the x- and y-values. So $y = f(x)$ has as its inverse $x = g(y)$. Symbolically, rather than using a function $g(y)$, a more meaningful symbol can be used: $f^{-1}(x)$.

MATH FACTS

In order for $f(x)$ and $f^{-1}(x)$ to be inverse functions, each of the following must be true:

1. $(f \circ f^{-1})(x) = f(f^{-1}(x)) = x$
2. $(f^{-1} \circ f)(x) = f^{-1}(f(x)) = x$

Example 1

Use the composition rules to show that $f(x) = 6x$ and $f^{-1}(x) = \frac{x}{6}$ are inverse functions.

Solution: $(f \circ f^{-1})(x) = f(f^{-1}(x)) = f\left(\frac{x}{6}\right) = 6 \cdot \frac{x}{6} = x$

$$(f^{-1} \circ f)(x) = f^{-1}(f(x)) = f^{-1}(6x) = \frac{6x}{6} = x$$

Therefore, functions $f(x)$ and $f^{-1}(x)$ are inverse functions.

Basically, two questions remain. First, how can the inverse of a function be determined? Second, does every function have an inverse?

Finding the Inverse of a Function

Consider the function $f(x) = 3x + 4$. How can the inverse of this function be determined? Notice that this is a linear equation. Its graph is shown below along with a partial table of values.

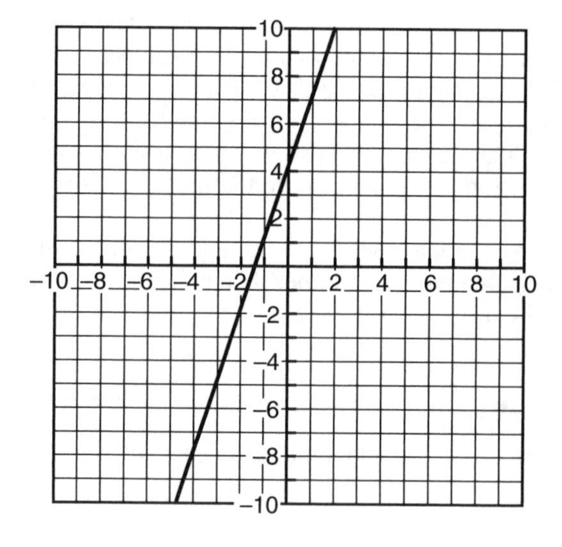

x	$3x + 4$
–5	–11
–4	–8
–3	–5
–2	–2
–1	1
0	4
1	7
2	10
3	13
4	16
5	19
6	22
7	25
8	28

The relationship between this function $f(x) = 3x + 4$ and its inverse is that the y-values for this function need to map to the x-values in its inverse. In fact, what really happens when the inverse is found is that the x-values for the function $f(x)$ become the y-values for its inverse $f^{-1}(x)$ and the y-values from $f(x)$ become the x-values in $f^{-1}(x)$. To find the inverse of function $f(x)$, rewrite the function in the format $y = 3x +4$ and switch the x's and y's in this function so that $x = 3y + 4$. Now, solve $x = 3y + 4$ for y. Therefore, $x - 4 = 3y$ and $y = \frac{x-4}{3}$ or $y = \frac{1}{3}x - \frac{4}{3}$. Examine the graph below. It includes the original $y = 3x + 4$ (solid line) as well as the new equation $y = \frac{1}{3}x - \frac{4}{3}$ (dashed line). The graph of the line $y = x$ is also displayed (dotted line).

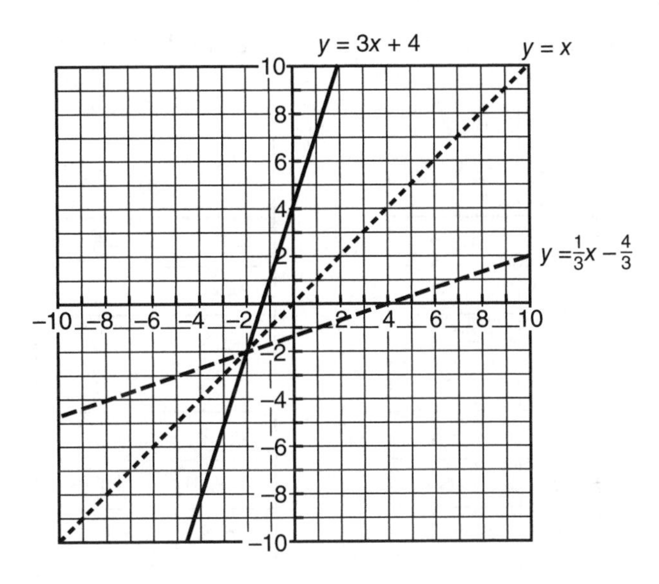

From this, it is possible to see that the inverse of a function is the reflection of that function over the line $y = x$.

Example 2

Determine the inverse: $f(x) = \frac{1}{2}x - 6$

Solution: $f(x) = \frac{1}{2}x - 6$

$$y = \frac{1}{2}x - 6$$

$$x = \frac{1}{2}y - 6$$

$$2x = y - 12$$

$$2x + 12 = y$$

$$f^{-1}(x) = 2x + 12$$

Check the validity of this by using the composition rules.

$$(f \circ f^{-1})(x) = f(f^{-1}(x)) = f(2x+12) = \frac{1}{2}(2x+12) - 6 = x + 6 - 6 = x$$

$$(f^{-1} \circ f)(x) = f^{-1}(f(x)) = f\left(\frac{1}{2}x - 6\right) = 2\left(\frac{1}{2}x - 6\right) + 12 = x - 12 + 12 = x$$

Let's also examine the sample table of values for both $f(x)$ and $f^{-1}(x)$.

x	$\frac{1}{2}x - 6$
–5	–8.5
–4	–8
–3	–7.5
–2	–7
–1	–6.5
0	–6
1	–5.5
2	–5
3	–4.5
4	–4
5	–3.5
6	–3
7	–2.5
8	–2

x	$2x + 12$
–8.5	–5
–8	–4
–7.5	–3
–7	–2
–6.5	–1
–6	0
–5.5	1
–5	2
–4.5	3
–4	4
–3.5	5
–3	6
–2.5	7
–2	8

Example 3

Determine the inverse: $f(x) = x^2$

Solution: $f(x) = x^2$

$$y = x^2$$

$$x = y^2$$

$$y = \pm\sqrt{x}$$

This means there is no function that can be named as the inverse of function $f(x)$ because with $y = \pm\sqrt{x}$, each x-value has two y-values. So $y = \pm\sqrt{x}$ is a relation, not a function. Examine the graph of the function and its mapping across the line $y = x$.

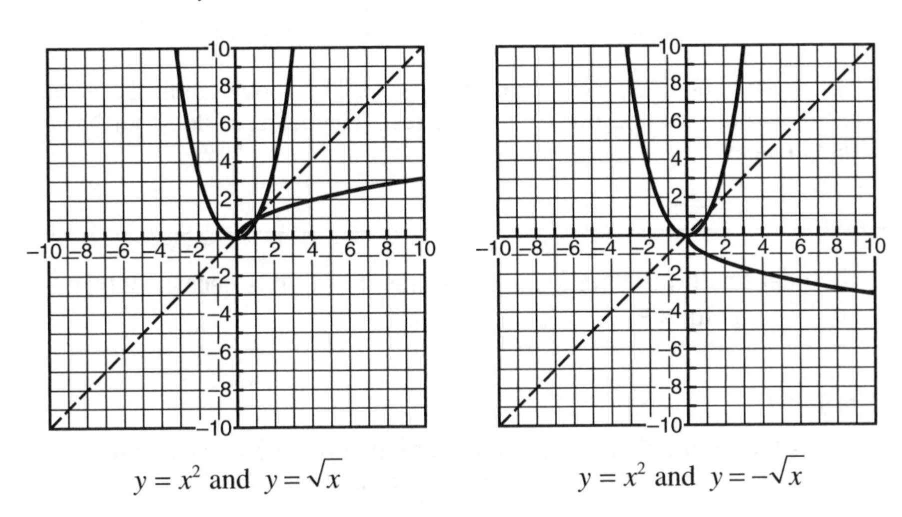

$y = x^2$ and $y = \sqrt{x}$ $y = x^2$ and $y = -\sqrt{x}$

Observe that $y = \sqrt{x}$ appears to be the reflection of $y = x^2$ on $(0,\infty)$ while $y = -\sqrt{x}$ appears to be the reflection of $y = x^2$ on $(-\infty, 0)$. This is not just a coincidence. Recall the horizontal line test for a one-to-one function. That is what is really affecting the function $y = x^2$ and causing the confusion seen here.

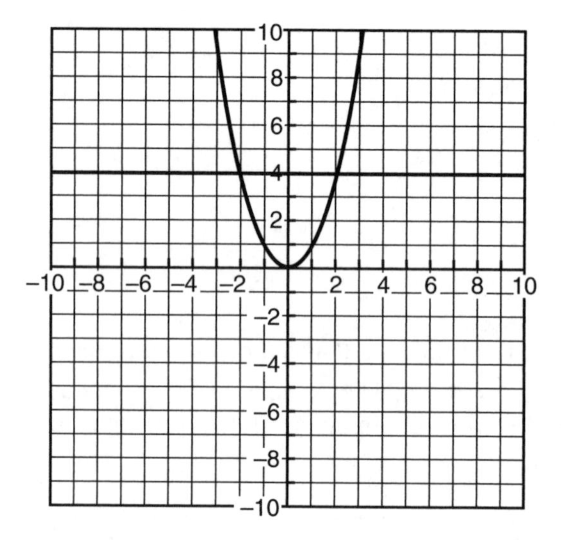

In fact, the function $f(x) = x^2$ is not one-to-one. The above graph shows that $y = x^2$ does not pass the horizontal line test. Therefore, it does not have a unique inverse.

TIP

In order for $f(x)$ to have a unique inverse, $f^{-1}(x)$, it is necessary that $f(x)$ be one-to-one.

Example 4

Which of the following functions has a unique inverse? In each case where there is an inverse, determine the inverse of that function.

$f(x) = x^3$, $g(x) = |x|$, $h(x) = 5x - 7$, and $j(x) = \sqrt{x+2}$

Solution: $f(x) = x^3$

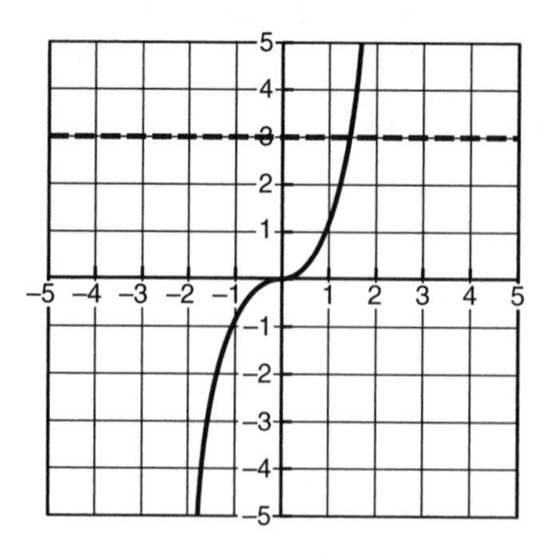

$g(x) = |x|$

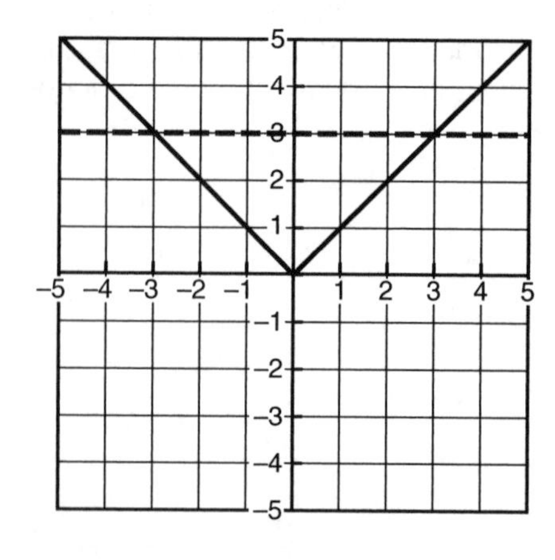

$h(x) = 5x - 7$ $\qquad\qquad$ $j(x) = \sqrt{x+2}$

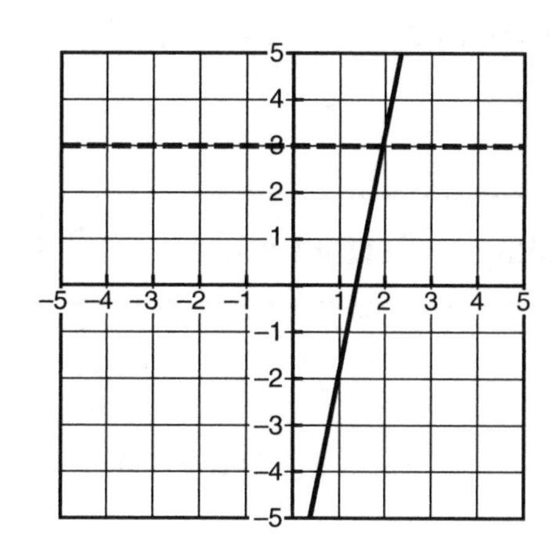

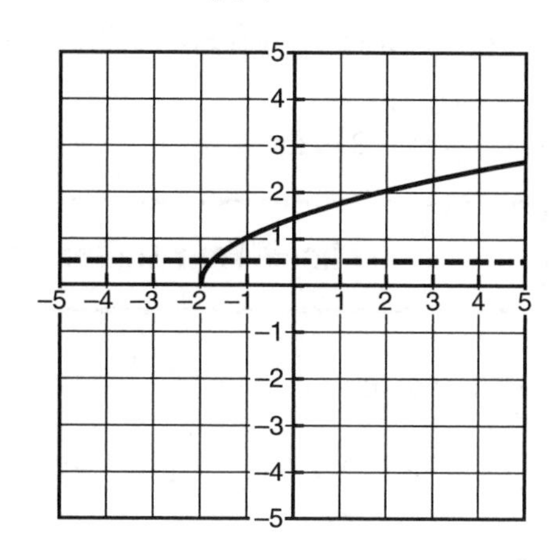

So $f(x) = x^3$, $h(x) = 5x - 7$, and $j(x) = \sqrt{x+2}$ pass the horizontal line test and have unique inverses. The development of these inverse functions follows.

$f(x) = x^3$

$y = x^3$

$x = y^3$

$y = \sqrt[3]{x}$

$f^{-1}(x) = \sqrt[3]{x}$

$h(x) = 5x - 7$

$y = 5x - 7$

$x = 5y - 7$

$5y = x + 7$

$y = \dfrac{x+7}{5}$

$y = \dfrac{1}{5}x + \dfrac{7}{5}$

$h^{-1}(x) = \dfrac{1}{5}x + \dfrac{7}{5}$

$j(x) = \sqrt{x+2}$

$y = \sqrt{x+2}$

$x = \sqrt{y+2}$

$x^2 = y + 2$

$y = x^2 - 2$

$j^{-1}(x) = x^2 - 2$

Test the inverses to see if they are correct.

$(f \circ f^{-1})(x) = f(f^{-1}(x)) = f\left(\sqrt[3]{x}\right) = \left(\sqrt[3]{x}\right)^3 = x$

$(f^{-1} \circ f)(x) = f^{-1}(f(x)) = f^{-1}(x^3) = \sqrt[3]{x^3} = x$

$f^{-1}(x) = \sqrt[3]{x}$ is the inverse.

$(h \circ h^{-1})(x) = h(h^{-1}(x)) = h\left(\dfrac{1}{5}x + \dfrac{7}{5}\right) = 5\left(\dfrac{1}{5}x + \dfrac{7}{5}\right) - 7 = x + 7 - 7 = x$

$(h^{-1} \circ h)(x) = h^{-1}(h(x)) = h^{-1}(5x - 7) = \dfrac{1}{5}(5x - 7) + \dfrac{7}{5} = x - \dfrac{7}{5} + \dfrac{7}{5} = x$

$h^{-1}(x) = \dfrac{1}{5}x + \dfrac{7}{5}$ is the inverse.

$$(j \circ j^{-1})(x) = j(j^{-1}(x)) = j(x^2 - 2) = \sqrt{(x^2 - 2) + 2} = \sqrt{x^2} = x$$
$$(j^{-1} \circ j)(x) = j^{-1}(j(x)) = j^{-1}\left(\sqrt{x+2}\right) = \left(\sqrt{x+2}\right)^2 - 2 = (x+2) - 2 = x$$

$j^{-1}(x) = x^2 - 2$ is the inverse.

Check Your Understanding of Section 4.2

1. The inverse of $g(x) = \frac{-1}{x}$ is equal to:

(1) $g^{-1}(x) = \frac{1}{x}$ (2) $g^{-1}(x) = \frac{-1}{x}$ (3) $g^{-1}(x) = x$ (4) $g^{-1}(x) = -x$

2. The inverse of $p(x) = 3x - 1$ is equal to:

(1) $p^{-1}(x) = \frac{1}{3}x - 1$
(2) $p^{-1}(x) = \frac{-1}{3}x + 1$
(3) $p^{-1}(x) = \frac{1}{3}x - \frac{1}{3}$
(4) $p^{-1}(x) = \frac{1}{3}x + \frac{1}{3}$

3. The inverse of $f(x) = x^3 + 5$ is equal to:

(1) $f^{-1}(x) = \sqrt[3]{x-5}$
(2) $f^{-1}(x) = \sqrt[3]{x-5}$
(3) $f^{-1}(x) = \sqrt[3]{x+5}$
(4) $f^{-1}(x) = x^3 - 5$

4. The inverse of $g(x) = \frac{1}{\sqrt{x}}$ is equal to:

(1) $g^{-1}(x) = \sqrt{x}$ (2) $g^{-1}(x) = \frac{1}{x}$ (3) $g^{-1}(x) = \frac{1}{x^2}$ (4) $g^{-1}(x) = x^2$

5. $f(x)$ is defined by the set of ordered pairs $\{(2, 5), (3, 7), (-1, 4), (6, -2), (5, 3)\}$. $f^{-1}(x) =$

(1) $\{(5, 2), (7, 3), (-1, 4), (6, -2), (3, 5)\}$
(2) $\{(5, 3), (6, -2), (-1, 4), (3, 7), (2, 5)\}$
(3) $\{(5, 2), (7, 3), (4, -1), (-2, 6), (3, 5)\}$
(4) $f^{-1}(x)$ does not exist

6. Which of the following functions does not have a unique inverse function?

(1) $f(x) = x^5$ (2) $g(x) = -4x + 6$ (3) $h(x) = \sqrt{x+4}$ (4) $j(x) = x^{\frac{2}{3}}$

7. $r(x) = x^2 - 2$ does not have a unique inverse. On which subset of the domain does it have a unique inverse?

(1) $\{x: x < 2\}$ (2) $\{x: x < 0\}$ (3) $\{x: x > 2\}$ (4) $\{x: x \neq 2\}$

8. Based on the graph of function $f(x)$ displayed below, which graph could represent the inverse of $f(x)$, $f^{-1}(x)$?

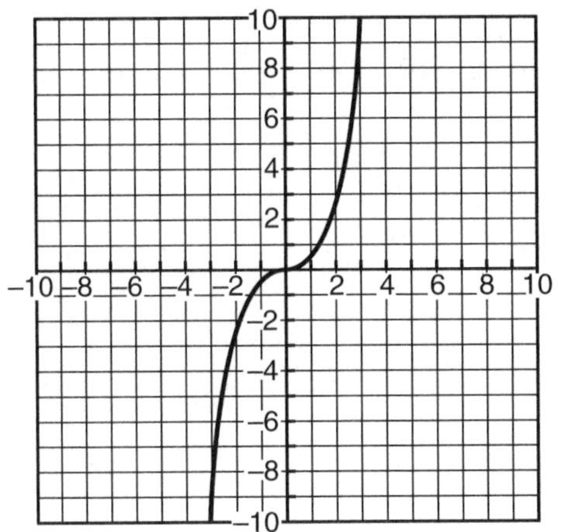

(1)

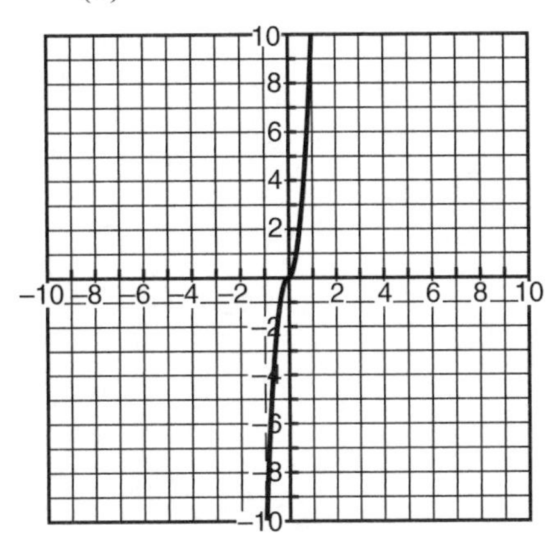

(2)

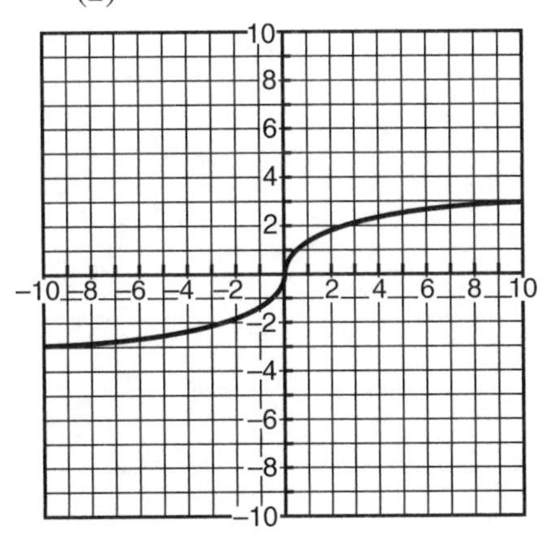

(3)

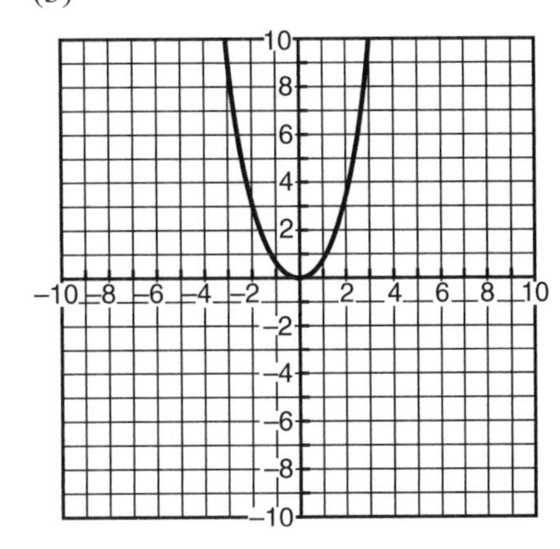

(4)

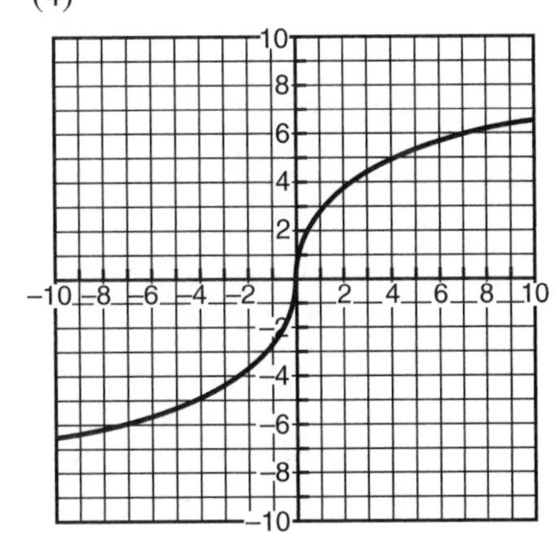

9. Determine the inverse of $g(x) = -7x + 3$ and use composition to justify your answer.

10. Determine the inverse of $h(x) = \frac{x}{x-2}$ and use composition to justify your answer.

Chapter Five TRANSFORMATIONS OF FUNCTIONS

5.1 PARENT FUNCTIONS AND TRANSFORMATIONS

Do you remember transformations, such as line reflections, translations, and rotations? These same transformations can be applied to basic functions. For instance, a translation on $f(x) = x^2$ can be used to determine the graph of $f(x + a)$.

Parent Functions and Transformations

There are certain basic functions whose graphs should now be easily recognizable. Transformations will be applied to the graphs of these functions. Because the graphs of the transformations will depend on the graph of the original functions, it is convenient to call the original functions parent functions.

Here is a library of parent functions that will be used with transformations of the plane. In subsequent chapters, other functions will be added to this library of parent functions. Also, make note of the specific ordered pairs associated with each parent function as written under its graph.

$f(x) = x^2$

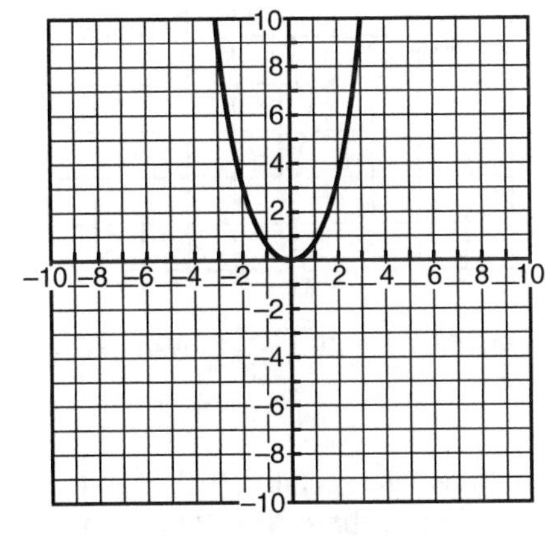

passes through (0, 0), (1, 1) and (–1, 1)

$f(x) = x^3$

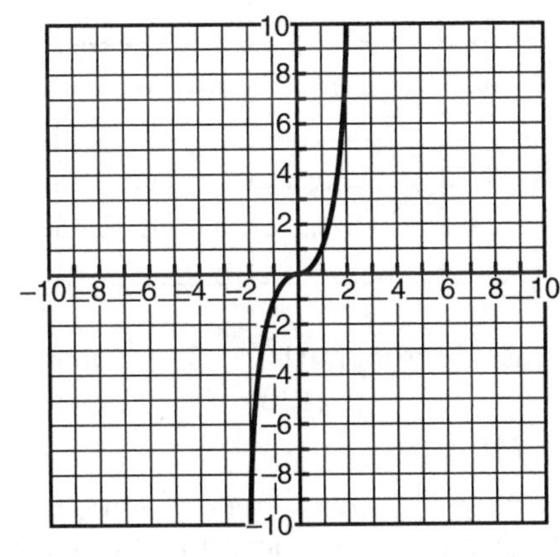

passes through (0, 0), (1, 1) and (–1, –1)

$f(x) = \sqrt{x}$

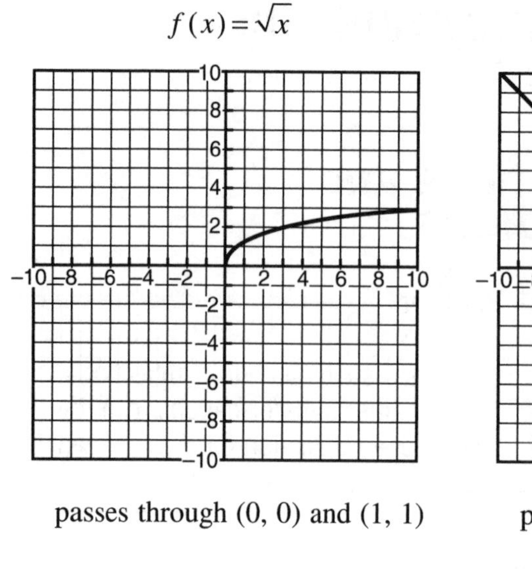

passes through (0, 0) and (1, 1)

$f(x) = |x|$

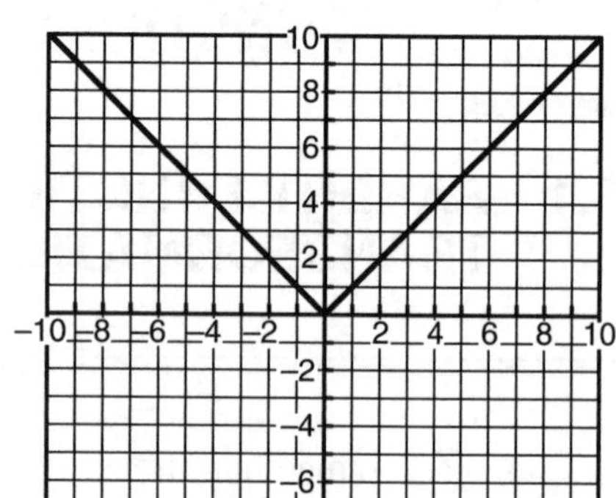

passes through (0, 0), (1, 1), and (–1, 1)

In addition to these functions, two other functions should be included in this library of parent functions. Their graphs may not be as familiar as the previous ones.

$f(x) = \sqrt[3]{x}$

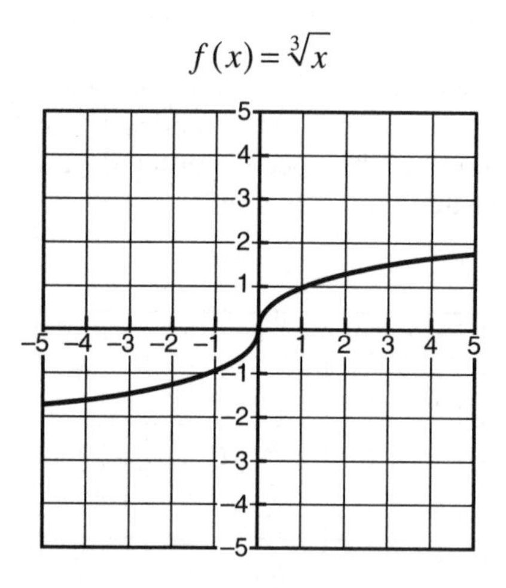

passes through (0, 0), (1, 1), and (–1, –1)

$f(x) = \dfrac{1}{x}$

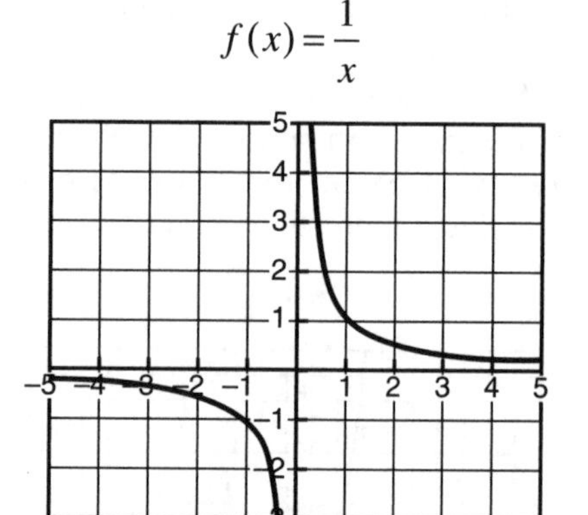

passes through (1, 1) and (–1, –1)

Now that this library of parent functions has been established, it is time to examine how transformations can be applied to them. For this purpose, the function $f(x) = x^2$ will be examined first. It is generally the most familiar function to everyone because it is the most basic parabola.

Two different translations can be applied to a function—a horizontal shift and a vertical shift. Of course, both types of shifts can be applied to the same function simultaneously. A vertical shift takes place when a constant is added to (or subtracted from) a function. Examine the two vertical shifts shown below:

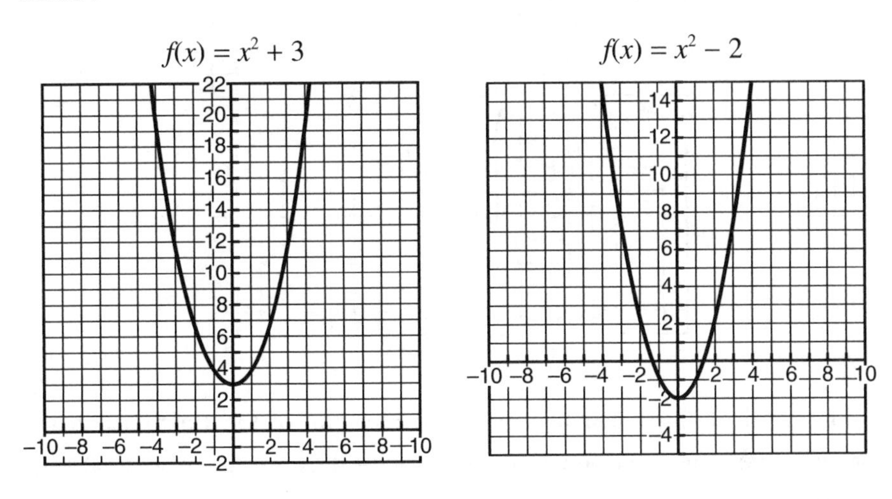

Notice that these two curves are the same shape as the curve graphed for $f(x) = x^2$ except that the location of the curve on the grid is different. In fact, examine the three ordered pairs that were listed in the library of parent functions for the graph of $f(x) = x^2$. Those ordered pairs were (0, 0), (1, 1) and (−1, 1). For the graph of $f(x) = x^2 + 3$, these ordered pairs were moved (or transformed) to (0, 3), (1, 4), and (−1, 4). Notice that the number 3 that was added in the function notation was added to the *y*-values of the original three ordered pairs. Similarly, when 2 is subtracted from the *y*-values of the original ordered pairs, the results are the corresponding ordered pairs found in the graph of $f(x) = x^2 - 2$. These ordered pairs are (0, −2), (1, −1), and (−1, −1). They are indeed found in the graph of $f(x) = x^2 - 2$.

Example 1

Sketch a graph of the function $f(x) = x^2 - 5$

Solution: The ordered pairs (0, 0), (1, 1), and (−1, 1) must be transformed by subtracting 5 from the *y*-values in each of the ordered pairs, resulting in (0, −5), (1, −4), and (−1, −4). Using these ordered pairs, the graph of $g(x) = x^2$ will be moved with a vertical shift of −5. The graph below shows the final result of this transformation.

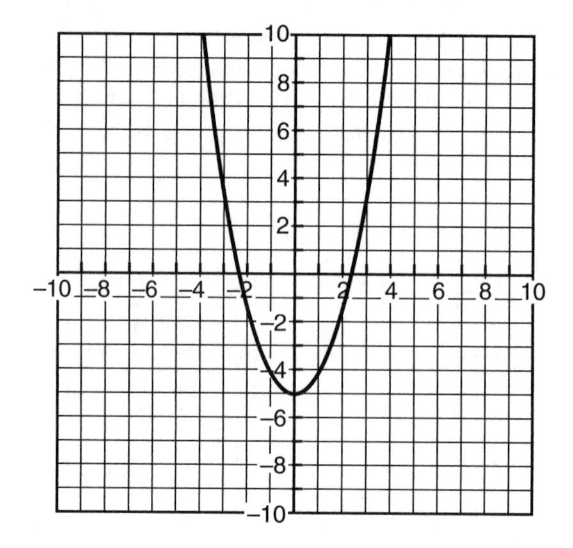

Similar vertical shifts can be applied to any one of the functions in the library. For instance, the function $h(x)=\sqrt[3]{x}+4$ is a positive vertical shift of 4 units on the function $f(x)=\sqrt[3]{x}$. It shifts the ordered pairs (0, 0), (1, 1), and (−1, −1) to (0, 4), (1, 5), and (−1, 3).

A sketch of $h(x)=\sqrt[3]{x}+4$ is shown below:

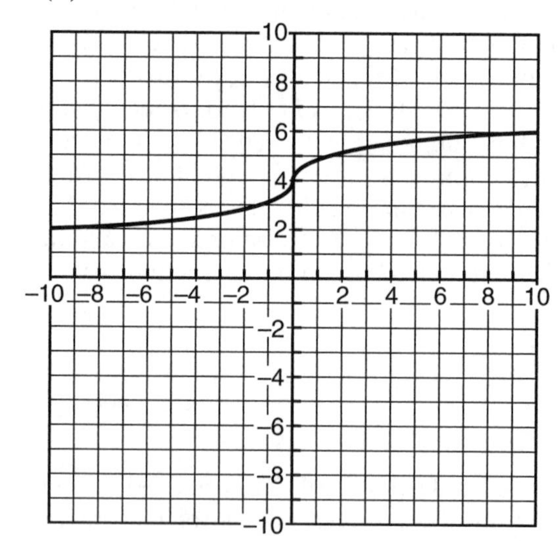

MATH FACTS

A vertical shift of a units on function $f(x)$ is written as $f(x) + a$ and adds a to the y-values of the ordered pairs that satisfy $f(x)$.

Example 2

Sketch a graph of the function $g(x) = x^3 + 2$

Solution: The ordered pairs (0, 0), (1, 1), and (−1, −1) must be transformed by adding 2 to the y-values in each of the three ordered pairs, resulting in (0, 2), (1, 3), and (−1, 1).

The resulting graph is shown below:

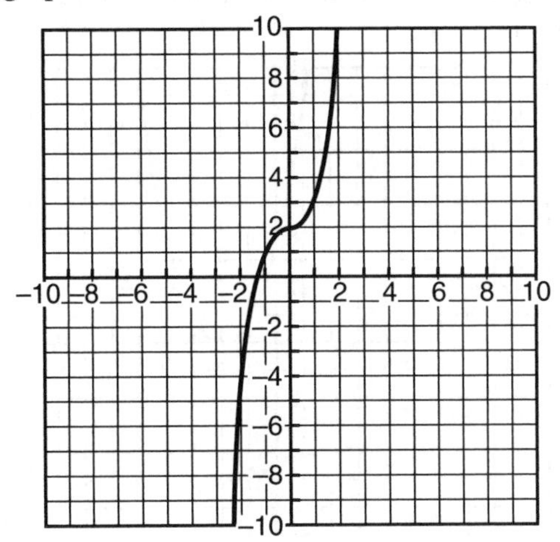

Example 3

Identify the function whose graph is shown below:

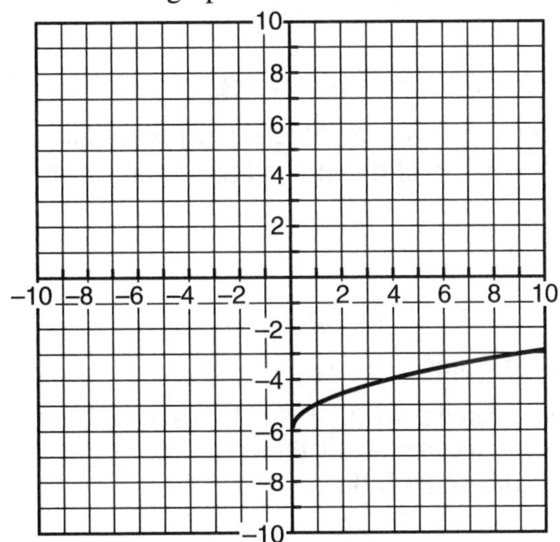

Solution: The graph shown appears to be similar to the graph of $f(x) = \sqrt{x}$. However, the graph has been translated vertically downward by 6 units. The function $f(x) = \sqrt{x}$ passes through the ordered pairs (0, 0) and

(1, 1). The graph on page 151 passes through (0, –6) and (1, –5). Therefore, the function is $g(x) = \sqrt{x} - 6$.

A horizontal shift occurs when a number is added or subtracted to the *x*-value within the function rule. For instance, a horizontal shift of 2 units on $f(x) = x^2$ has the functional rule $g(x) = (x - 2)^2$. Notice that a positive horizontal shift of 2 units subtracts a 2 within the functional rule. Below is the graph of $g(x) = (x - 2)^2$.

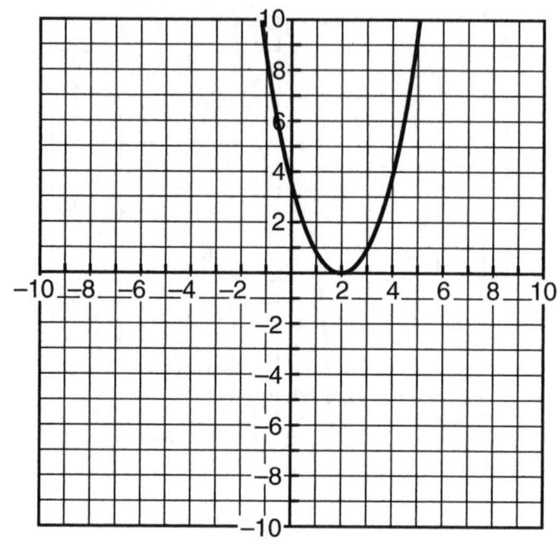

MATH FACTS

A horizonal shift of *a* units on function $f(x)$ is written as $f(x - a)$ and adds *a* to the *x*-values of the ordered pairs that satisfy $f(x)$.

Example 4

Sketch a graph of the function $g(x) = |x - 3|$.

Solution: The ordered pairs (0, 0), (1, 1), and (–1, 1) must be transformed by adding 3 to the *x*-values of all three ordered pairs, resulting in (3, 0), (4, 1), and (2, 1). The graph follows.

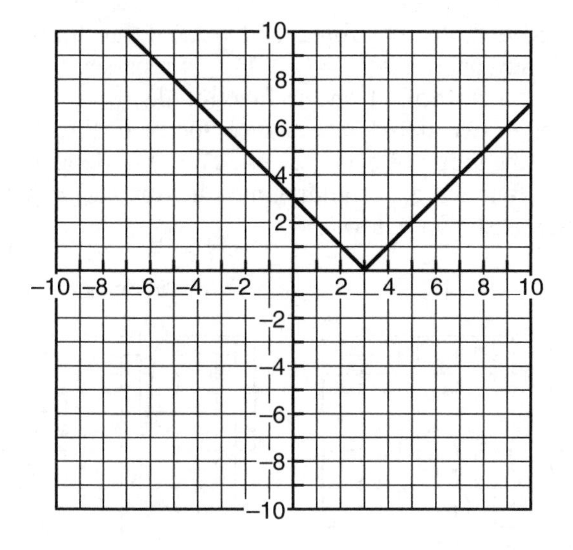

Example 5

Sketch a graph of the function $g(x) = \dfrac{1}{x+2}$.

Solution: The ordered pairs (1, 1) and (–1, –1) from $f(x) = \dfrac{1}{x}$ must be transformed by subtracting 2 from the x-values of each of these ordered pairs resulting, in (–1, 1) and (–3, –1). The graph is shown below:

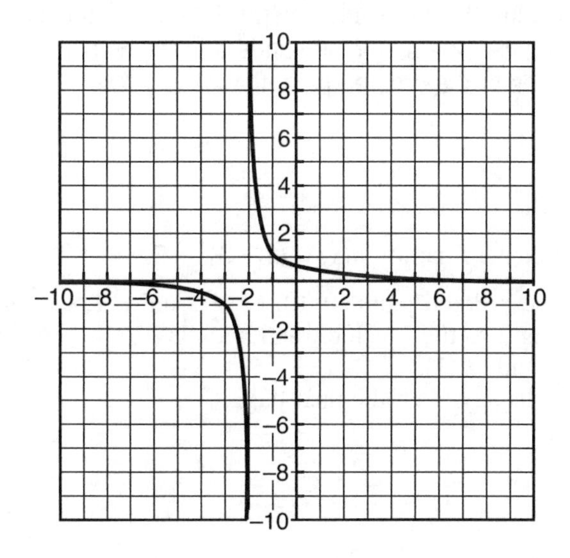

A function can have both a horizontal and vertical transformation applied to it. For instance, the function $h(x) = (x + 1)^2 - 4$ represents a vertical shift of -4 and a horizontal shift of -1 on the function $f(x) = x^2$. It can be sketched in either one step or two steps. For instance, the vertical shift can be applied first, resulting in the graph of $g(x) = x^2 - 4$. Then a horizontal shift of -1 can be applied to the function $g(x)$, yielding the function $h(x) = (x + 1)^2 - 4$. This two-stage sketch is shown below:

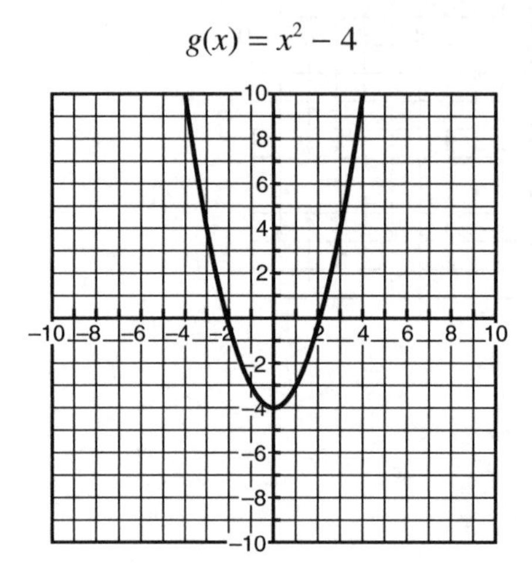

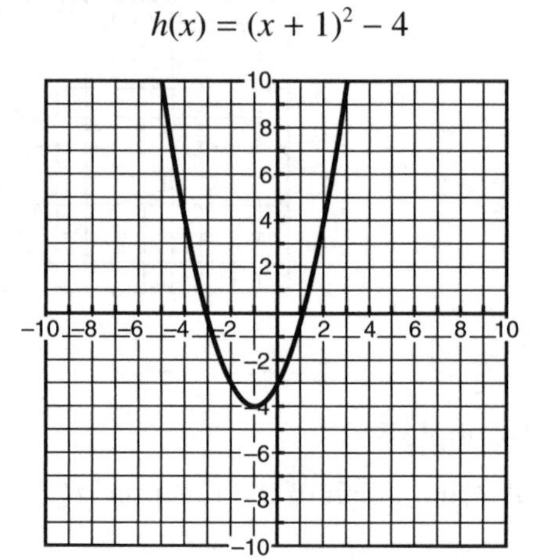

MATH FACTS

A horizonal shift of a units and a vertical shift of b units on function $f(x)$ is written as $f(x - a) + b$ and adds a to the x-values and b units to the y-values of the ordered pairs that satisfy $f(x)$.

Example 6

Sketch a graph of the function $h(x) = (x - 3)^3 + 2$.

Solution: The ordered pairs (0, 0), (1, 1), and (−1, −1) must be transformed by adding 3 to the x-values, and adding 2 to the y-values, resulting in (3, 2), (4, 3), and (2, 1).

The graph is displayed on the next page.

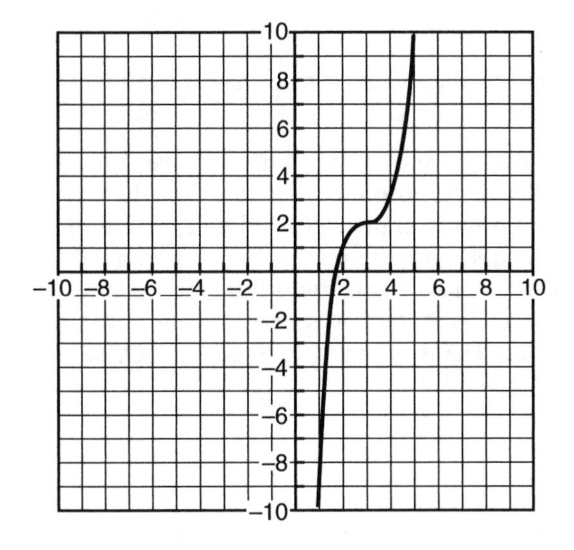

Example 7

Sketch a graph of the function $f(x) = \sqrt{x+3} - 1$.

Solution: The ordered pairs (0, 0) and (1, 1) must be transformed by subtracting 3 from the *x*-values and subtracting 1 from the *y*-values of these ordered pairs. The result is (–3, –1) and (–2, 0). A sketch of the graph of *f*(*x*) is displayed.

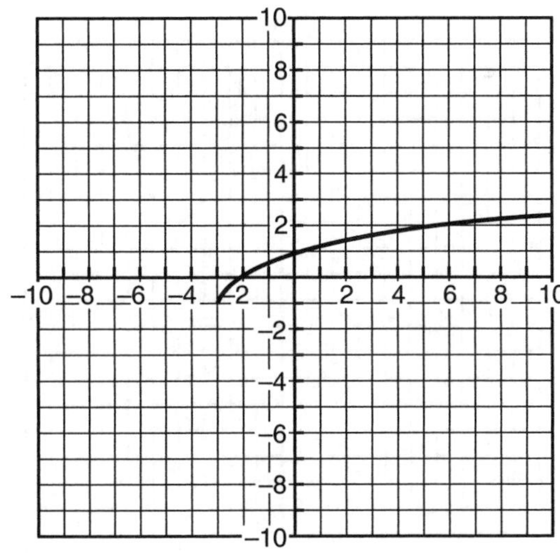

Example 8

Describe the transformations that map the function $f(x) = |x|$ to the function $g(x) = |x - 5| + 2$.

Solution: Since 5 is subtracted from the x inside the absolute value function and 2 is added to the absolute value function, the transformation involves a horizontal shift 5 units to the right and a vertical shift 2 units up.

This same process can be applied to sketch a graph of a quadratic function or parabola. Recall that all quadratic functions can be written in the form $f(x) = ax^2 + bx + c$. To do this, it is necessary to utilize the process of completing the square. The next example uses this method.

Example 9

Sketch a graph of the function $f(x) = x^2 - 8x + 19$.

Solution: Complete the square for the first two terms of $f(x) = x^2 - 8x + 19$. To do this, take one-half of the coefficient of x, -8, which is -4. Now square the number, $(-4)^2 = 16$. However, it is not valid just to add 16 because that changes the function rule. So, instead, add the 16 to the $x^2 - 8x$ and subtract the 16 from the rest of the function rule, 19. The result is $f(x) = x^2 - 8x + 16 + 19 - 16 = x^2 - 8x + 16 + 3 = (x - 4)^2 + 3$. This reveals that the function $f(x) = x^2 - 8x + 19$ is a horizontal shift 4 units to the right and a vertical shift 3 units up. Its graph can be displayed.

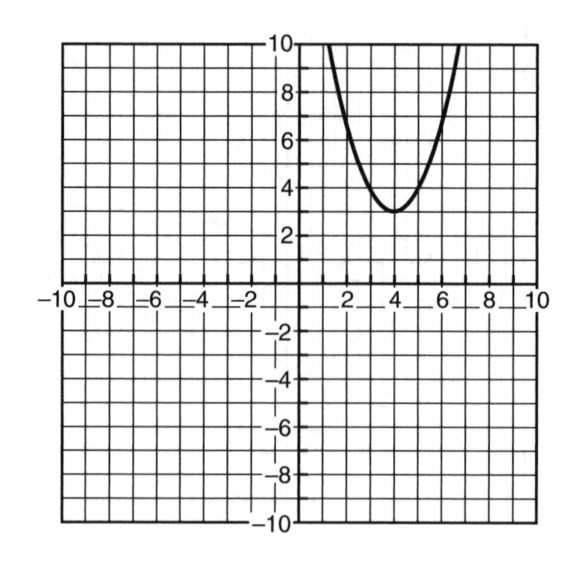

Reflections over the x-axis can be observed by comparing $f(x)$ with $-f(x)$. This can be seen in the following two examples.

$f(x) = x^2$ and $-f(x) = -x^2$

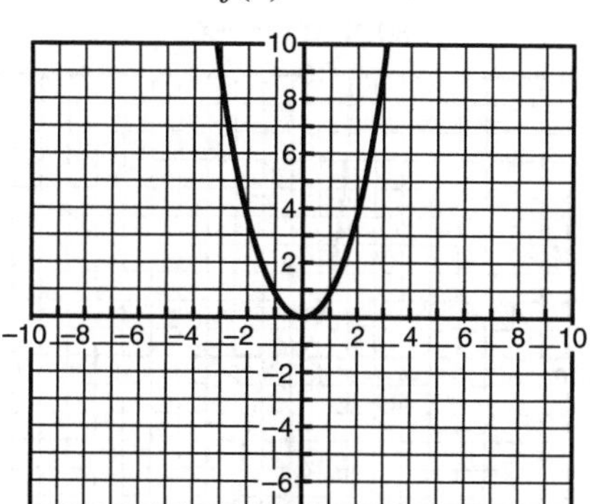

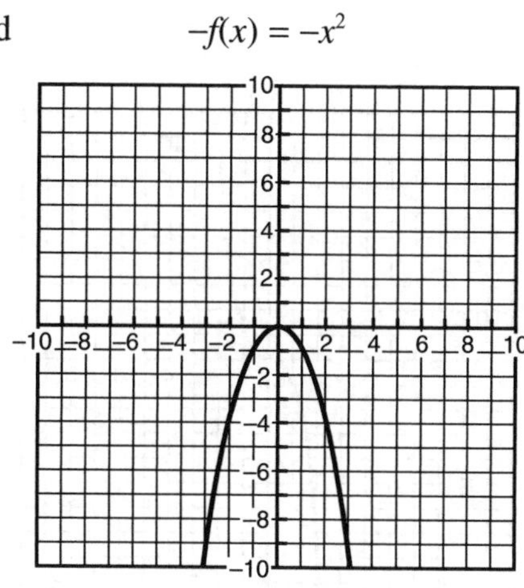

$f(x) = \sqrt{x}$ and $-f(x) = -\sqrt{x}$

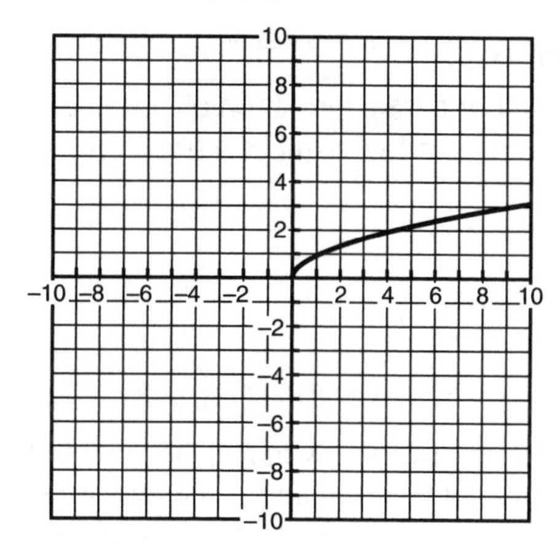

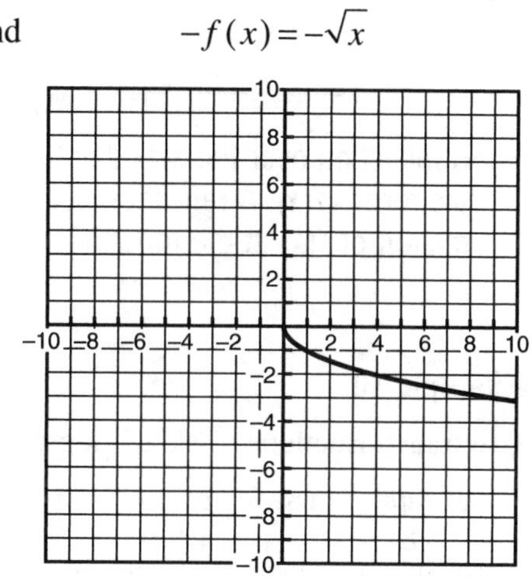

Reflections over the y-axis can be observed by comparing $f(x)$ with $f(-x)$. This can be seen in the following example.

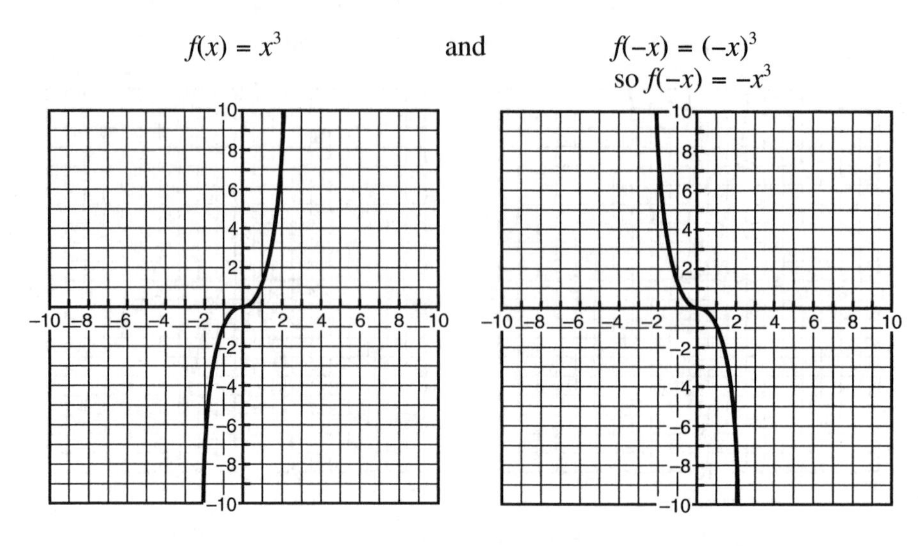

MATH FACTS

A reflection over the x-axis is the result of a transformation of $f(x)$ in the form $-f(x)$. A reflection over the y-axis is the result of a transformation of $f(x)$ in the form $f(-x)$.

Example 10

Use transformations to sketch a graph of the function $f(x) = -x^2 - 8x - 7$.

Solution: First factor out -1 from the first two terms in $f(x)$. That is, $f(x) = -x^2 - 8x - 7 = -1(x^2 + 8x) - 7$. Now complete the square for the two terms inside the parentheses, yielding $f(x) = -1(x^2 + 8x) - 7 = -1(x^2 + 8x + 16) - 7 + 16 = -1(x + 4)^2 + 9$. Notice that although 16 was added to the expression inside the parentheses, 16 was also added to the -7. That is because the 16 added to the $x^2 + 4x$ was multiplied by -1 so it was actually subtracted and, therefore, needed to be added to cancel out with the -16. At any rate, the graph can be sketched since the -1 indicates a reflection over the x-axis. That is combined with a horizontal shift of 4 units to the left and a vertical shift of 9 units upward.

The graph is displayed below.

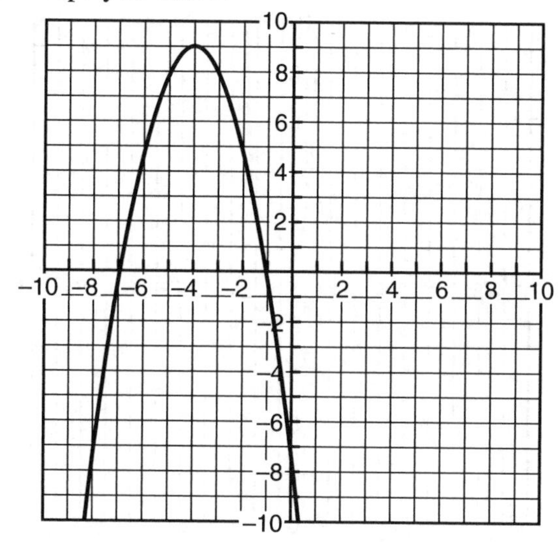

A final type of transformation occurs when the function rule is multiplied by a constant. Compare the graph of $f(x) = x^2$ with the graph of $g(x) = 2x^2$. How does the 2 in function $g(x)$ transform $f(x)$?

$f(x) = x^2$

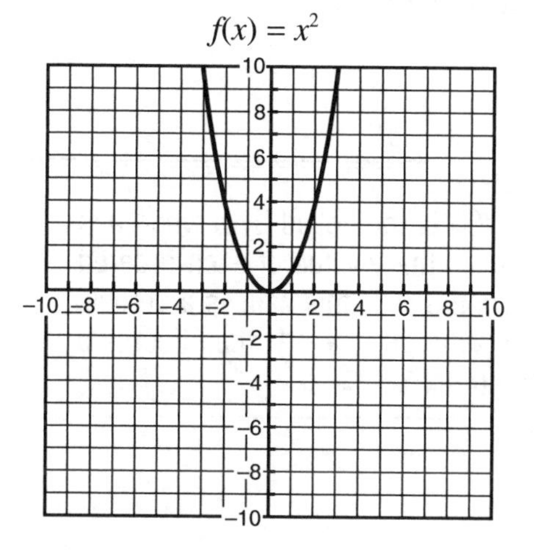

$g(x) = 2x^2$

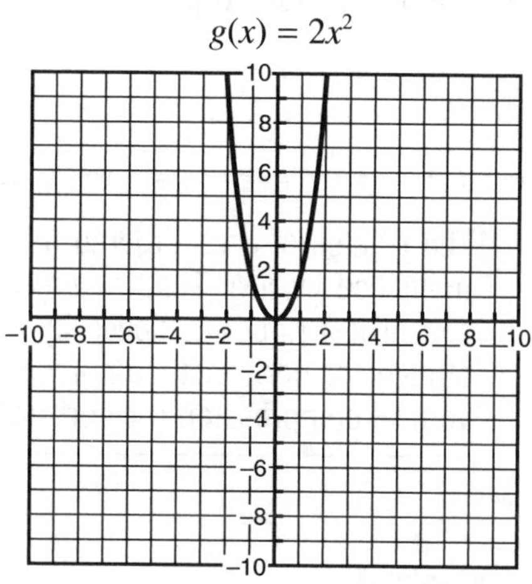

Notice how $g(x)$ appears to be narrower than $f(x)$. Instead of passing through (0, 0), (1, 1), and (−1, 1), $g(x)$ passes through (0, 0), (1, 2), and (−1, 2). Now see what happens if the function rule is multiplied by 4 instead of 2.

$h(x) = 4x^2$

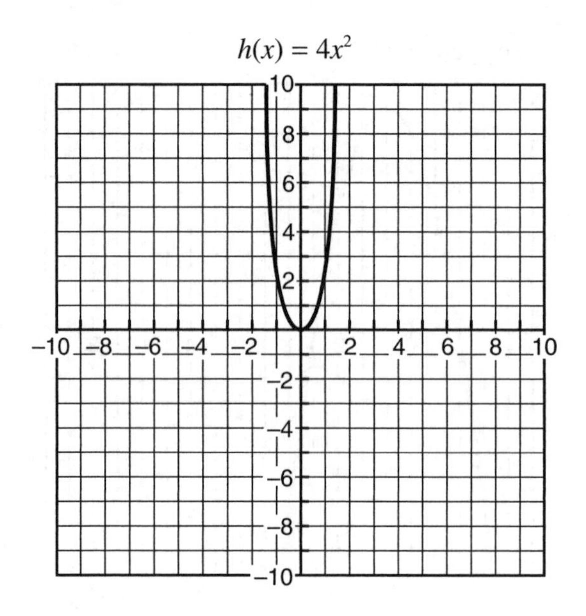

$h(x) = 4x^2$ passes through (0, 0), (1, 4), and (−1, 4). The relation between the function rule and the corresponding graph should be apparent. The constant that is multiplied by the x in the function rule also is multiplied by the y-value of each of the ordered pairs in the original function.

MATH FACTS

The y-value of each ordered pair of $f(x)$ is multiplied by a when a is multiplied by $f(x)$. That is, $a \cdot f(x)$ causes the y-value of each ordered pair of $f(x)$ to be multiplied by a. When $a > 1$, this causes a vertical stretch (or horizontal compression). When $0 < a < 1$, the effect is a vertical compression (or horizontal stretch).

Example 11

Use transformations to sketch a graph of the function $f(x) = 3x^3$.

Solution: Since $g(x) = x^3$ passes through (0, 0), (1, 1), and (−1, −1), $f(x) = 3x^3$ passes through (0, 0), (1, 3), and (−1, −3). A sketch of $f(x) = 3x^3$ is display on the next page.

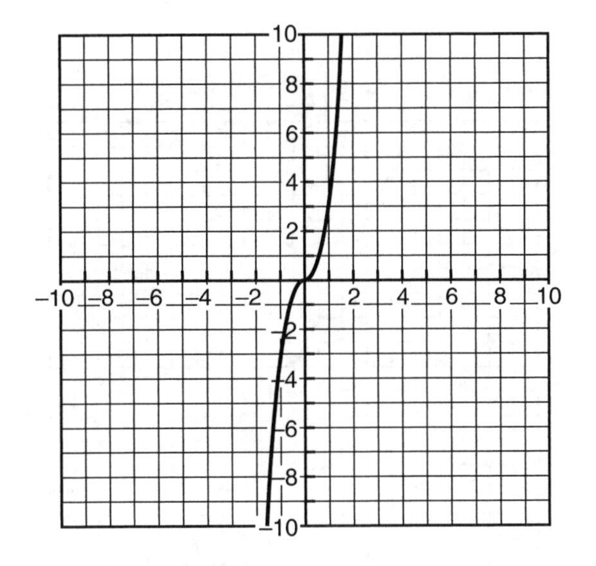

Example 12

Use transformations to sketch a graph of the function $f(x)=\frac{1}{3}x^3$.

Solution: Since $g(x) = x^3$ passes through (0, 0), (1, 1), and (−1, −1), $f(x)=\frac{1}{3}x^3$ passes through (0, 0), (1, $\frac{1}{3}$), and (−1, $-\frac{1}{3}$).

A sketch of $f(x)=\frac{1}{3}x^3$ is displayed below.

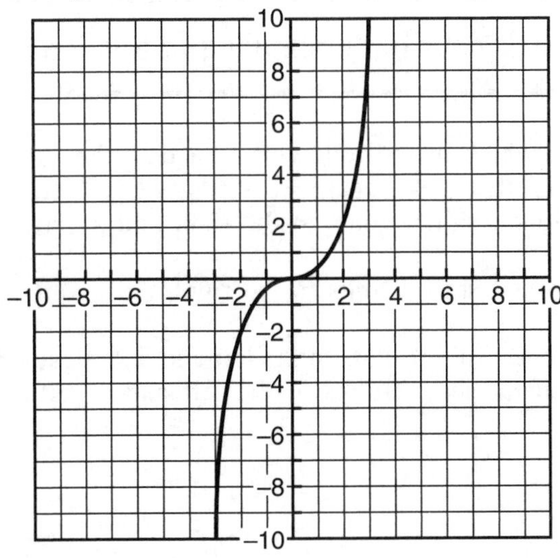

Example 13

Identify the function rule for the graph shown below.

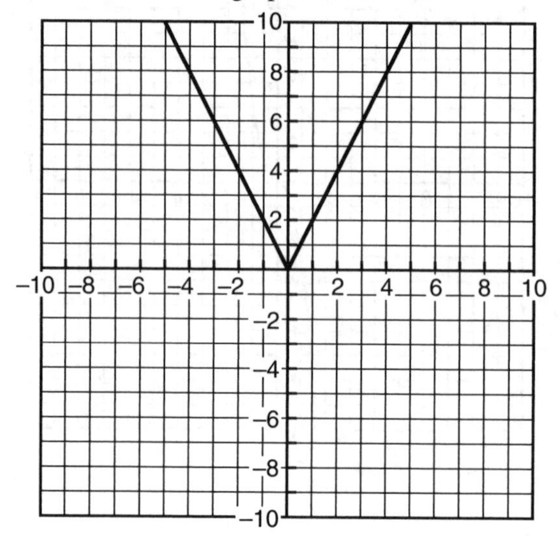

Solution: The basic shape of the graph appears to be an absolute value function. Rather than passing through (0, 0), (1, 1), and (–1, 1), this graph passes through (0, 0), (1, 2), and (–1, 2). Therefore, the function is $f(x) = |2x|$.

Check Your Understanding of Section 5.1

1. If $f(x) = x^3$, which of the following best describes the transformation applied to $f(x)$ that yields $g(x) = (x + 2)^3 - 4$?

(1) A horizontal shift 2 units to the right and a vertical shift 4 units down

(2) A horizontal shift 4 units to the right and a vertical shift 2 units up

(3) A horizontal shift 4 units to the left and a vertical shift 2 units up

(4) A horizontal shift 2 units to the left and a vertical shift 4 units down

2. If $f(x) = |x|$, which of the following best describes the transformation applied to $f(x)$ that yields $g(x)=\frac{1}{2}|x-1|$?

(1) A vertical compression scale factor of $\frac{1}{2}$ and a vertical shift 1 unit down

(2) A vertical expansion scale factor of $\frac{1}{2}$ and a vertical shift 1 unit down

(3) A vertical expansion scale factor of $\frac{1}{2}$ and a horizontal shift 1 unit to the right

(4) A vertical compression scale factor of $\frac{1}{2}$ and a horizontal shift 1 unit right

3. If $f(x)=\sqrt[3]{x}$, which of the following best describes the transformation applied to $f(x)$ that yields $g(x)=\frac{3}{2}\sqrt[3]{x+2}+1$?

(1) A horizontal compression of 1.5 with a horizontal shift 2 units to the left and a vertical shift up 1 unit
(2) A horizontal expansion of 1.5 with a horizontal shift 2 units to the left and a vertical shift up 1 unit
(3) A horizontal compression of 1.5 with a horizontal shift 2 units to the right and a vertical shift up 1 unit
(4) A horizontal expansion of 1.5 with a horizontal shift 2 units to the left and a vertical shift down 1 unit

4. Which of the following functions represents a transformation that moves the ordered pairs (0, 0) and (1, 1) from function $f(x)=\sqrt{x}$ to (–1, 1) and (0, 2), respectively?

(1) $g(x)=\sqrt{x-1}+1$
(2) $g(x)=\sqrt{x+1}+1$
(3) $g(x)=\sqrt{x+1}+2$
(4) $g(x)=\sqrt{x-1}+2$

5. Which of the following functions represents a transformation that moves the ordered pairs (0, 0), (1, 1), and (–1, 1) from function $f(x) = x^2$ to (–2, –3), (–1, –1), and (–3, –1), respectively?

(1) $g(x) = 2(x - 2)^2 - 3$
(2) $g(x) = 2(x - 2)^2 + 3$
(3) $g(x) = 2(x + 2)^2 - 3$
(4) $g(x) = 2(x + 2)^2 + 3$

6. The graph of $g(x) = \dfrac{1}{x}$ is shown below.

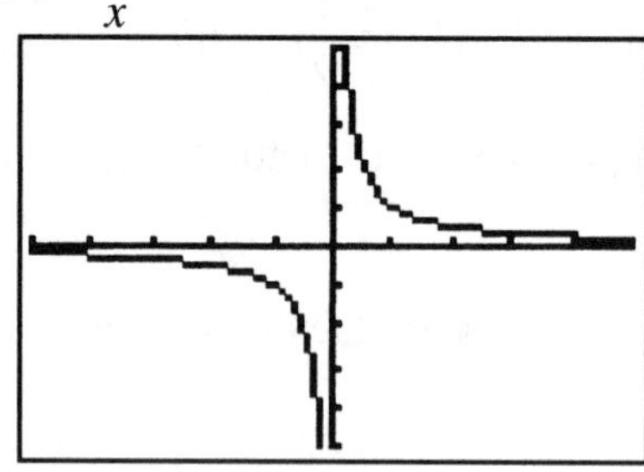

Which of the following graphs shows $g(x - 2) - 2$?

(1)

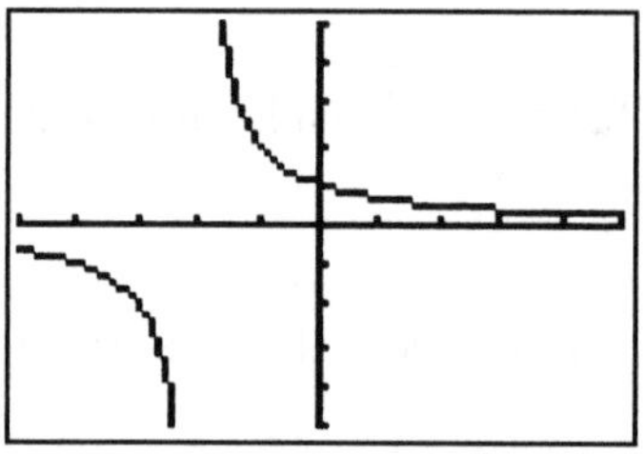

(3) 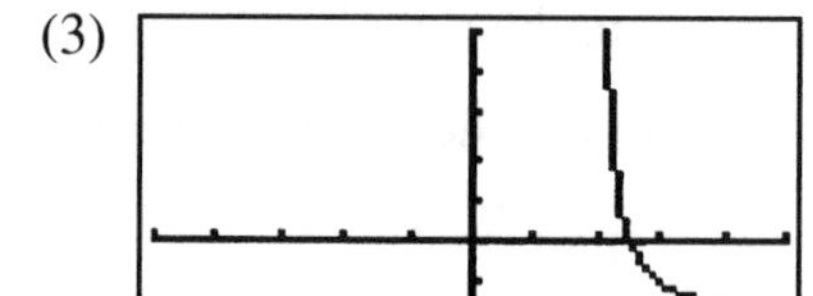

(2) 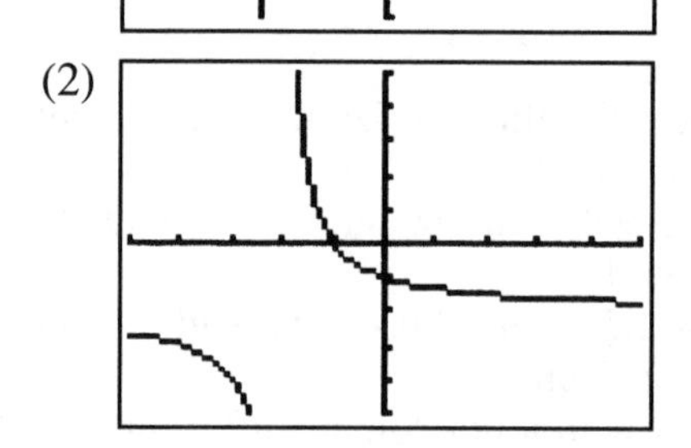

(4) 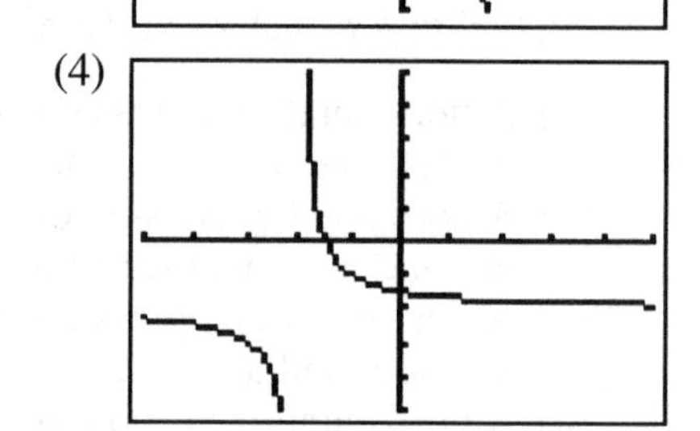

7. The graph of $f(x) = |x|$ has been moved through a transformation as displayed below. Which function results in this transformation?

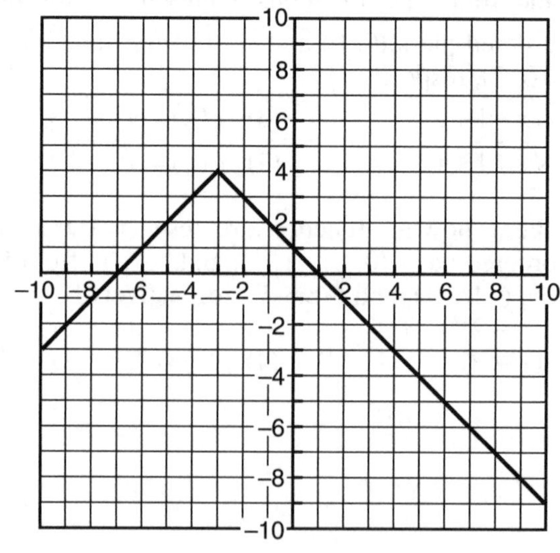

(1) $g(x) = -|x + 3| + 4$
(2) $g(x) = -|x - 3| + 4$
(3) $g(x) = |x + 3| + 4$
(4) $g(x) = -|x + 4| - 3$

8. Based on the graph of function $h(x)$ as shown below, sketch the graphs of the functions $h(x + 2)$, $h(x) - 1$ and $-h(x)$.

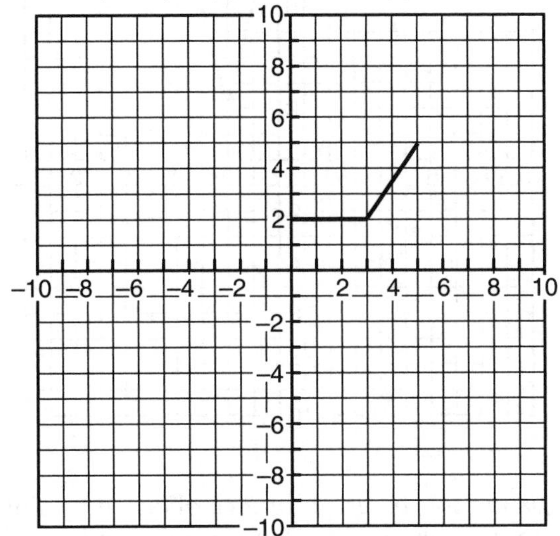

9. Use transformations to sketch a graph of the function $f(x) = 2x^2 + 12x + 17$.

10. Use transformations to sketch a graph of the function $f(x) = \frac{1}{2}x^2 - 4x + 3$.

11. Use transformations to sketch a graph of the function $g(x) = \frac{2}{x+3} + 1$.

12. Use transformations to sketch a graph of the function $h(x) = \frac{1}{3}(x-2)^3 - 4$.

13. A transformation on $g(x) = |x|$ results in the graph displayed below. Identify the function $f(x)$ for this graph.

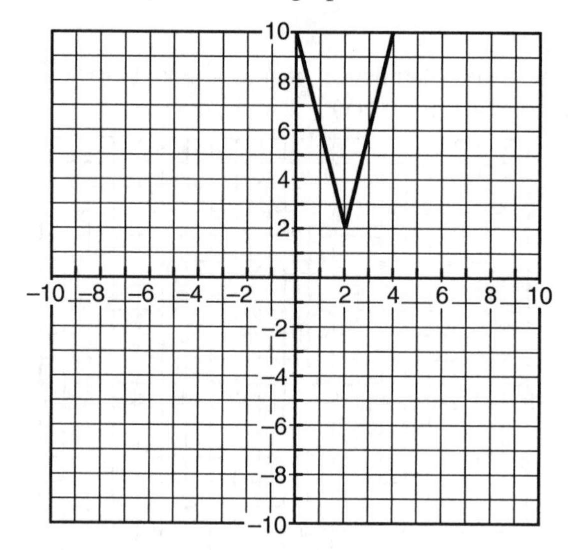

14. Write an equation for the graph of $f(x)$, obtained by shifting the graph of $h(x) = x^3$ to the right 2 units, down 3 units, and stretching the graph vertically by a factor of 2.

15. Describe the transformation that produces the graph of $h(x) = \frac{1}{2}\sqrt{x+1} - 3$

from the graph of the function $g(x) = \sqrt{x}$.

5.2 CIRCLES

KEY IDEAS

In Geometry, a circle was defined as the locus of points in a plane equidistant from a fixed point in the plane. This definition will be used to develop the general equation of a circle and to graph circles in the coordinate plane.

The Standard Form of the Equation of a Circle

Since a circle is the set of points at fixed distance from a point in a plane, it is possible to position a circle at the origin of the graph and determine its equation.

In the right triangle shown inside this circle with radius 1, the Pythagorean Theorem can be used to see that any ordered pair (x, y) on the circle is such that $x^2 + y^2 = 1$. Notice that a circle does not pass the vertical line test so, it is not a function.

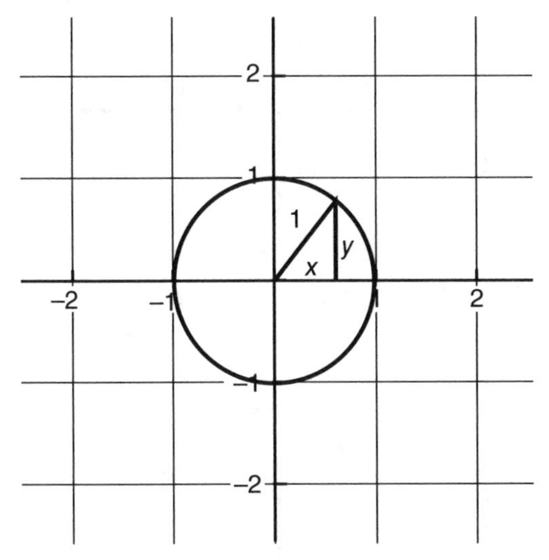

In fact, from this it is possible to realize that the equation of circle centered at the origin of the graph is $x^2 + y^2 = r^2$ because of the Pythagorean Theorem.

MATH FACTS

The equation of a circle centered at the origin with radius r is $x^2 + y^2 = r^2$.

Let's examine how to graph a circle such as $x^2 + y^2 = 25$.

PLEASE NOTE

Since a circle is not a function, it cannot be graphed on a graphing calculator by entering a rule in **Y =**. However, $x^2 + y^2 = r^2$ can be solved for y, resulting in $y = \pm\sqrt{r^2 - x^2}$. Therefore, enter $y = \sqrt{r^2 - x^2}$ in Y_1 and $y = -\sqrt{r^2 - x^2}$ in Y_2. Be careful to graph in ZSquare or the circle will look squashed.

Press **Y =** and enter $y = \sqrt{25 - x^2}$ in Y_1 and $y = -\sqrt{25 - x^2}$ in Y_2. In ZStandard, it looks like an ellipse.

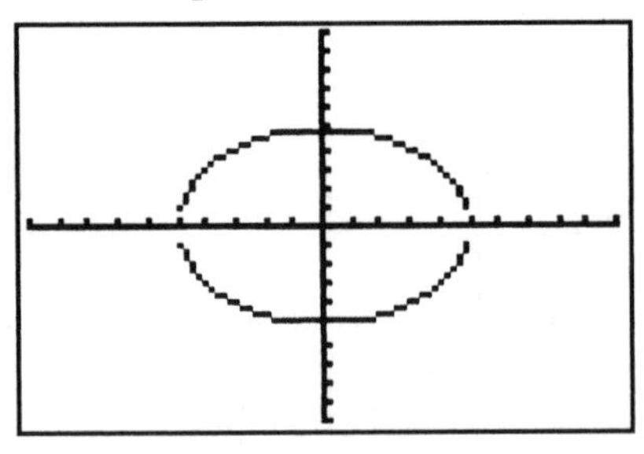

Press **ZOOM** **5** or ZSquare to get a better graph.

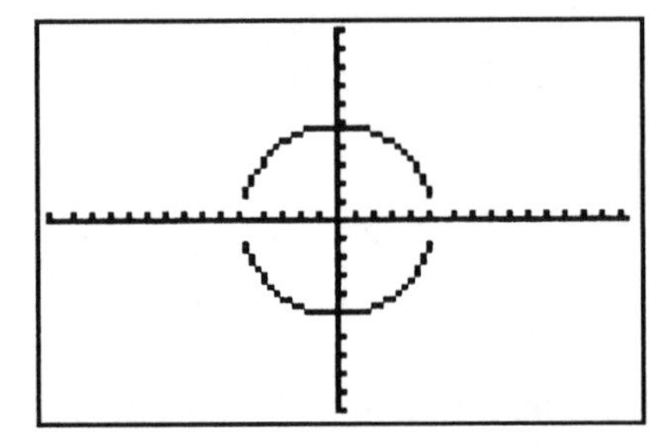

What if the graph of a circle is not centered at the origin? This, of course, would be the result of a translation applied to $x^2 + y^2 = r^2$. By the translation $T_{h,k}$ a circle whose equation is $x^2 + y^2 = r^2$ is mapped to $(x - h)^2 + (y - k)^2 = r^2$.

MATH FACTS

The equation of a circle whose center is (h, k) with radius r is $(x - h)^2 + (y - k)^2 = r^2$.

Writing the Equation of a Circle Given Its Center and Radius

Example 1

Write the equation of a circle whose center is at (2, 5) and whose radius is 3.

Solution: $h = 2$, $k = 5$, and $r = 3$. Replacing these values in the standard form of an equation of a circle with center (h, k), $(x - h)^2 + (y - k)^2 = r^2$, yields $(x - 2)^2 + (y - 5)^2 = 3^2$ or $(x - 2)^2 + (y - 5)^2 = 9$.

Example 2

Write the equation of a circle whose center is at (–4, 1) and passes through (3, –2).

Solution: $h = -4$, $k = 1$, but r is not given. However, r can be determined by using the distance formula from the center, (–4, 1), and a point on the circle, (3, –2). The distance formula is $d = \sqrt{(x_2 - x_1)^2 + (y_2 - y_1)^2}$. Replacing the values from these two ordered pairs into the formula results in $= \sqrt{((3-(-4))^2 + (-2) - 1)^2} = \sqrt{7^2 + (-3)^2} = \sqrt{49+9} = \sqrt{58}$. So $r = \sqrt{58}$. Replacing the values of h, k, and r into the standard form of an equation of a circle with center (h, k), $(x - h)^2 + (y - k)^2 = r^2$, yields $(x-(-4))^2 + (y-1)^2 = \left(\sqrt{58}\right)^2$ or $(x + 4)^2 + (y - 1)^2 = 58$.

Identifying the Center and Radius of a Circle Given Its Equation

Example 3

If the equation of a circle is $(x - 2)^2 + (y + 8)^2 = 36$, identify the center and radius of the circle. Then sketch its graph.

Solution: By comparing $(x - 2)^2 + (y + 8)^2 = 36$ with the standard form of the equation of a circle, $(x - h)^2 + (y - k)^2 = r^2$, the values of h, k, and r can be determined as $h = 2$, $k = -8$, and $r = \sqrt{36} = 6$. So the center is at (2, –8) and its radius is 6. Here is a sketch of this circle.

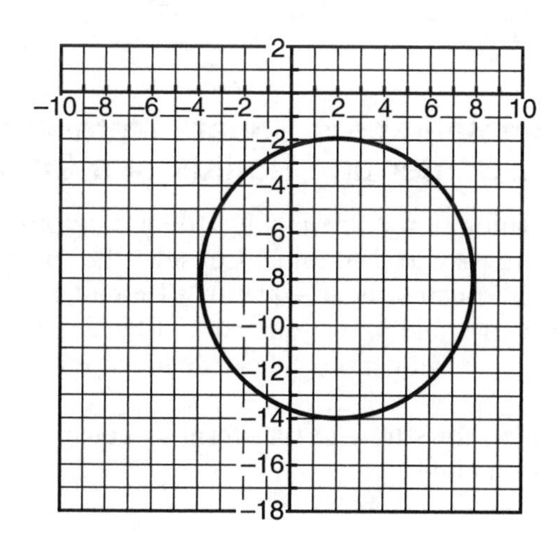

Writing the Equation of a Circle Given Its Graph

It has already been shown that once the center and radius of a circle are known, the equation of the circle can be written. That means that if a graph of a circle is provided, then the center and radius can be determined and, therefore, the equation can also be written.

Example 4

Determine the equation of the circle in the graph.

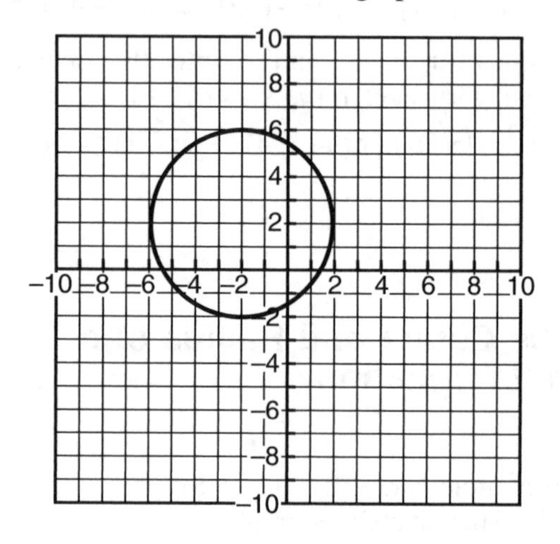

Solution: The center of this circle appears to be $(-2, 2)$ with a radius of 4 so $h = -2$, $k = 2$, and $r = 4$.

Replacing these values in the standard form of an equation of a circle with center (h, k), $(x - h)^2 + (y - k)^2 = r^2$, yields $(x - (-2))^2 + (y - 2)^2 = 4^2$ or $(x + 2)^2 + (y - 2)^2 = 16$.

Using the Completing the Square Method to Determine the Center and Radius of a Circle

Sometimes the equation of a circle is not written in a form from which the center and radius are easy to determine. Consider the equation $x^2 + 6x + y^2 - 8y - 39 = 0$. This equation is an equation of a circle. The reason it is a circle is because both x and y are squared and have the same coefficient. That is an indication of an equation that can be transformed into the standard form of a circle.

The next example shows how to transform an equation to determine the center and radius of a circle.

Example 5

Determine the center and radius of the circle whose equation is $x^2 + 6x + y^2 - 8y - 39 = 0$.

Solution: The process to transform this equation into the standard form of the equation of a circle involves completing the square of a trinomial. The square of a trinomial will be completed for the first two terms, $x^2 + 6x$, and also for $y^2 - 8y$. In $x^2 + 6x$, the coefficient of the middle term is 6. Taking half of 6 and squaring results in 9. Similarly, taking half of -8 and squaring results in 16. The perfect square trinomials thus found are $x^2 + 6x + 9$ and $y^2 - 8y + 16$. However, adding 9 and 16 to these terms changes the equation. In order not to change the equation, first add the 39 to both sides of the equation. The resulting equation is $x^2 + 6x + y^2 - 8y = 39$. If 9 plus 16 are added to the left side of the equation, their sum, or 25, must be added to the right side of the equation. The equation is now $x^2 + 6x + 9 + y^2 - 8y + 16 = 39 + 25$ or $x^2 + 6x + 9 + y^2 - 8y + 16 = 64$. Factoring the two perfect square trinomials puts the equation into standard form, $(x + 3)^2 + (y - 4)^2 = 64$. Therefore, the center is $(-3, 4)$ and the radius is 8.

Example 6

Determine the center and radius of the circle whose equation is $3x^2 + 12x + 3y^2 + 6y - 12 = 0$.

Solution: Similar to Example 5, both x and y are squared but their coefficients are both 3. That still indicates a circle. To complete the square requires first factoring out the 3.

$3x^2 + 12x + 3y^2 + 6y - 12 = 0$
$3(x^2 + 4x) + 3(y^2 + 2y) = 12$
$3(x^2 + 4x + 4) + 3(y^2 + 2y + 1) = 12 + 3 \cdot 4 + 3 \cdot 1$

Note: Adding 4 and 1 is not sufficient.

$3(x + 2)^2 + 3(y + 1)^2 = 12 + 12 + 3$
$3(x + 2)^2 + 3(y + 1)^2 = 27$
$3(x + 2)^2 + 3(y + 1)^2 = 3 \cdot 9$

Now divide by 3 to get

$(x + 2)^2 + (y + 1)^2 = 9$

The center is at $(-2, -1)$ and the radius is 3.

Check Your Understanding of Section 5.2

1. What is the equation of a circle with its center at (4, –5) and a radius of 7?

(1) $(x + 4)^2 + (y - 5)^2 = 49$
(2) $(x - 4)^2 + (y + 5)^2 = 7$
(3) $(x + 4)^2 + (y - 5)^2 = 7$
(4) $(x - 4)^2 + (y + 5)^2 = 49$

2. The equation of a circle is $(x - 1)^2 + (y - 3)^2 = 25$. What are the coordinates of its center and the length of its radius?

(1) (1, 3) and 5
(2) (–1, –3) and 5
(3) (1, 3) and 25
(4) (–1, –3) and 25

3. The center of a circular flower has a diameter of 6 centimeters and is located on a map of a garden (with units measured in centimeters) at the coordinates (4, –2). Which equation represents the flower?

(1) $(x + 4)^2 + (y - 2)^2 = 3$
(2) $(x + 4)^2 + (y - 2)^2 = 9$
(3) $(x - 4)^2 + (y + 2)^2 = 3$
(4) $(x - 4)^2 + (y + 2)^2 = 9$

4. Which equation represents the equation of the circle shown in the graph below?

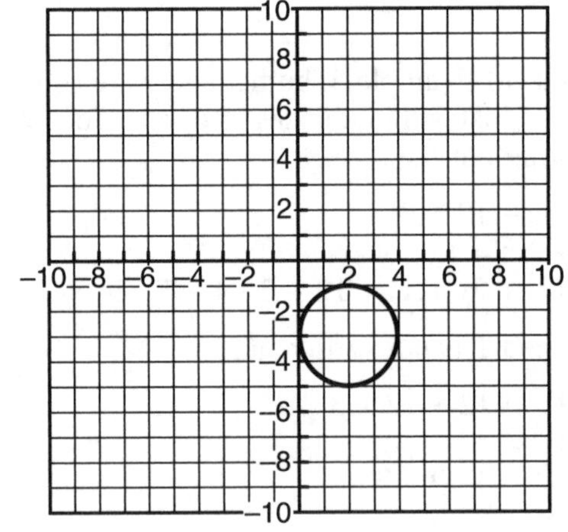

(1) $(x - 2)^2 + (y + 3)^2 = 2$
(2) $(x - 2)^2 + (y + 3)^2 = 4$
(3) $(x + 2)^2 + (y - 3)^2 = 2$
(4) $(x + 2)^2 + (y - 3)^2 = 4$

5. What are the coordinates for the center and the length of the radius of a circle whose equation is $2(x + 3)^2 + 2(y - 2)^2 = 50$?

(1) (3, –2) and 25
(2) (3, –2) and $\sqrt{50}$
(3) (–3, 2) and 5
(4) (3, –2) and 5

6. Which is the graph of $(x - 1)^2 + (y - 4)^2 = 16$?

(1)

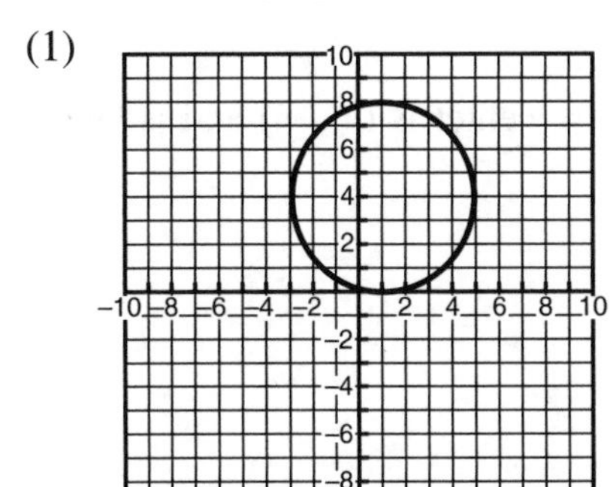

(3)

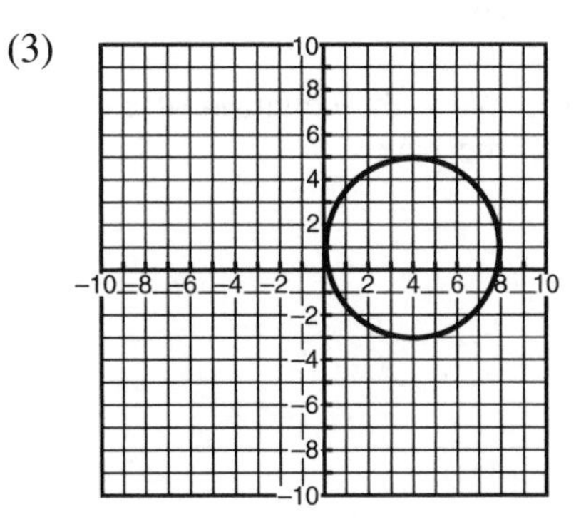

(2)

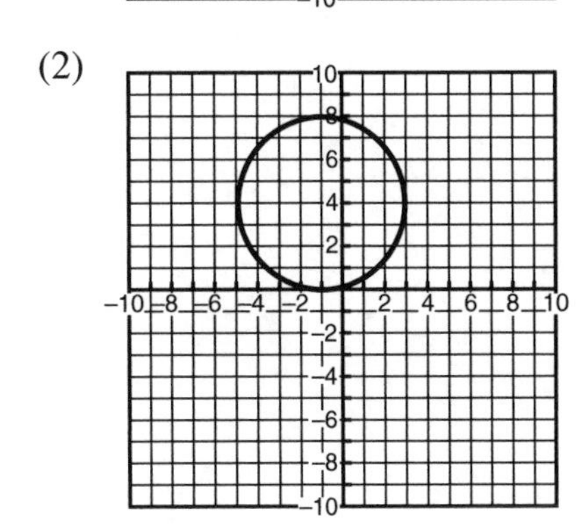

(4)

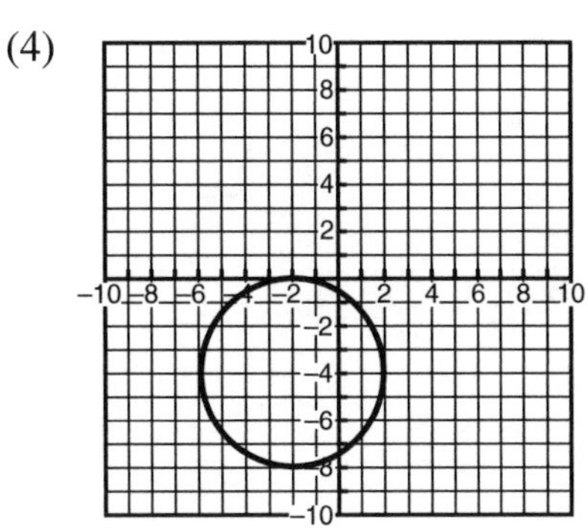

7. What is the equation of the circle $x^2 + y^2 = 16$ after a translation $T_{-2,1}$?

(1) $(x + 2)^2 + (y - 1)^2 = 16$
(2) $(x - 2)^2 + (y + 1)^2 = 16$
(3) $(x - 1)^2 + (y + 2)^2 = 16$
(4) $(x + 1)^2 + (y - 2)^2 = 16$

8. What are the coordinates of the center and the length of the radius of a circle whose equation is $x^2 + 12x + y^2 - 4y + 31 = 0$?

9. Sketch the graph of a circle whose equation is $x^2 - 10x + y^2 - 6y + 18 = 0$.

10. Sketch the graph of a circle whose equation is $4x^2 + 24x + 4y^2 - 16y = 48$.

11. Write the equation of a circle whose center is at (2, 3) and passes through (−1, −1).

12. Write the equation of a circle whose center is at (5, −2) and passes through (−6, 1).

13. Write the equation of a circle whose diameter has endpoints $(-2, 1)$ and $(6, 7)$.

14. Write the equation of a circle whose center is $(4, -3)$ and is tangent to the x-axis.

Chapter Six

IMAGINARY AND COMPLEX NUMBERS

6.1 SQUARE ROOTS OF NEGATIVE NUMBERS

KEY IDEAS

Remember that taking the square root of a number determines what positive number occurs as two equal factors of the number inside the radical (the radicand). What happens if the radicand is a negative number?

Negative Radicands

The product of two positive numbers is a positive number, and the product of two negative numbers is also a positive number. So can two equal factors of a negative number be found? Clearly, the answer is no!

How can an answer for the square root of a number, such as −12, be determined? With some insight and a lot of imagination, there is a way to work with such numbers as $\sqrt{-12}$. To start, notice that $\sqrt{-12} = \sqrt{-1 \cdot 4 \cdot 3} = 2\sqrt{-3}$. In fact, for the square root of any negative number, it is possible to take out the perfect square factors of the radicand and simplify the expression in a similar fashion. Of course, that still does not really solve the problem at hand.

The Imaginary Unit

The real problem that needs to be addressed is what is $\sqrt{-1}$? Here is where our imagination comes into play. Everyone knows that there is no real number that can satisfy this expression. What if a special symbol is defined for this imaginary number $\sqrt{-1}$?

MATH FACTS

Definition: $i^2 = -1$

Based on this definition, it can be seen that $i = \sqrt{-1}$. The number i is sometimes referred to as the imaginary unit.

Simplifying, Adding, and Subtracting Imaginary Numbers

Let's reinvestigate $\sqrt{-12}$

$$\sqrt{-12} = \sqrt{-1} \cdot \sqrt{4} \cdot \sqrt{3} = i \cdot 2\sqrt{3} = 2i\sqrt{3}$$

Notice that it is easier to read this expression if the 2 is written as the coefficient of the imaginary number, i.

Example 1

Simplify $\sqrt{-50}$

Solution: –50 can be factored as $-1 \cdot 25 \cdot 2$ where 25 is a perfect square factor.

$$\sqrt{-50} = \sqrt{25 \cdot -1 \cdot 2} = \sqrt{25} \cdot \sqrt{-1} \cdot \sqrt{2} = 5i\sqrt{2}$$

All of the operations for radical expressions previously learned apply to radicals of negative expressions. The next example shows how addition can be performed.

Example 2

Add and simplify $\sqrt{-48} + \sqrt{-108}$

Solution: $\sqrt{-48} + \sqrt{-108} = \sqrt{-1 \cdot 16 \cdot 3} + \sqrt{-1 \cdot 36 \cdot 3} = 4i\sqrt{3} + 6i\sqrt{3} = 10i\sqrt{3}$

Check Your Understanding of Section 6.1

1. Simplify: $\sqrt{-72}$

(1) $3i\sqrt{2}$ (2) $6i\sqrt{2}$ (3) $6 + i\sqrt{2}$ (4) $3i\sqrt{8}$

2. Simplify: $\sqrt{-112}$

(1) $4i\sqrt{7}$ (2) $16i\sqrt{7}$ (3) $4i\sqrt{28}$ (4) $7i\sqrt{4}$

3. Add and simplify: $\sqrt{-2} + 5i\sqrt{2}$

(1) $4i\sqrt{2}$ (2) $10i\sqrt{2}$ (3) $4i\sqrt{-2}$ (4) $6i\sqrt{2}$

4. Subtract and simplify: $4\sqrt{-75} - 2\sqrt{-27}$

(1) $26i\sqrt{3}$ (2) $2i\sqrt{3}$ (3) $10i\sqrt{3}$ (4) $14i\sqrt{3}$

5. Simplify: $\sqrt{-56a^4b^3}$

(1) $4a^2bi\sqrt{7b}$ (2) $4a^2bi\sqrt{14b}$ (3) $2a^2bi\sqrt{14b}$ (4) $4ab^2i\sqrt{14b}$

6. Subtract and simplify: $6x\sqrt{-24x} - \sqrt{-600x^3}$

(1) $24xi\sqrt{2x}$ (2) $2xi\sqrt{6x}$ (3) $12xi\sqrt{6x}$ (4) $2xi\sqrt{2x}$

7. Simplify: $-5\sqrt{-125s^7t^4}$

(1) $25s^3t^2\sqrt{5s}$ (2) $-25s^3t^2\sqrt{5s}$ (3) $-25s^3t^2i\sqrt{5s}$ (4) $-125s^3t^2i\sqrt{5s}$

8. Perform the indicated operations, and express your answer in simplest radical form (without negative radicands):

$3a\sqrt{-12ab^2} + 2b\sqrt{-75b} - 2b\sqrt{-27a^3} + \sqrt{-48b^3}$

9. Subtract $2\sqrt{-80}$ from the sum of $4\sqrt{-45}$ and $2\sqrt{-180}$.

10. Determine the sum of the difference between $\sqrt{-96}$ and $2\sqrt{-6}$ and the difference between $3\sqrt{-24}$ and $\sqrt{-54}$.

6.2 POWERS OF *i* AND OPERATIONS ON IMAGINARY NUMBERS

Since $i = \sqrt{-1}$, how does that affect multiplication of imaginary numbers?

Multiplication of Imaginary Numbers

In order to multiply $\sqrt{-6}$ by $\sqrt{-15}$, it is important to understand how to determine powers of the imaginary number i. Let's investigate the powers of i and then examine the above product.

$i^2 = i \cdot i = \sqrt{-1} \cdot \sqrt{-1} = -1$

$i^3 = i^2 \cdot i = -1 \cdot i = -i$

$i^4 = i^3 \cdot i = -i \cdot i = -i^2 = -(-1) = 1$

$i^5 = i^4 \cdot i = 1 \cdot i = i$

$i^6 = i^5 \cdot i = i \cdot i = i^2 = -1$

$i^7 = i^6 \cdot i = -1 \cdot i = -i$

$i^8 = i^7 \cdot i = -i \cdot i = -i^2 = -(-1) = 1$

$i^9 = i^8 \cdot i = 1 \cdot i = i$

$i^{10} = i^9 i = i \cdot i = i^2 = -1$

$i^{11} = i^{10} \cdot i = -1 \cdot i = -i$

Notice that there is a pattern. Once it was established that $i^4 = 1$, a repeating pattern of $i, -1, -i, 1$ occurs. For instance, to evaluate i^{13}, factor to see that $i^{13} = i^4 \cdot i^4 \cdot i^4 \cdot i = 1 \cdot 1 \cdot 1 \cdot i = i$. This suggests that the same problem can be evaluated by dividing the exponent by 4 and recognizing that if the remainder is 1, the result is i. If the remainder is 2, the result is -1. If the remainder is 3, the result is $-i$. Finally, if there is no remainder, the result is 1.

Example 1

Simplify each of the following: a. i^{19} b. i^{48} c. i^{34} d. i^{73}

Solution: a. $i^{19} = -i$ because 19 divided by 4 has a remainder of 3.

b. $i^{48} = 1$ because there is no remainder when 48 is divided by 4.

c. $i^{34} = -1$ because 34 divided by 4 has a remainder of 2.

d. $i^{73} = i$ because 73 divided by 4 has a remainder of 1.

So how does this information help find the product of $\sqrt{-6}$ and $\sqrt{-15}$?

$$\sqrt{-6} \cdot \sqrt{-15} = i\sqrt{6} \cdot i\sqrt{15} = i^2\sqrt{90} = -1 \cdot \sqrt{9 \cdot 10} = -3\sqrt{10}$$

Example 2

Determine the product and simplify: $2\sqrt{-35} \cdot 4\sqrt{-40}$

Solution: $2\sqrt{-35} \cdot 4\sqrt{-40} = 2i\sqrt{5 \cdot 7} \cdot 4i\sqrt{5 \cdot 4 \cdot 2} = 5 \cdot 2 \cdot 8i^2\sqrt{2 \cdot 7}$

$$80(-1)\sqrt{14} = -80\sqrt{14}$$

Division of Imaginary Numbers

Since imaginary numbers are similar to radical expressions, division by imaginary numbers is sometimes performed by rationalizing the denominator. However, rules of exponentiation may apply to a specific example instead of the process of rationalizing a denominator. For instance, if an example involves dividing a real number by an imaginary number, the denominator must be rationalized. The example that follows shows how this is done. Remember, just because there is no square root sign, the imaginary unit, i, represents $\sqrt{-1}$. So i behaves as if there is a square root present.

Example 3

Determine the quotient: $\frac{3}{i}$

Solution: $\frac{3}{i} = \frac{3}{i} \cdot \frac{i}{i} = \frac{3i}{i^2} = \frac{3i}{-1} = -3i$

Example 4

Determine the quotient: $\frac{2i^3}{i^2}$

Solution: $\frac{2i^3}{i^2} = \frac{2i \cdot \cancel{i^2}}{\cancel{i^2}} = 2i$

Of course, this example can be approached by subtracting exponents as if the example were $\frac{2x^3}{x^2}$ instead of $\frac{2i^3}{i^2}$.

Check Your Understanding of Section 6.2

1. Simplify: i^{235}
(1) i (2) $-i$ (3) 1 (4) -1

2. Simplify: $i^{17} \cdot i^{23} \cdot i^{46}$
(1) i (2) $-i$ (3) 1 (4) -1

3. Simplify: $(i^{27})^3$
(1) i (2) $-i$ (3) 1 (4) -1

4. Determine the product and simplify: $5\sqrt{-30} \cdot 2\sqrt{-20}$
(1) $-100i\sqrt{6}$ (2) $-100\sqrt{6}$ (3) $-10i\sqrt{6}$ (4) $50i\sqrt{6}$

5. If $5\sqrt{-4}$ is multiplied by $3\sqrt{-9}$, what is the result?
(1) 90 (2) $90i$ (3) $-15i$ (4) -90

6. Multiply and simplify the result: $(3i)^3 \cdot (-2i)^2$
(1) $-108i$ (2) 108 (3) -108 (4) $108i$

7. Multiply and simplify the result: $(-5i)(2i)(-4i)$

(1) $-40i$ (2) $40i$ (3) $11i$ (4) $-40i^3$

8. Divide: $\frac{12}{2i}$

(1) $6i$ (2) 6 (3) $-6i$ (4) $6i$

9. Divide and simplify the result: $\frac{28i^7}{4i^3}$

(1) -7 (2) $-7i$ (3) 7 (4) $7i$

6.3 THE SET OF COMPLEX NUMBERS

Now that the imaginary number has been introduced, it is easy to extend the concept to numbers that are part imaginary and part real. Such numbers are called complex numbers and take the general form $a + bi$.

Complex Numbers

A complex number is a number that can be written in the form $a + bi$, where a and b are real numbers and b is the coefficient of an imaginary number. In other words, a complex number is the sum of a real number and an imaginary number. Notice that any real number is actually a complex number where $b = 0$. Any imaginary number is a complex number where $a = 0$.

Operations on Complex Numbers

Let's examine how to perform the basic operations on the set of complex numbers. Addition and subtraction of complex numbers is very similar to addition and subtraction of binomial expressions because addition requires that only like terms be combined.

Example 1

Express the sum in $a + bi$ form: $(3 - 5i) + (-7 + 2i)$

Solution: $(3 - 5i) + (-7 + 2i) = 3 + (-7) + (-5 + 2)i = -4 - 3i$

Example 2

Express the difference in $a + bi$ form: $(-6 + 8i) - (-2 - 3i)$

Solution: $(-6 + 8i) - (-2 - 3i) = -6 - (-2) + (8i - (-3i)) = -4 + 11i$

Multiplication of complex numbers can also be compared to multiplication of binomial expressions. It can be performed by the FOIL method, box method, or distributive method, as discussed in section 1.1 of this book.

Example 3

Express the product in $a + bi$ form: $(3 + 2i)(2 - 5i)$

FOIL Method	Distributive Method	Box Method
$(3 + 2i)(2 - 5i) =$ $6 - 15i + 4i - 10i^2 =$ $6 - 11i - (10 \cdot -1) =$ $6 - 11i + 10 =$ $16 - 11i$	$(3 + 2i)(2 - 5i) =$ $(3 + 2i)2 - (3 + 2i)5i =$ $6 + 4i - 15i - 10i^2 =$ $6 - 11i + 10 =$ $16 - 11i$	$3 + 2i$ $\underline{2 - 5i}$ $-15i - 10i^2$ $\underline{6 + 4i}$ $6 - 11i - 10i^2 =$ $6 - 11i + 10 =$ $16 - 11i$

Solution: $16 - 11i$

In order to perform division with complex numbers, the denominator must be rationalized. Let's examine dividing a complex number by an imaginary number.

Example 4

Express the quotient in $a + bi$ form: $\frac{2+3i}{5i}$

Solution: $\frac{2+3i}{5i} = \frac{(2+3i)}{5i} \cdot \frac{i}{i} = \frac{2i+3i^2}{5i^2}$

Since $i^2 = -1$, this is equal to $\frac{2i-3}{-5} = \frac{3}{5} - \frac{2}{5}i$

Notice that in order to place this answer in $a + bi$ form, it was necessary to express the real number part and the coefficient of the imaginary part of this complex number in fractional form.

In division, recall that to rationalize an irrational binomial denominator requires multiplying by the conjugate expression. The conjugate for the complex number $a + bi$ is $a - bi$. For both of the examples in the table below, similar binomial expressions is solved along with the complex number examples. This lets the reader compare processes.

Examples	Similar Binomial Example	Solution for the Complex Number Example
Example 5 $\frac{2}{1+3i}$	$\frac{2}{1+3\sqrt{2}}=\frac{2}{(1+3\sqrt{2})}\cdot\frac{(1-3\sqrt{2})}{(1-3\sqrt{2})}$ $\frac{2-6\sqrt{2}}{1-9\cdot 2}=\frac{2-6\sqrt{2}}{1-18}=\frac{2-6\sqrt{2}}{-17}$ $=\frac{-2}{17}+\frac{6}{17}\sqrt{2}$	$\frac{2}{1+3i}=\frac{2}{(1+3i)}\cdot\frac{(1-3i)}{(1-3i)}$ $\frac{2-6i}{1-9i^2}=\frac{2-6i}{1+9}=\frac{2-6i}{10}$ $=\frac{1}{5}-\frac{3}{5}i$
Example 6 $\frac{4+7i}{2-4i}$	$\frac{4+7\sqrt{3}}{2-4\sqrt{3}}=\frac{(4+7\sqrt{3})}{(2-4\sqrt{3})}\cdot\frac{(2+4\sqrt{3})}{(2+4\sqrt{3})}$ $=\frac{8+16\sqrt{3}+14\sqrt{3}+28\cdot 3}{4-16\cdot 3}$ $=\frac{8+30\sqrt{3}+84}{4-48}$ $=\frac{92+30\sqrt{3}}{-44}$ $=\frac{-23}{11}-\frac{15}{22}\sqrt{3}$	$\frac{4+7i}{2-4i}=\frac{(4+7i)}{(2-4i)}\cdot\frac{(2+4i)}{(2+4i)}$ $=\frac{8+16i+14i+28i^2}{4-16i^2}$ $=\frac{8+30i-28}{4+16}$ $=\frac{-20+30i}{20}$ $=-1+\frac{3}{2}i$

Check Your Understanding of Section 6.3

1. Add and express the answer in $a + bi$ form: $(2 - 5i) + (-7 + 2i)$
(1) $5 + 3i$ (2) $-5 + 3i$ (3) $-5 - 3i$ (4) $5 + 3i$

2. Determine the product of $(4 - 3i)(1 + 2i)$ and express the answer in the standard form for a complex number.
(1) $-2 + 5i$ (2) $5 - i$ (3) $10 - 5i$ (4) $10 + 5i$

3. Express the sum of $\sqrt{64}+\sqrt{-50}-3.1+\sqrt{-162}$ in simplest $a + bi$ form.
(1) $4.9+14i\sqrt{2}$ (2) $11.1+14i\sqrt{2}$ (3) $4.9-4i\sqrt{2}$ (4) $11.1-14i\sqrt{2}$

4. From the sum of $6 - 2i$ and $3 + 5i$ subtract $-4 + 7i$.
(1) $9 + 3i$ (2) $4 - 5i$ (3) $13 + 10i$ (4) $13 - 4i$

5. Simplify $\frac{6+2i}{3-4i}$ and express the answer in simplest $a + bi$ form.
(1) $\frac{26}{25}-\frac{12}{25}i$ (2) $\frac{-26}{7}+\frac{12}{7}i$ (3) $\frac{2}{5}+\frac{6}{5}i$ (4) $\frac{-10}{7}-\frac{36}{7}i$

6. Divide 12 by $5 + 2i$ and express the answer in simplest $a + bi$ form.
(1) $\frac{60}{29}-\frac{24}{29}i$ (2) $\frac{60}{29}+\frac{24}{29}i$ (3) $\frac{60}{21}-\frac{24}{21}i$ (4) $\frac{60}{21}+\frac{24}{21}i$

7. Josh and Tanya are playing a game with complex numbers. Josh picked the number $3 - 5i$. Tanya has to select the complex number needed to multiply by Josh's number to get a result of $21 - i$. Which of the following numbers should Tanya select for this game?
(1) $2 + 3i$ (2) $2 + i$ (3) $2 - i$ (4) $2 - 3i$

8. Divide $\sqrt{-49}$ by $\left(\sqrt{16}+\sqrt{-64}\right)$ and express the answer in the standard form for a complex number.
(1) $\frac{-7}{10}-\frac{7}{20}i$ (2) $\frac{7}{10}+\frac{7}{20}i$ (3) $\frac{7}{20}+\frac{7}{10}i$ (4) $\frac{-7}{10}+\frac{7}{20}i$

9. Perform the indicated operations and express the answer in simplest $a + bi$ form: $3i(2 - 5i) - (6 - 2i)$
(1) $-6 + 23i$ (2) $9 + 8i$ (3) $9 - 8i$ (4) $9 + 4i$

10. In an electrical circuit, the equation $E = IZ$ is used to calculate the voltage, E, in volts, using the current, I, in amps, and the opposition to the flow of current, called impedance, Z. A circuit has a current of $(5 - 2i)$ amps and an impedance of $(4 + 3i)$ ohms. Determine the voltage in $a + bi$ form.
(1) $14 + 7i$ (2) $26 + 23i$ (3) $26 + 7i$ (4) $14 + 23i$

11. From the product of $3 + 6i$ and $-2 + 4i$ subtract the product of $1 - 3i$ and $5 + 2i$.

12. Perform the indicated operations and express the answer in $a + bi$ form:

$$\frac{2i}{3-2i}+\frac{5}{3+2i}$$

13. Determine the product of $5 - 4i$ and its conjugate. Express the answer in $a + bi$ form.

14. Subtract the conjugate of $2 - 7i$ from the conjugate of $6 + 3i$.

15. Solve for x: $\sqrt{5-10x} = 5i$

16. If $f(x) = x^3 - 2x$, evaluate $f(6 + 3i)$

6.4 QUADRATIC FUNCTIONS AND THE DISCRIMINANT

What are the roots for a quadratic equation whose graph does not intersect the x-axis? In this section, the relationship between the roots of a quadratic equation will be examined in the system of complex numbers.

The Nature of the Roots of a Quadratic Function

Remember that the quadratic formula $x=\dfrac{-b\pm\sqrt{b^2-4ac}}{2a}$ is used to solve a quadratic equation in the form $y = ax^2 + bx + c$. This formula can used to solve for the roots of any quadratic equation, even if the quadratic expression is factorable. Notice that the formula includes a square root. What would happen if the radicand is a negative number? Obviously, the root would be a complex number. The following example is of a quadratic function that does have complex roots.

Example 1

Determine the roots of $f(x) = x^2 + 2x + 3$.

Solution: $a = 1$, $b = 2$, and $c = 3$

$$x = \frac{-2 \pm \sqrt{2^2 - 4 \cdot 1 \cdot 3}}{2 \cdot 1}$$

$$= \frac{-2 \pm \sqrt{4-12}}{2} = \frac{-2 \pm \sqrt{-8}}{2} = \frac{-2 \pm 2i\sqrt{2}}{2} = -1 \pm i\sqrt{2}$$

The roots are $-1 + i\sqrt{2}$ and $-1 - i\sqrt{2}$.

It is easy to recognize from this example that when the radicand is negative, the roots are complex conjugates. Furthermore, the radicand inside the quadratic formula actually determines the nature of the roots of any quadratic function. This is because the radicand can cause the equation to have complex roots, irrational roots, rational roots, and even equal roots. Let's examine the radicand more carefully to see why this is so. The radicand inside the quadratic formula is the expression $b^2 - 4ac$. Since this expression actually determines the nature of the roots of the quadratic equation, it is called the discriminant.

If the discriminant $b^2 - 4ac$ equals 0, then the quadratic equation yields the fraction $\frac{-b}{2a}$. If $b^2 - 4ac$ is a perfect square, the quadratic equation yields two different rational expressions. If $b^2 - 4ac$ is irrational, the result has two irrational values. Finally, if $b^2 - 4ac$ is a negative number, the quadratic equation has two complex conjugates. Clearly, the discriminate determines the nature of the roots of the quadratic function.

MATH FACTS

The discriminant, $b^2 - 4ac$, determines the nature of the roots of a quadratic equation.

1. If $b^2 - 4ac$ equals 0, the roots are real, rational, and equal.
2. If $b^2 - 4ac$ is equal to a perfect square, the roots are real, rational, and unequal.
3. If $b^2 - 4ac$ is equal to a positive number that is not a perfect square, the roots are real, irrational, and unequal.
4. If $b^2 - 4ac$ is equal to a negative number, the roots are complex conjugates.

The following examples explore each of these situations.

Examples	$a =$	$b =$	$c =$	Discriminant $b^2 - 4ac =$	Solution $x = \dfrac{-b \pm \sqrt{b^2 - 4ac}}{2a}$
Example 2 $f(x) = x^2 + 6x + 9$	1	6	9	$6^2 - 4 \cdot 1 \cdot 9 =$ $36 - 36 = 0$	$x = \dfrac{-6 \pm 0}{2} = \dfrac{-6}{2} = -3$
Example 3 $f(x) = x^2 - x - 20$	1	−1	−20	$(-1)^2 - 4(1)(-20)$ $= 1 + 80 = 81$	$x = \dfrac{1 \pm \sqrt{81}}{2} = \dfrac{1 \pm 9}{2}$ $= \dfrac{10}{2}$ or $\dfrac{-8}{2} = 5$ or -4
Example 4 $f(x) = 2x^2 + 3x - 4$	2	3	−4	$3^2 - 4(2)(-4) =$ $9 + 32 = 41$	$x = \dfrac{-3 \pm \sqrt{41}}{4}$ $= -\dfrac{3}{4} \pm \dfrac{\sqrt{41}}{4}$
Example 5 $f(x) = 2x^2 + 3x + 4$	2	3	4	$3^2 - 4(2)(4) =$ $9 - 32 = -23$	$x = \dfrac{-3 \pm \sqrt{-23}}{4}$ $= \dfrac{-3 \pm i\sqrt{23}}{4}$ $= -\dfrac{3}{4} \pm \dfrac{\sqrt{23}}{4}i$

Let's examine the graphs of the same four functions from the examples in the chart.

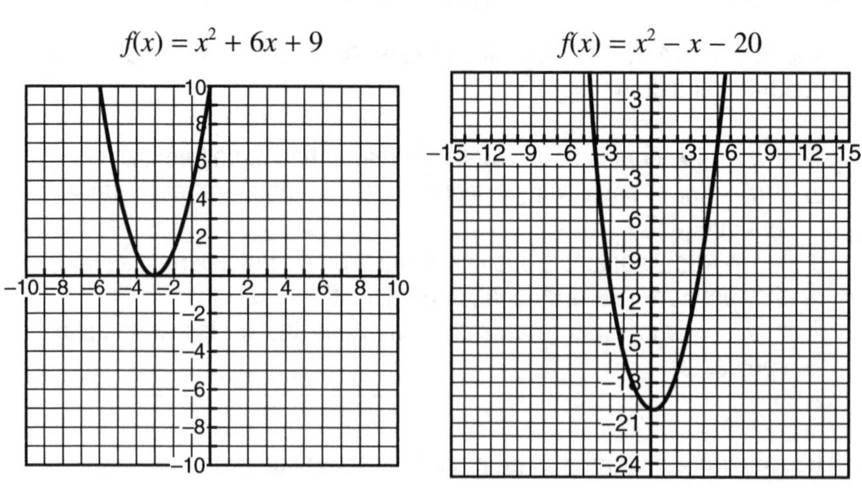

$f(x) = 2x^2 + 3x - 4$

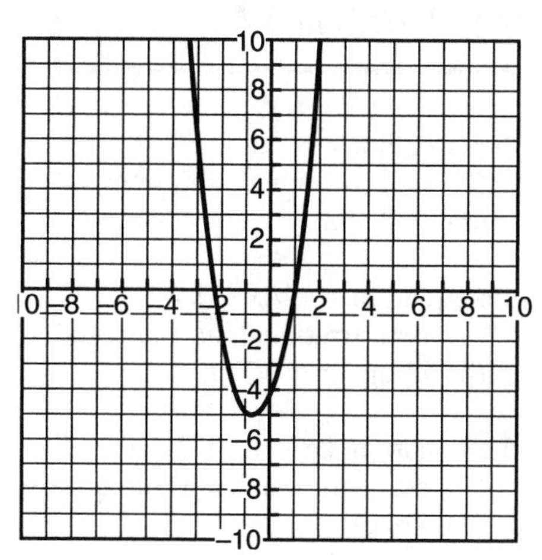

$f(x) = 2x^2 + 3x + 4$

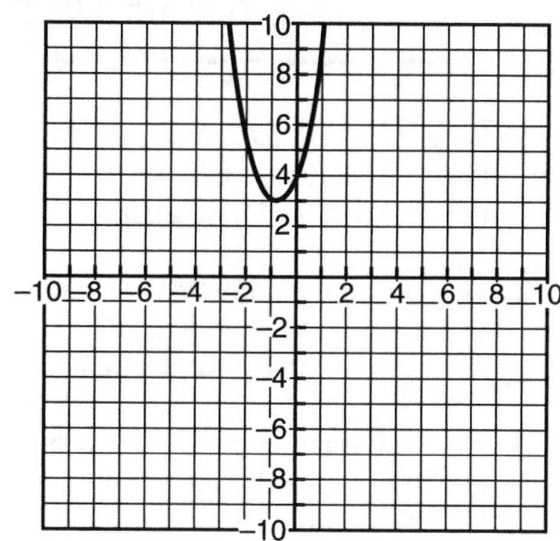

Notice that the one root of $f(x) = x^2 + 6x + 9$, which is -3, is easy to read off the graph of this function. The roots of -4 and $+5$ can be read off the graph of $f(x) = x^2 - x - 20$. The roots for $f(x) = 2x^2 + 3x - 4$ are not clear from the graph but appear to be close to -2.5 and 1. However, the graph of $f(x) = 2x^2 + 3x + 4$ does not intersect the x-axis. This suggests that there are no real numbers that are roots of the equation.

The roots for $f(x) = 2x^2 + 3x - 4$ can be approximated with a graphing calculator using **2ND** **TRACE** **2** after entering the function in **Y =**. Scroll over to the left side of the root on the left and press **ENTER**. Then scroll to the right of the root and press **ENTER** **ENTER**. The result is shown below.

Zero
X=-2.350781 Y=0

Repeat this process to determine the numerical value of the other root.

Check Your Understanding of Section 6.4

1. Determine the nature of the roots of the equation $x^2 + 5x + 7 = 0$.
(1) real, rational, and equal
(2) real, rational, and unequal
(3) real, irrational, and unequal
(4) imaginary

2. Determine the nature of the roots of the equation $x^2 - 4x - 2 = 0$.
(1) real, rational, and equal
(2) real, rational, and unequal
(3) real, irrational, and unequal
(4) imaginary

3. Determine the nature of the roots of the equation $4x^2 - 8x + 3 = 0$.
(1) real, rational, and equal
(2) real, rational, and unequal
(3) real, irrational, and unequal
(4) imaginary

4. Determine the nature of the roots of the equation $3x^2 + 2x - 5 = 0$.
(1) real, rational, and equal
(2) real, rational, and unequal
(3) real, irrational, and unequal
(4) imaginary

5. Determine the nature of the roots of the equation $2x^2 + 3 = 6x$.
(1) real, rational, and equal
(2) real, rational, and unequal
(3) real, irrational, and unequal
(4) imaginary

6. Determine the nature of the roots of the equation $x^2 = 6 - 2x$.
(1) real, rational, and equal
(2) real, rational, and unequal
(3) real, irrational, and unequal
(4) imaginary

7. For which positive value of c will the equation $9x^2 + cx + 4 = 0$ have roots that are real, equal, and rational?
(1) 12 (2) 9 (3) 4 (4) 3

8. The equation $3x^2 + 10x + k = 0$ has imaginary roots when k is equal to what number?
(1) 1 (2) 3 (3) 6 (4) 9

9. Which equation has roots that are real, rational, and unequal?
(1) $x^2 - 9 = 0$
(2) $x^2 - 5x + 9 = 0$
(3) $x^2 - 5x + 1 = 0$
(4) $x^2 + x + 2 = 0$

10. The equation $px^2 + 8x + 16 = 0$ has imaginary roots when
(1) $p < 1$ (2) $p > 1$ (3) $-1 < p < 1$ (4) $p < -1$

11. Which is a true statement about the graph of the equation $y = 2x^2 + 3x + 5$?
(1) It is tangent to the x-axis.
(2) It does not intersect the x-axis.
(3) It intersects the x-axis in two distinct points that have irrational coordinates.
(4) It intersects the x-axis in two distinct points that have rational coordinates.

12. Which graph represents a function whose discriminate is equal to 0?

discriminant

(1)

(2)

(3)

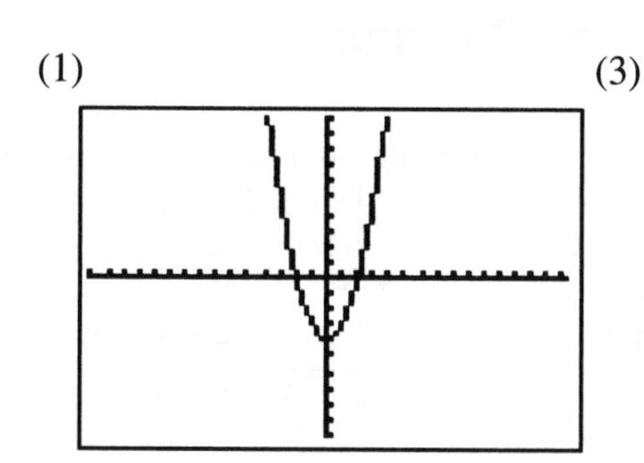

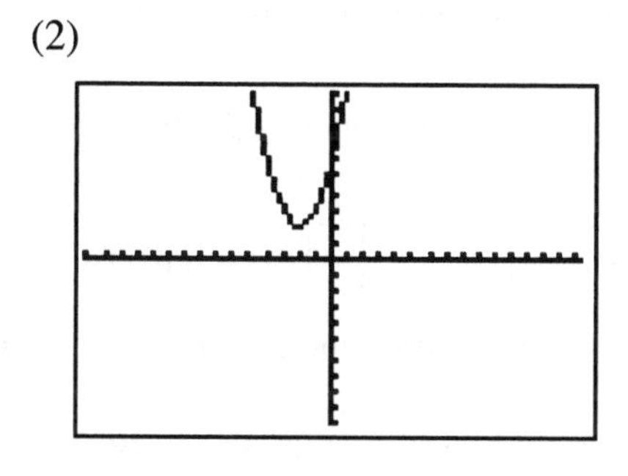

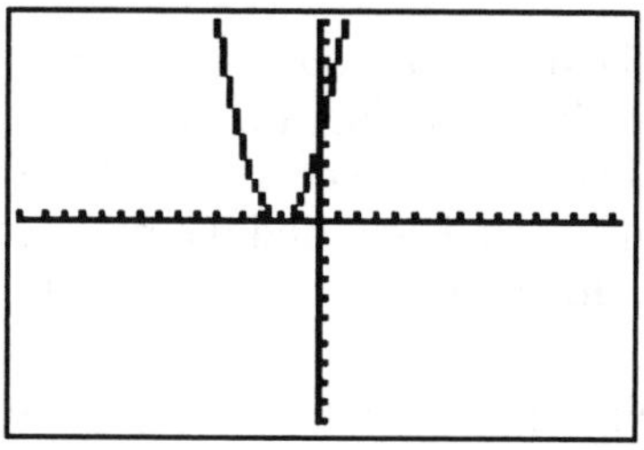

(4)

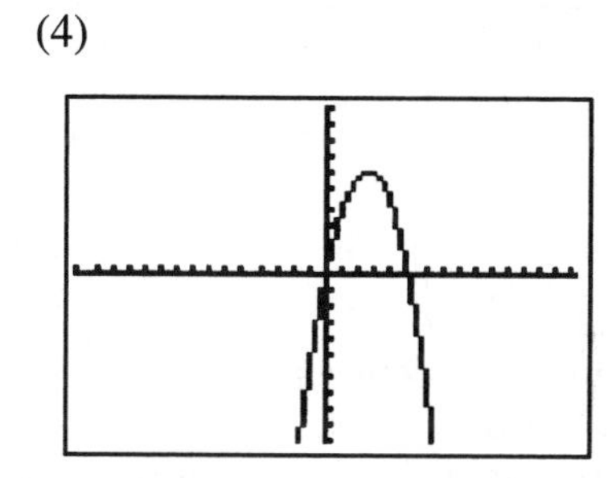

13. Solve for c: $c^2 + 3c + 5 = 0$

14. Solve for t: $4t^2 + 3 = 2t$

15. Solve for x: $4x = x^2 + 3$

16. Solve for m: $2m^2 + 8 = 5m$

6.5 COEFFICIENTS AND ROOTS OF A QUADRATIC FUNCTION

The roots of the quadratic equation $x^2 + 3x - 4 = 0$ are $x = -4$ and $x = 1$. What is the relationship between the coefficients and the roots of a quadratic equation?

The Sum and Product of the Roots of a Quadratic Function

The quadratic equation $x^2 + 3x - 4 = 0$ is factorable, so it can be solved using the Zero Product Property. Factoring yields $(x + 4)(x - 1) = 0$. Therefore, either $x + 4 = 0$ or $x - 1 = 0$. So $x = -4$ or $x = 1$. Notice that the sum of these two roots is -3, and the product of these two roots is -4.

Examine $3x^2 + 2x + 1 = 0$. In this equation, $a = 3$, $b = 2$, and $c = 1$. Solving by the quadratic equation yields $x = \frac{-2 \pm \sqrt{4 - 4 \cdot 3 \cdot 1}}{2 \cdot 3} = \frac{-2 \pm \sqrt{4 - 12}}{6}$ $= \frac{-2 \pm \sqrt{-8}}{6} = \frac{-2 \pm 2i\sqrt{2}}{6} = -\frac{1}{3} \pm \frac{\sqrt{2}}{3}i$. The sum of the two roots is $\left(-\frac{1}{3} + \frac{\sqrt{2}}{3}i\right) + \left(-\frac{1}{3} - \frac{\sqrt{2}}{3}i\right)$ or $-\frac{2}{3}$. The product of the roots is $\left(-\frac{1}{3} + \frac{\sqrt{2}}{3}i\right)\left(-\frac{1}{3} - \frac{\sqrt{2}}{3}i\right) = \frac{1}{9} - \frac{2}{9}i^2 = \frac{1}{9} + \frac{2}{9} = \frac{3}{9} = \frac{1}{3}$. There appears to be a relationship between the coefficients of a quadratic function and the sum and product of its roots.

To establish an actual formula, use the roots of any quadratic equation as expressed in the quadratic formula, $x = \frac{-b \pm \sqrt{b^2 - 4ac}}{2a}$. The sum of the two general roots is $\frac{-b + \sqrt{b^2 - 4ac}}{2a} + \frac{-b - \sqrt{b^2 - 4ac}}{2a}$. Since these have common denominators, just add the numerators to get $\frac{-2b}{2a} = \frac{-b}{a}$. Finding the product of these general roots is also easy because they are conjugate expressions. So $\left(\frac{-b + \sqrt{b^2 - 4ac}}{2a}\right)\left(\frac{-b - \sqrt{b^2 - 4ac}}{2a}\right) = \frac{b^2 - (b^2 - 4ac)}{4a^2}$ $= \frac{4ac}{4a^2} = \frac{c}{a}$.

MATH FACTS

The sum of the roots of the quadratic equation $ax^2 + bx + c = 0$ is $\frac{-b}{a}$.

The product of the roots of the quadratic equation $ax^2 + bx + c = 0$ is $\frac{c}{a}$.

Example 1

Determine the sum and product of the roots of $f(x) = x^2 + 4x + 7$.

Solution: $a = 1$, $b = 4$, and $c = 7$. The sum of the roots is $\frac{-b}{a} = \frac{-4}{1} = -4$. The product of the roots is $\frac{c}{a} = \frac{7}{1} = 7$.

Example 2

Determine the sum and product of the roots of $f(x) = 4x^2 - 5x + 2$.

Solution: $a = 4$, $b = -5$, and $c = 2$. The sum of the roots is $\frac{-b}{a} = \frac{-(-5)}{4} = \frac{5}{4}$. The product of the roots is $\frac{c}{a} = \frac{2}{4} = \frac{1}{2}$.

Determining a Quadratic Equation From Its Roots

Two approaches can be used to determine the equation of a quadratic function from its roots. Suppose the roots of a quadratic function are 3 and 7. The first method is to reverse the steps taken with the Zero Product Property. The final result gave $x = 3$ and $x = 7$. These roots resulted from the equations $x - 3 = 0$ and $x - 7 = 0$.

Reversing steps again, multiply the factors that cause these two equations, yielding $(x - 3)(x - 7) = 0$ or $x^2 - 10x + 21 = 0$.

A second approach to forming the equation of a quadratic function with roots 3 and 7 uses the sum and product formulas investigated in the first part of this section. The sum of the roots is $r_1 + r_2 = 3 + 7 = 10$. Therefore, $\frac{-b}{a} = 10 = \frac{10}{1}$. The product of the roots is $r_1 \cdot r_2 = 3 \cdot 7 = 21$ and $\frac{c}{a} = 21 = \frac{21}{1}$. By inspection, it appears as though $a = 1$, $b = -10$, and $c = 21$. So the equation, $x^2 - 10x + 21 = 0$ is once again established.

Example 3

Determine a quadratic function whose roots are $\frac{2}{3}$ and $\frac{-3}{5}$.

Solution: Both methods are shown below for this example.

Solution by Reversing Steps	Solution Using the Sum and Product of the Roots
$x=\frac{2}{3}$ yields $3x = 2$ or $3x - 2 = 0$, $x=\frac{-3}{5}$ yields $5x = -3$ or $5x + 3 = 0$ $(3x - 2)(5x + 3) = 15x^2 - x - 6 = 0$	$r_1+r_2=\frac{2}{3}+\frac{-3}{5}=\frac{10}{15}-\frac{9}{15}=\frac{1}{15}$ $r_1 \cdot r_2=\frac{2}{3}\cdot\frac{-3}{5}=\frac{-6}{15}$ By inspection, $a = 15$, $b = -1$, and $c = -6$. $15x^2 - x - 6 = 0$

Example 4

Determine a quadratic function whose roots are $2+\sqrt{5}$ and $2-\sqrt{5}$.

Solution: The method using the sum and the product of the roots is illustrated below.

$$r_1+r_2=\left(2+\sqrt{5}\right)+\left(2-\sqrt{5}\right)=4$$

$$r_1 \cdot r_2=\left(2+\sqrt{5}\right)\left(2-\sqrt{5}\right)=4-5=-1$$

$a = 1$, $b = -4$, and $c = -1$

The function is $f(x) = x^2 - 4x - 1$.

Example 5

Determine a quadratic function whose roots are $\frac{1}{3}+\frac{2}{9}i$ and $\frac{1}{3}-\frac{2}{9}i$.

Solution: The method using the sum and the product of the roots is illustrated below.

$$r_1+r_2=\left(\frac{1}{3}+\frac{2}{9}i\right)+\left(\frac{1}{3}-\frac{2}{9}i\right)=\frac{2}{3}=\frac{54}{81}$$

$$r_1 \cdot r_2=\left(\frac{1}{3}+\frac{2}{9}i\right)\left(\frac{1}{3}-\frac{2}{9}i\right)=\frac{1}{9}-\frac{4}{81}i^2=\frac{1}{9}+\frac{4}{81}=\frac{9}{81}+\frac{4}{81}=\frac{13}{81}$$

$a = 81$, $b = -54$, and $c = 13$.

The function is $f(x) = 81x^2 - 54x + 13$.

Notice how the quadratic equation insures that if there is an irrational root, the conjugate of that root is also a root.

MATH FACTS

Irrational roots of a quadratic equation, $ax^2 + bx + c = 0$, occur in conjugate pairs. If $a+\sqrt{b}$ is a root of a quadratic equation, then $a-\sqrt{b}$ is also a root. If $a + bi$ is a root of a quadratic equation, then $a - bi$ is also a root.

Example 6

Determine a quadratic function, one of whose roots is $3 - 5i$.

Solution: If $3 - 5i$ is a root, then $3 + 5i$ also occurs as a root of the equation.

$$r_1 + r_2 = (3 - 5i) + (3 + 5i) = 6$$

$$r_1 \cdot r_2 = (3 - 5i)(3 + 5i) = 9 - 25i^2 = 9 + 25 = 34$$

$$a = 1,\ b = -6, \text{ and } c = 34$$

The function is $f(x) = x^2 - 6x + 34$

Check Your Understanding of Section 6.5

1. Determine the sum and the product of the roots of $x^2 - 8x + 3 = 0$.
(1) sum = −8, product = 3
(2) sum = 8, product = 3
(3) sum = −3, product = 8
(4) sum = 8, product = −3

2. Write an equation of a quadratic function the sum of whose roots is $\frac{2}{3}$ and the product of whose roots is $\frac{-5}{9}$.
(1) $f(x) = x^2 - 2x - 5$
(2) $f(x) = 9x^2 - 6x - 5$
(3) $f(x) = x^2 + 2x - 5$
(4) $f(x) = 9x^2 + 6x - 5$

3. What is the sum of the roots of the quadratic equation $5x^2 + 2x - 3 + 0$?

(1) $\frac{2}{5}$ (2) -2 (3) $\frac{-2}{5}$ (4) $\frac{-3}{5}$

4. What is the product of the roots of the quadratic equation $6x^2 - 3x + 1 = 0$?

(1) $\frac{1}{6}$ (2) $-\frac{1}{6}$ (3) $\frac{1}{2}$ (4) $-\frac{1}{2}$

5. If the roots of the quadratic equation $2x^2 - x - 5 = 0$ are named r_1 and r_2, what is $\frac{r_1 \cdot r_2}{r_1 + r_2}$?

(1) $\frac{5}{2}$ (2) $\frac{-5}{2}$ (3) 5 (4) -5

6. Write an equation of a quadratic function one of whose roots is $3 - 4i$.

7. Write an equation of a quadratic function one of whose roots is $\frac{2}{3} + \frac{\sqrt{3}}{5}$.

8. Write a quadratic equation such that the sum of its roots is $\frac{3}{5}$ and the product of its roots is $\frac{2}{25}$. What are the roots of this equation?

9. Determine the roots of the quadratic equation $2x^2 + 5x - 12 = 0$. If 2 is added to the positive root from this original quadratic equation and 3 is subtracted from the negative root of the original quadratic equation, write the equation of the resulting new quadratic equation.

Chapter Seven

EXPONENTIAL AND LOGARITHMIC FUNCTIONS

7.1 EXPONENTIAL FUNCTIONS

Some functions contain exponents as variables with constants as bases. Compare $f(x) = x^2$ with $g(x) = 2^x$. $f(x)$ is a polynomial function, and $g(x)$ is an exponential function.

Evaluating Exponential Expressions, Including Base *e*

What does 2^3 mean? The symbol 2^3 indicates 3 factors of the base number 2. So $2^3 = 2 \cdot 2 \cdot 2$. The exponent indicates the number of factors of the base number that are multiplied together.

Example 1

Evaluate 4^2 and 2^4. Explain why they result in the same answer.

Solution: $4^2 = 4 \cdot 4 = 16$ and $2^4 = 2 \cdot 2 \cdot 2 \cdot 2 = 16$. They are both equal to 16 because $2^4 = (2^2)^2 = 4^2$.

The irrational number e is called the natural base because it is used to model exponential growth and decay. Similar to the irrational number π, e can only be approximated. Expressions are often left in terms of e rather than using an approximation. $e \approx 2.71828\ldots$

To evaluate the powers of e, a graphing calculator can be used to find an approximate value. For instance, to evaluate e^5, press **2ND** **+** **^** **5** **ENTER**. The calculator will now display 148.4131591.

Exponential equations are used in many common applications. The example that follows is a typical application resulting in an exponential equation.

Example 2

A population model was created for a town that predicts its school population will grow according to the function $P(t) = 1350e^{0.0241t}$, where t is time in years and $t = 0$ corresponds to 2005. Use this model to approximate the population of the town in (a) 2005 (b) 2010 (c) 2020.

Solution: On a graphing calculator, press **Y=**. In Y_1 enter **1** **3** **5** **0** **2ND** **÷** **^** **(** **0** **.** **0** **2** **4** **1** **X** **)**. Now press **2ND** **WINDOW**. Scroll down and over to Indpnt: Ask **ENTER** and **2ND** **GRAPH**. Finally, 0, 5, and 15 can be entered to get the results as shown below:

X	Y1	
0	1350	
5	1522.9	
15	1937.9	

X=0

Graphing Exponential Equations

MATH FACTS

Exponential functions are functions that have a constant as a base and a variable as an exponent. The function $f(x) = b^x$ such that $b > 0$, $b \neq 1$, and x represents any real number is an exponential function.

Examine the table below for three different exponential functions for integral values of x ranging from -2 to $+3$.

x	-2	-1	0	1	2	3
$f(x) = 2^x$	0.25	0.5	1	2	4	8
$g(x) = 3^x$	0.1111 . . .	0.3333 . . .	1	3	9	27
$h(x) = 10^x$	0.01	0.1	1	10	100	1000

Notice that when $x = 0$, each of these functions map 0 to 1. The graphs are displayed in the following graph with $f(x) = 2^x$ as the narrow black curve, $g(x) = 3^x$ as the medium black curve, and $h(x) = 10^x$ as the wide black curve.

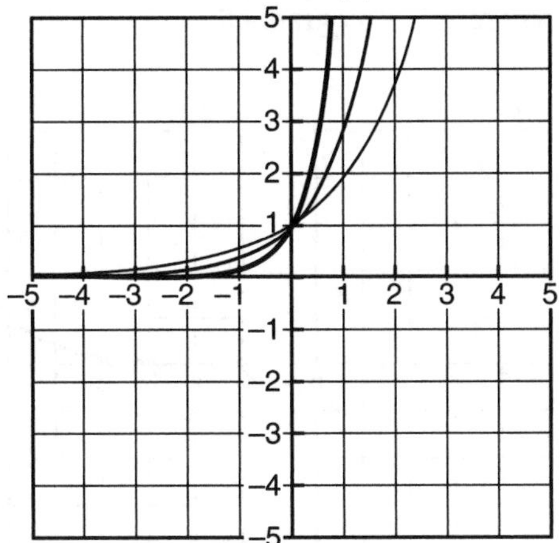

Example 3

On separate graph grids, sketch the graphs of $f(x) = \left(\frac{1}{2}\right)^x$ and $g(x) = 2^{-x}$.

Solution:

$f(x) = \left(\frac{1}{2}\right)^x$ $\qquad$ $g(x) = 2^{-x}$

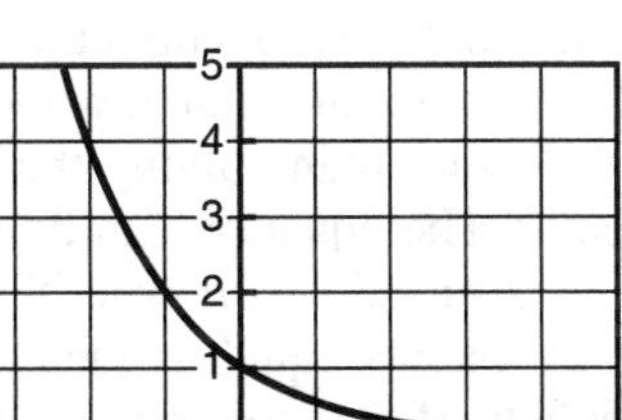

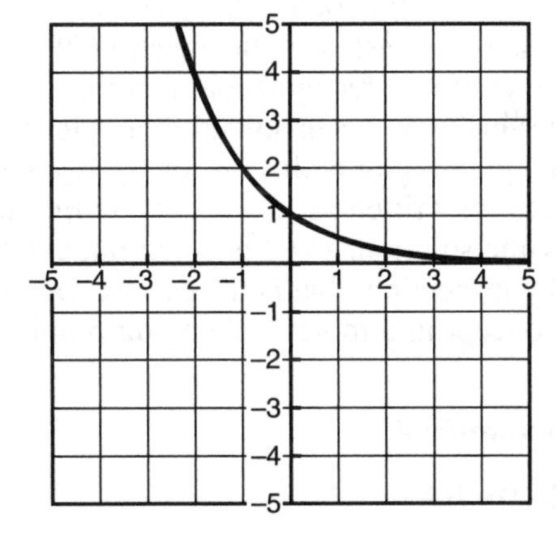

It is probably not surprising that these graphs are exactly the same since $2^{-x} = \frac{1}{2^x} = \left(\frac{1}{2}\right)^x$.

Since e is a real number between 2 and 3, its graph lies between that of $y = 2^x$ and $y = 3^x$. Since e is very close to 3, graphs of both $y = 3^x$ and $y = e^x$ are displayed side by side for comparison. Notice that at $x = 1$, $y = e^x$ is plotted just below where $y = 3^x$ is graphed.

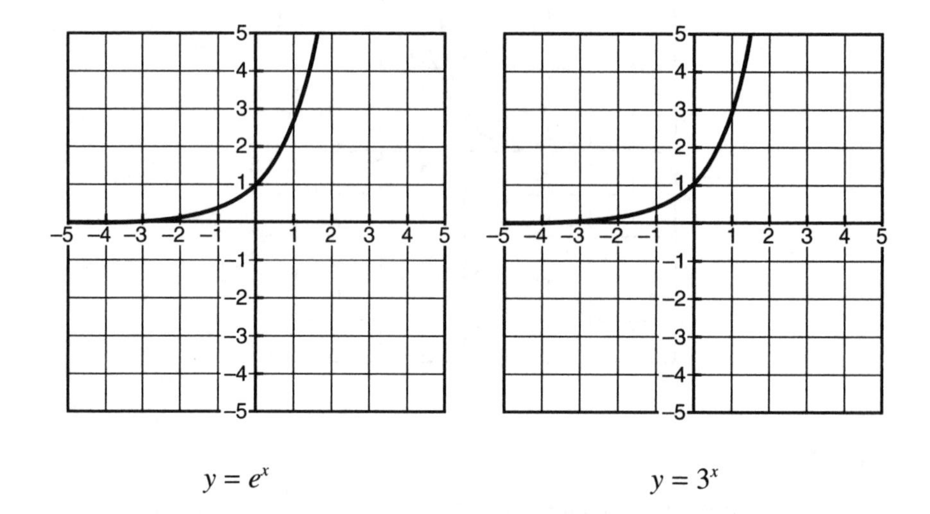

$y = e^x$ $\qquad$ $y = 3^x$

Solving Exponential Equations with Common Bases

When both sides of an equation can be expressed as powers of the same base, the equation can be solved by setting their exponents equal to each other. For instance, the equation $2^x = 4$ does not require much work to solve since it is a recognizable fact that $2^2 = 4$. Rather than solving the equation in this way, it can solved more algebraically. The equation can be rewritten by expressing both sides of the equation as 2 raised to a power. Now $2^x = 4$ can be written as $2^x = 2^2$. By setting the exponents equal to each other, the same solution, $x = 2$ is determined. Obviously, if an equation can be solved by inspection, that is the easiest way to proceed. Now, let's look at the same concept in a more challenging problem.

Example 4

Solve for x: $3^{x+2} = 9^x$

Solution: $3^{x+2} = 9^x$

Recognizing that $3^2 = 9$ yields $3^{x+2} = (3^2)^x$

Using rules of exponents results in $3^{x+2} = 3^{2x}$

Since both sides of the equation are now base 3, this equation can be solved by setting the exponents equal to each other.

$$x + 2 = 2x$$
$$2 = x$$

$$\begin{aligned} \text{Check: } 3^{x+2} &= 9^x \\ 3^{2+2} &= 9^2 \\ 3^4 &= 81 \\ 81 &= 81 \end{aligned}$$

Example 5

Solve for x: $5^{x+5} = 125^{3-x}$

Solution: $5^{x+5} = 125^{3-x}$

$$5^{x+5} = (5^3)^{3-x}$$
$$5^{x+5} = 5^{9-3x}$$
$$x + 5 = 9 - 3x$$
$$4x + 5 = 9$$
$$4x = 4$$
$$x = 1$$

Example 6

Solve for x: $4^{2x-7} = \dfrac{1}{64^{4x}}$

Solution: $4^{2x-7} = \dfrac{1}{64^{4x}}$

$$4^{2x-7} = (4^{-3})^{4x}$$
$$4^{2x-7} = 4^{-12x}$$
$$2x - 7 = -12x$$
$$-7 = -14x$$
$$x = \frac{1}{2}$$

Check Your Understanding of Section 7.1

1. Solve for x: $4^x = 32$

(1) $x = \frac{2}{5}$ (2) $x = 2$ (3) $x = \frac{5}{2}$ (4) $x = 3$

2. Solve for a: $3^{2a-9} = 243$

(1) $a = -2$ (2) $a = 5$ (3) $a = -7$ (4) $a = 7$

3. Solve for s: $6^{2-s} = \frac{1}{216}$

(1) $s = -3$ (2) $s = -5$ (3) $s = 6$ (4) $s = 5$

4. Solve for z: $9^{z+8} = 27^{3z+10}$

(1) $z = 7$ (2) $z = 2$ (3) $z = -7$ (4) $z = -2$

5. Solve for k: $4^{k-2} = 8^{12-k}$

(1) $k = 7$ (2) $k = 10$ (3) $k = 5$ (4) $k = 8$

6. Solve for c: $\left(\frac{1}{125}\right)^{c-3} = 5^{3c-15}$

(1) $c = 2$ (2) $c = 4$ (3) $c = -2$ (4) $c = -4$

7. Solve for y: $3125^{3y-5} = 625^{y+2}$

(1) $y = 3$ (2) $y = 11$ (3) $y = \frac{7}{2}$ (4) $y = \frac{-7}{2}$

8. Solve for y in terms of x if $16^x = 8^{y+5}$.

9. The number bacteria that grow in a gel plate is approximated by the function $A(t) = 28e^{0.216t}$, where t is time in minutes, with $t = 0$ corresponding to the starting time. Use this model to approximate the number of bacteria present after t minutes when t equals (a) 0 minutes, (b) 1 minute, (c) 5 minutes, (d) 15 minutes, and (e) 1 hour.

10. Solve for all values of x: $25^{x^2-5} = 5^x$

11. Complete the table of values below, and sketch a graph of $f(x)=\left(\frac{2}{3}\right)^x$.

x	$f(x)$
−2	
−1	
0	
1	
2	
3	

12. Sketch the graph of $g(x)=\sqrt{e^x}$.

7.2 LOGARITHMIC FUNCTIONS

Any exponential function is one-to-one. Therefore, all exponential functions have unique inverses. A logarithmic function is the inverse function of an exponential function.

The Graph of the Inverse of an Exponential Function

Remember that inverse functions are reflections of each other over the line $y = x$. The following graph shows the exponential function $f(x) = 2^x$. Next to this is the same graph reflected over $y = x$.

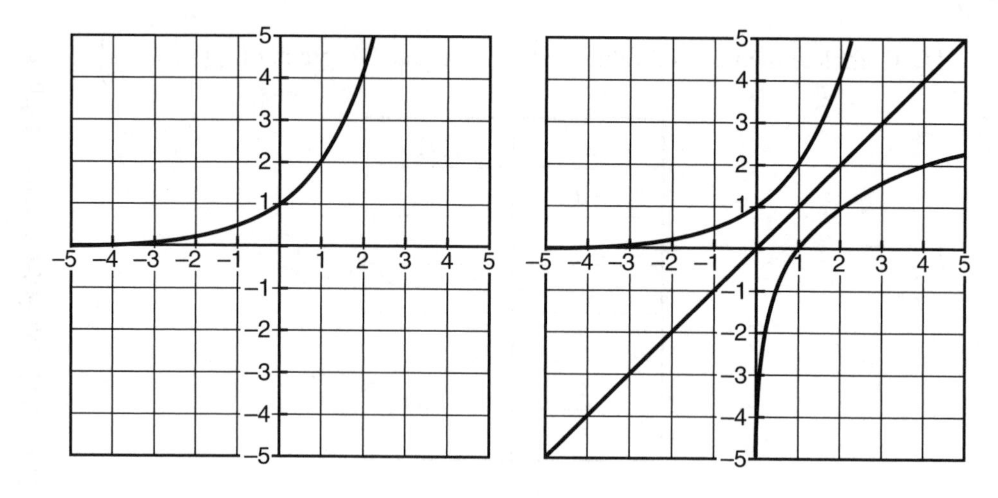

Examine these two graphs carefully. Notice that for the function $f(x) = 2^x$, for very small negative numbers, the graph of $f(x)$ is very close to the x-axis. In other words, as x approaches $-\infty$, y approaches 0. In other words, the x-axis is an asymptote for the function as x approaches $-\infty$. Similarly, in the graph of the inverse of $f(x) = 2^x$, for very small negative values of y, the graph of the inverse function is very close to the y-axis. In other words, as y approaches $-\infty$, x approaches 0. The y-axis is an asymptote for the function. Also notice that $f(x) = 2^x$ passes through the ordered pair (0, 1), as do all exponential functions. The inverse function passes through the ordered pair (1, 0) since an inverse switches x and y for each ordered pair.

As always, to find the function rule for the inverse of a function, set $y = f(x)$. In other words, switch x and y and solve for y or $f^{-1}(x)$. If $f(x) = 2^x$, $y = 2^x$, and $x = 2^y$. However, there is no way to solve $x = 2^y$ for y. Therefore, a new type of function was invented to provide a function rule for this inverse—a logarithmic function. In general, if $x = b^y$ then $y = \log_b x$. This clearly indicates that a logarithm is the exponent that the base is raised to that yields a value of x.

Note that when a logarithm has a base equal to 10, the base 10 is not actually written. Base 10 is assumed to be there. This is similar to how a variable that has no coefficient, such as x, assumes that the coefficient is 1. In fact, the logarithmic function with base 10 is called a common logarithm. A logarithm whose base is e is called the natural logarithm, and it is written as $\ln x$ rather than $\log_e x$.

MATH FACTS

A logarithmic function is the inverse function of an exponential function. The equation $x = b^y$ can be converted to $y = \log_b x$. This relationship can be used to convert an exponential equation into its equivalent logarithmic equation and vice versa.

Example 1

Convert the equation $y = 3^x$ to an equivalent logarithmic equation.

Solution: If $y = 3^x$, then $x = \log_3 y$.

Example 2

Convert the equation $y = \log_6 x$ to an equivalent exponential equation.

Solution: If $y = \log_6 x$, then $x = 6^y$.

This same conversion rule can be used to evaluate logarithmic expressions. For instance, to evaluate $\log_3 27$, set this expression equal to what could be called a dummy variable. Let $\log_3 27 = n$. Now convert this logarithmic equation to its equivalent exponential equation and solve.

$$3^n = 27$$

Since $27 = 3^3$, $3^n = 3^3$

$$n = 3$$

Example 3

Evaluate each logarithmic expression:

a. $\log_5 625$ b. $\log_2 \dfrac{1}{128}$ c. $\log_7 7$ d. $\ln e^3$ e. $\log_{\frac{1}{4}} 64$

Solution: a. Set $\log_5 625 = n$

Convert that to $5^n = 625$ or $5^n = 5^4$

$$n = 4$$

b. Set $\log_2 \dfrac{1}{128} = n$

$$2^n = \frac{1}{128} = \frac{1}{2^7} = 2^{-7}$$

$$n = -7$$

c. Set $\log_7 7 = n$

$7^n = 7 = 7^1$

$n = 1$

Notice that this was not dependent on the base. For all values of b, $\log_b b = 1$.

d. Since $\ln e^3$ stands for $\log_e e^3$, $\log_e e^3 = n$

Set $\log_e e^3 = n$

$e^n = e^3$

$n = 3$

e. Set $\log_{\frac{1}{4}} 64 = n$

$$\left(\frac{1}{4}\right)^n = 64$$

$$\left((4)^{-1}\right)^n = 4^{-n} = 4^3$$

$n = -3$

Example 4

Solve for x: $\log_x 49 = 2$

Solution: If $\log_x 49 = 2$, then $x^2 = 49$ and $x = 7$. Remember that the domain of all logarithmic functions is the set of positive numbers, so $x = -7$ is not a solution.

Example 5

Solve for x: $\log x = 3$

Solution: If $\log x = 3$ implies that $10^3 = x$. So $x = 1000$.

A graphing calculator can be used to evaluate a logarithmic expression. For instance, an approximate numerical value for the expression $\log 6$ can be found by pressing **LOG** **6** **)** **ENTER**. The result .7781512504 is now displayed and can be rounded to any degree of accuracy desired.

Graphing calculators can also evaluate an expression such as $\ln 6$. To do this, press **LN** **6** **)** **ENTER**. The result is displayed as 1.791759469.

Notice that a graphing calculator does not have a button for logarithms with bases other than 10 or e. There is a very useful formula that can be used to change the base of a logarithm so that the calculator can be used. Conveniently, the formula is called the change of base formula. This formula is $\log_b x = \frac{\log_a x}{\log_a b}$.

MATH FACTS

The change of base formula for logarithms is:

$$\log_b x = \frac{\log_a x}{\log_a b}$$

To evaluate a logarithm such as $\log_5 8$, the expression can be changed so that it is expressed in base 10 or base e. For example, $\log_5 8 = \frac{\log 8}{\log 5}$ or $\frac{\ln 8}{\ln 5}$. Either of these yield an answer of 1.292029674.

Check Your Understanding of Section 7.2

1. The inverse of the function $g(x) = 8^x$ is:

(1) $g^{-1}(x) = \log_x 8$
(2) $g^{-1}(x) = \log_{3x} 8$
(3) $g^{-1}(x) = \log_8 x$
(4) $g^{-1}(x) = \log_2 3x$

2. Solve for z: $\log_z 125 = 3$

(1) $z = 5$ (2) $z = 25$ (3) $z = -5$ (4) $z = 10$

3. Evaluate: $\log_{\frac{1}{3}} 27$

(1) 3 (2) −3 (3) 9 (4) −9

4. Evaluate: $\log_4 2$

(1) 4 (2) 2 (3) 1 (4) $\frac{1}{2}$

5. Solve for t: $\log t = -2$

(1) $t = \frac{1}{2}$ (2) $t = \frac{1}{10}$ (3) $t = \frac{1}{100}$ (4) $t = \frac{1}{5}$

6. Evaluate: $\ln \frac{1}{e^3}$

(1) -3 (2) $\frac{1}{3}$ (3) 3 (4) e

7. Evaluate: $\log_6 12$

(1) 2 (3) approximately 1.901708569

(2) approximately 1.386852807 (4) $\frac{1}{2}$

8. Evaluate: $\log_c c^5$

9. Evaluate: $\ln e^8$

10. Use the change of base formula to approximate the value of $\log_3 7$ to the nearest hundredth.

11. What is the domain of $\log_2(x^2 - 4)$?

12. Write as an equivalent expression involving x without logs:

$\log_3 9^{x+1} - \log_3 27^x$

13. Sketch the graph of $h(x) = \log_5 x$.

7.3 PROPERTIES OF LOGARITHMS

Since a logarithm is the exponent that a base is raised to, operational rules for exponents can be modified to provide operational rules for logarithms.

Properties of Logarithms

The following table lists several exponent rules and the properties for logarithms that are based on them.

Exponentiation Rules	Logarithm Properties
To multiply like bases raised to exponents, simply add their exponents: $x^a \cdot x^b = x^{a+b}$	The logarithm of the product of two numbers is the sum of their logarithms: $\log_b(u \cdot v) = \log_b u + \log_b v$
To divide like bases raised to exponents, simply subtract their exponents: $x^a \div x^b = x^{a-b}$	The logarithm of the quotient of two numbers is the difference of their logarithms: $\log_b\left(\frac{u}{v}\right) = \log_b u - \log_b v$
To raise a base that is raised to an exponent to another exponent, simply multiply their exponents: $\left(x^a\right)^b = 4^{a \cdot b}$	The logarithm of a number raised to an exponent is the exponent multiplied by the logarithm of the number: $\log_b u^n = n \cdot \log_b u$

These properties are used to simplify logarithmic expressions. The following examples show how they are applied.

Example 1

Use the properties of logarithms to rewrite as the sum or difference of logarithms: $\log(x^2)(3x + 1)$

Solution: $\log(x^2)(3x + 1) = \log x^2 + \log(3x + 1) = 2\log x + \log(3x + 1)$.

Example 2

Use the properties of logarithms to rewrite as the sum or difference of logarithms: $\log_3 \frac{\sqrt{x-4}}{2x}$

$$\textit{Solution}: \log_3 \frac{\sqrt{x-4}}{2x} = \log_3 \sqrt{x-4} - \log_3 2x = \log_3 (x-4)^{\frac{1}{2}} - \log_3 2x$$

$$\frac{1}{2}\log_3(x-4) - \log_3 2x = \frac{1}{2}\log_3(x-4) - \log_3 2 - \log_3 x$$

Example 3

Use the properties of logarithms to rewrite as the sum or difference of logarithms: $\ln(5e^x)^3$

Solution: $\ln(5e^x)^3 = 3\ln(5e^x) = 3(\ln 5 + \ln e^x) = 3(\ln 5 + x) = 3\ln 5 + 3x$

Check Your Understanding of Section 7.3

1. If $\log 3 = t$, then $\log 2700$ can be expressed as:
(1) $900t$ (2) $100 + 3t$ (3) $3t + 2$ (4) $90t$

2. Which expression is equivalent to $\ln 2.5$?
(1) $2\ln 5$ (2) $2\ln 5 - \ln 10$ (3) $\dfrac{\ln 25}{\ln 10}$ (4) $\dfrac{1}{2}\ln 5$

3. If $\log_b u = r$, $\log_b v = s$, and $\log_b w = t$, then $\log_b \dfrac{\sqrt[3]{uv}}{u^2 w^4}$ is equivalent to:
(1) $\frac{1}{3}r + \frac{1}{3}s - 2r - 4t$
(2) $\frac{1}{3}(r + s) - 2r + 4t$
(3) $\frac{1}{3}r + \frac{1}{3}s - 2r + 4t$
(4) $\frac{1}{3}(r + t) - 2r - 4s$

4. Which of the following expressions is equivalent to $\log_b 108$?
(1) $\log_b 27 \cdot \log_b 4$
(2) $3\log_b 3 \cdot 2\log_b 2$
(3) $3\log_b 3 + 2\log_b 2$
(4) $2\log_b 3 + 3\log_b 2$

5. Which of the following expressions is equal to x?
(1) $\ln x$ (2) $\log_2 2^x$ (3) $\log e^x$ (4) $\ln x^x$

6. If $\log_b 2 = r$ and $\log_b 3 = s$, then express in simplest form the number whose logarithm in base b is equal to $\frac{1}{3}r + \frac{2}{3}s - 5r - 3s$.

7. Use the properties of logarithms to rewrite as the sum or difference of logarithms: $\log_b \dfrac{5x^3y}{\sqrt[4]{2z}}$

8. Use the properties of logarithms to rewrite as the sum or difference of logarithms: $\log_b \sqrt{\dfrac{13(x+5)}{(2x-3)^3}}$

7.4 SOLVING LOGARITHMIC EQUATIONS

Simply put, a logarithmic equation is an equation involving logarithms. To solve these, use the properties of logarithms from Section 7.3 and convert the equation to an equivalent exponential equation.

Logarithmic Equations

A simple logarithmic equation such as $\log_3 x = 2$ can be solved by converting the equation to the equivalent exponential equation using the conversion formula $x = b^y$ implies that $y = \log_b x$.

Example 1

Solve for x: $\log_3 x = 2$

Solution: If $\log_3 x = 2$, then $3^2 = x$, therefore $x = 9$.

Many logarithmic equations will require applying the properties of logarithms from Section 7.3 to simplify the logarithmic expression within the equation before converting it to the equivalent exponential equation. The following examples illustrate this.

Example 2

Solve for x: $\log 2 + \log x = 3$

Solution: If $\log 2 + \log x = 3$, then $\log (2 \cdot x) = 3$. Since this equation involves a common logarithm, the base of the logarithm is 10.

$$10^3 = 2x$$
$$1000 = 2x$$
$$x = 500$$

Example 3

Solve for x: $\log_2 x + \log_2(x - 2) = 3$

Solution: If $\log_2 x + \log_2(x - 2) = 3$, then $\log_2 x(x - 2) = 3$ and $\log_2(x^2 - 2x) = 3$. So, $x^2 - 2x = 2^3 = 8$ and $x^2 - 2x - 8 = 0$. This is a factorable quadratic equation. Factoring yields the equation $(x - 4)(x + 2) = 0$. The answers are therefore $x = 4$ and $x = -2$. However, notice that $x = -2$ is not possible since -2 is not in the domain of either $\log_2 x$ or $\log_2(x - 2)$. So the solution set is $\{4\}$.

MATH FACTS

If $\log_b x = \log_b y$, then $x = y$.

Example 4

Solve for x: $\log_5(x + 1) - \log_5 4 = \log_5 6$

Solution: If $\log_5(x + 1) - \log_5 4 = \log_5 6$, then $\log_5 \frac{x+1}{4} = \log_5 6$. Therefore, $\frac{x+1}{4} = 6$ and $x + 1 = 24$. Subtracting 1 from each side yields $x = 23$.

Sometimes it is necessary to use a calculator to solve the exponential equation equivalent to the logarithmic equation.

Example 5

Solve for x to the nearest hundredth: $\log x = 2.1$

Solution: If $\log x = 2.1$, then $x = 10^{2.1} \approx 125.89$. The diagram below shows the display from a calculator used to determine this answer.

```
10^2.1
            125.8925412
```

Check Your Understanding of Section 7.4

1. Solve for z: $\log_z 8 = -3$

(1) $z = 2$ (2) $z = \frac{1}{2}$ (3) $z = 8$ (4) $z = \frac{1}{8}$

2. Solve for x: $\log_7 x + \log_7 2 = \log_7 8$

(1) $x = 2$ (2) $x = 4$ (3) $x = 6$ (4) $x = 8$

3. Solve for x: $2\ln x - \ln 4 = \ln 25$

(1) $x = 5$ (2) $x = 20$ (3) $x = 10$ (4) $x = \sqrt{104}$

4. Solve for x: $3\log x + \log 7 = \log 56$

(1) $x = 1$ (2) $x = 7$ (3) $x = 8$ (4) $x = 2$

5. Solve for x: $\log_6 4 + \log_6 x = 3$

(1) $x = \frac{3}{2}$ (2) $x = \frac{3}{4}$ (3) $x = 54$ (4) $x = 9$

6. Solve for t: $\log_3(t + 1) - \log_3 t = 4$

(1) $t = \frac{1}{80}$ (2) $t = \frac{1}{81}$ (3) $t = -80$ (4) $t = -81$

7. Solve for n: $\log_{16}(n+30)+\log_{16}n = 1\frac{1}{2}$

(1) $n = 2$
(2) $\{n : n = 2 \text{ or } n = 32\}$
(3) $n = 32$
(4) $\{n : n = 2 \text{ or } n = -32\}$

8. Solve for k: $\log_3(k + 3) + \log_3(k - 5) = 2$

(1) $k = 4$ (2) $k = 5$ (3) $k = 6$ (4) no solution

9. Solve for v: $\log_v 8 + \log_v 16 = 7$

(1) $v = 2$ (2) $v = 49$ (3) $v = \frac{128}{7}$ (4) $v = 128$

10. The formula for distance modulus M is $M = \log r - 5$, where r is the distance of a star from Earth as measured in parsecs, where 1 parsec is approximately 3.26 light-years. The distance modulus of the star Cygni is approximately −2.32. Use this formula to determine the distance from the Earth to Cygni in parsecs. Express your answer to the nearest hundredth. Then convert this distance to light-years. Express your answer to the nearest tenth.

11. The power of a sound, I, is measured in watts and can be converted to decibels by the formula $D = 10(\log I + 16)$. If the sound made by an animal is measured as 86 decibels, what is the power of this sound to the nearest hundredth?

7.5 SOLVING EXPONENTIAL EQUATIONS WITH UNLIKE BASES

Exponential equations were solved in the first section of this chapter, but not all exponential equations can be simplified as a power of the same base. When the same base cannot be used to solve an exponential equation, then the equation is solved by using logarithms.

Exponential Equations with Different Bases

An exponential equation such as $2^x = 8$ was easy to solve since 8 is 2^3. What happens if the equation is changed to $2^x = 7$? Since 7 cannot be written as a power of 2, just take the logarithm of each side of the equation, resulting in $\log 2^x = \log 7$. Now apply the logarithm rule $\log_b u^n = n \cdot \log_b u$ to get $x \log 2 = \log 7$. Now divide both sides of this equation by $\log 2$. The resulting equation is $x = \frac{\log 7}{\log 2}$. Use the calculator and press LOG 7) ÷ LOG 2) ENTER. The display on the calculator now provides the answer, 2.807354922. The same result could be found by taking the natural log of both sides of the equation instead of the common logarithm. The calculator displays for both approaches are shown.

Example 1

Solve for x to the nearest thousandth: $5^{4x} = 12$

Solution: If $5^{4x} = 12$, then $\log 5^{4x} = \log 12$.

$$4x\log 5 = \log 12$$

$$4x = \frac{\log 12}{\log 5} \approx 1.543959311$$

$$x = 0.3859898277 \approx 0.386$$

Example 2

Solve for x to the nearest thousandth: $9^{x-6} = 7$

Solution: If $9^{x-6} = 7$, then $\log 9^{x-6} = \log 7$.

$$(x - 6)\log 9 = \log 7$$

$$x - 6 = \frac{\log 7}{\log 9} = .8856218746$$

$$x = .8856218746 + 6 = 6.8856218746 \approx 6.886$$

Example 3

Solve for x to the nearest thousandth: $6^{2x-1} = 4^{x+3}$

Solution: If $6^{2x-1} = 4^{x+3}$, then $\log 6^{2x-1} = \log 4^{x+3}$

$$(2x - 1)\log 6 = (x + 3)\log 4$$

$$\frac{2x-1}{x+3} = \frac{\log 4}{\log 6} = .7737056145$$

$$2x - 1 = .7737056145(x + 3)$$

$$2x - 1 = .7737056145x + 2.321116843$$

$$1.226294386x = 3.321116843$$

$$x = \frac{3.321116843}{1.226294386} = 2.708254138 \approx 2.708$$

Now that this process to solve an exponential equation has been investigated, it offers a second approach that can be used to solve certain logarithmic equations. In order to solve an equation such as $\log_2 5 = x$, the change of base formula can be used to result in $\frac{\log 5}{\log 2} = x$ or approximately 2.3219.

This can also be solved by converting the logarithmic equation to an exponential equation as illustrated in the next example.

Example 4

Solve for x to the nearest thousandth: $\log_2 5 = x$

Solution: $\log_2 5 = x$ converts to the exponential equation $2^x = 5$. Take the log of both sides, $\log 2^x = \log 5$

$$x\log 2 = \log 5$$

$$x = \frac{\log 5}{\log 2} \approx 2.3219$$

Check Your Understanding of Section 7.5

1. Solve for x and round the answer to the nearest hundredth: $3^x = 6$
(1) $x = 2$ (2) $x = 1.63$ (3) $x = 0.61$ (4) $x = 0.95$

2. Solve for t and round the answer to the nearest thousandth: $4^x = 5$
(1) $t = 3.010$ (2) $t = 0.861$ (3) $t = 1.1.61$ (4) $t = 1.25$

3. Solve for s and round the answer to the nearest thousandth: $9^s = 2$
(1) $s = 512$ (2) $s = 81$ (3) $s = 3.170$ (4) $s = 0.3155$

4. Solve for x and round the answer to the nearest hundredth: $8^{2x+5} = 17$
(1) $x = -1.82$ (2) $x = 1.82$ (3) $x = -2.13$ (4) $x = 2.13$

5. Solve for t and round the answer to the nearest hundredth: $14 = 3^{2t-7}$
(1) $t = -3.71$ (2) $t = 3.71$ (3) $t = -4.70$ (4) $t = 4.70$

6. Solve for x and round the answer to the nearest hundredth: $3^{4x-1} = 7^{x+1}$

7. Solve for x and round the answer to the nearest thousandth: $\log_8 15 = x$

8. Solve for y and round the answer to the nearest hundredth: $6^{2-y} - 4^y = 0$

9. Solve for m and round the answer to the nearest hundredth:

$$12^{m+3} = \left(\frac{1}{3}\right)^{2m}$$

10. The number of bacteria present in a culture is modeled by the formula $N = 100e^{5t}$, where N is the number of bacteria present after t hours. To the nearest tenth, in how many hours will N equal 20,000? **Hint**: Use the natural logarithm.

Chapter Eight

TRIGONOMETRIC FUNCTIONS

8.1 RADIAN MEASURE AND ARC LENGTH

KEY IDEAS

Previous to this course, angles and arcs were measured in degrees. This measure is based on an arbitrary 360 degrees around a point. In this section, another type of measure is used. It is based on a circle with radius 1 whose circumference is then 2π. It is called radian measure.

What Is a Radian?

Start with an angle whose initial side is a ray positioned on the positive half of the x-axis with its endpoint at the origin. Rotate this ray in a counterclockwise direction to find the terminal side of the angle. Begin with the ordered pair (1, 0) in this rotation. The distance this point travels as a result of the rotation is called s. When a rotation causes the distance, s to be equal to 1 (the radius), then the measure of the rotation is called 1 radian. This is shown in the following diagram.

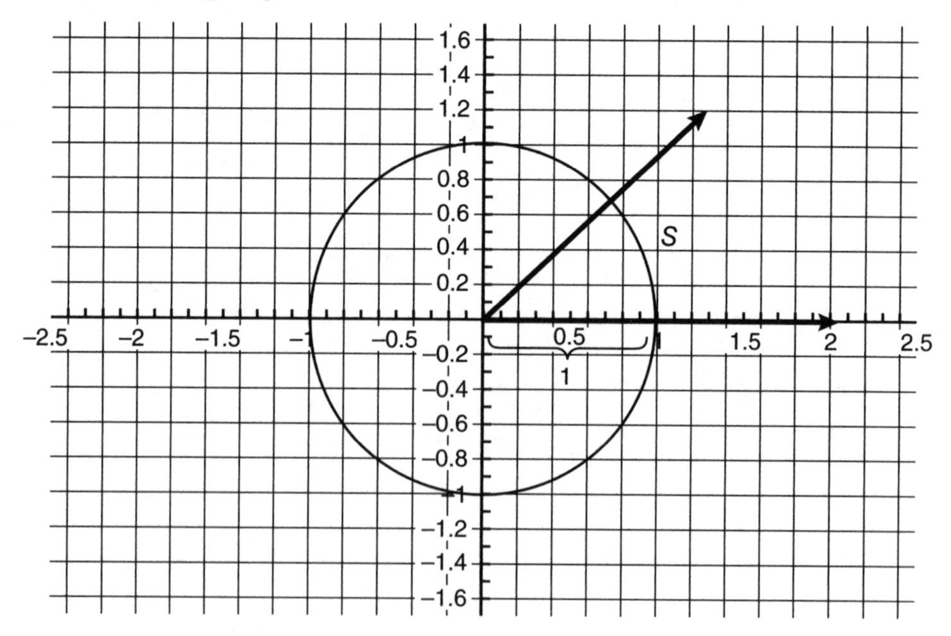

Converting Between Radian and Degree Measures

In a circle whose radius is 1, the circumference is $2\pi(1) = 2\pi$. This is the measure of the rotation of the ordered pair (1, 0) back to (1, 0). That suggests a proportion between the degree measure of the whole circle of 360° to the length of the arc of the entire circle 2π.

The proportion is $\frac{\text{radian measure}}{2\pi} = \frac{\text{degree measure}}{360}$. Dividing each denominator by 2 yields the formula that is used to convert between radian and degree measures: $\frac{\text{radian measure}}{\pi} = \frac{\text{degree measure}}{180}$.

MATH FACTS

To convert from radian measure to degree measure and vice versa, use the formula $\frac{\theta}{\pi} = \frac{d}{180}$, where θ is the radian measure of an angle (or rotation) and d is the degree measure of the same angle (or rotation).

Example 1

Convert 30° to radian measure.

Solution: Since $\frac{\theta}{\pi} = \frac{d}{180}$, $\frac{\theta}{\pi} = \frac{30}{180} = \frac{1}{6}$. Using the product of the means is equal to the product of the extremes on $\frac{\theta}{\pi} = \frac{1}{6}$ results in $6\theta = \pi$ or $\theta = \frac{\pi}{6}$.

Several additional examples appear in the table on the next page.

Degree Measure	Using $\frac{\theta}{\pi} = \frac{d}{180}$	Radian Measure
Example 2 $d = 75°$	$\frac{\theta}{\pi} = \frac{75}{180} = \frac{5}{12}$ $12\theta = 5\pi$	$\theta = \frac{5\pi}{12}$
Example 3 $d = 252°$	$\frac{\theta}{\pi} = \frac{252}{180} = \frac{7}{5}$ $5\theta = 7\pi$	$\theta = \frac{7\pi}{5}$
Example 4 $d = 140°$	$\frac{\theta}{\pi} = \frac{140}{180} = \frac{7}{9}$ $7\pi = 9\theta$	$\theta = \frac{7\pi}{9}$
Example 5 $d = 270°$	$\frac{\theta}{\pi} = \frac{270}{180} = \frac{3}{2}$ $3\pi = 2\theta$	$\theta = \frac{3\pi}{2}$

The same formula can be used to convert radian measures of angles (or rotations) to equivalent degree measure(s).

Example 6

Convert $\frac{3\pi}{4}$ radians to degree measure.

Solution: Since $\frac{\theta}{\pi} = \frac{d}{180}$, $\frac{\frac{3\pi}{4}}{\pi} = \frac{\text{d}}{180}$ and $\frac{\frac{3\pi}{4}}{\pi} = \frac{3\not{\pi}}{4} \cdot \frac{1}{\not{\pi}} = \frac{3}{4}$. Therefore, $\frac{3}{4} = \frac{d}{180}$ and $4d = 540$. So $d = 135°$.

An alternate method can be used to solve Example 6. Since $2\pi = 360°$, $\pi = 180°$. Substitute 180° for π in the expression $\frac{3\pi}{4}$ to yield $\frac{3 \cdot 180}{4} = \frac{540}{4} = 135°$.

The chart that follows uses the alternate approach of replacing 180° for π in the radian form for the measure of an angle (or rotation).

Example	Radian Measure	Replace 180° for π	Degree Measure
Example 7	$\frac{\pi}{2}$	$\frac{\pi}{2}=\frac{180}{2}$	90°
Example 8	$\frac{2\pi}{3}$	$\frac{2\pi}{3}=\frac{2\cdot180}{3}=\frac{360}{3}$	120°
Example 9	$\frac{5\pi}{6}$	$\frac{5\pi}{6}=\frac{5\cdot180}{6}=\frac{5\cdot30}{1}$	150°
Example 10	$\frac{7\pi}{12}$	$\frac{7\pi}{12}=\frac{7\cdot180}{12}=\frac{7\cdot15}{1}$	105°

What if a radian measure does not include the symbol π? Then divide this number by π. See the example below.

Example 11

How many degrees are in 1 radian?

Solution: Use the formula $\frac{\theta}{\pi}=\frac{d}{180}$ to get $\frac{1}{\pi}=\frac{d}{180}$ or $d=\frac{180}{\pi}$ Calculate $\frac{180}{\pi}$ using a calculator. Press **1** **8** **0** **÷** **π** **ENTER**. The calculator display is shown.

```
180/π
          57.29577951
```

Arc Length

The length of an arc in radian measure is S. The following diagram shows that the proportion $\frac{1}{r} = \frac{\theta}{S}$ can be formed. Manipulating this proportion results in the equation $S = r\theta$.

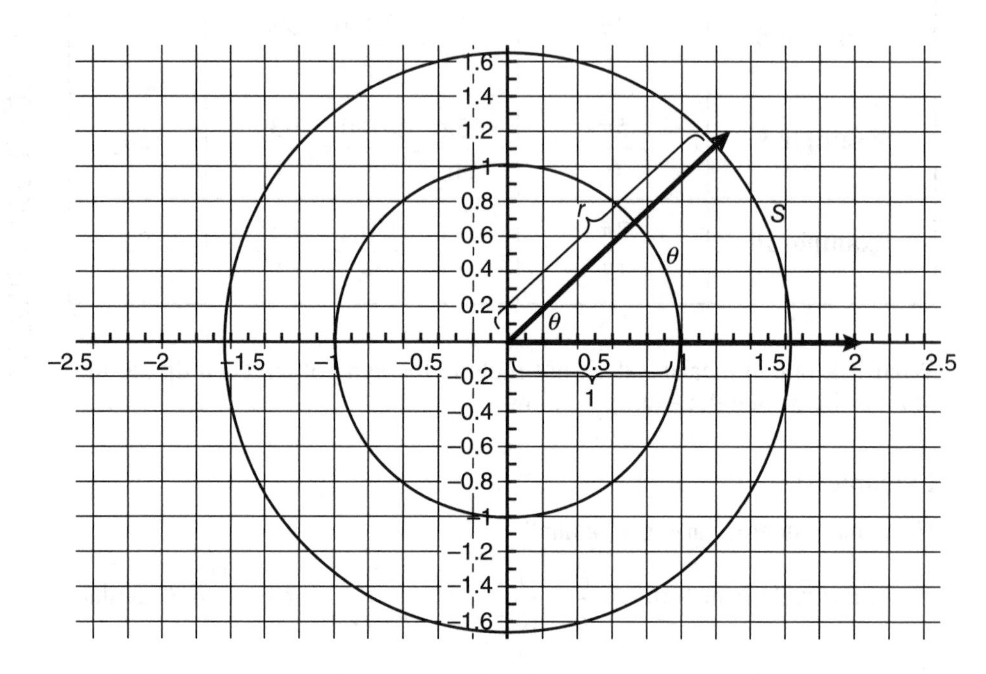

MATH FACTS

The formula to determine the length of an arc, S, on a circle with radius r intercepted by a central angle expressed in radian measure θ is $S = r\theta$.

Example 12

On a circle with radius 5 inches, an arc is intercepted by a central angle whose measure is $\frac{\pi}{3}$. How many inches long is this arc?

Solution: $S = r\theta = 5 \cdot \frac{\pi}{3} = \frac{5\pi}{3}$ inches.

Example 13

On a circle with radius 4 cm, an arc 5π cm long is intercepted by a central angle.

a. What is the radian measure of the central angle?
b. What is the degree measure of the central angle?

Solution: a. $S = r\theta$

$$5\pi = 4\theta$$

$$\theta = \frac{5\pi}{4} \text{ cm}$$

b. $$\frac{5\pi}{4} = \frac{5\cdot 180}{4}$$
$$= 5\cdot 45$$
$$= 225°$$

Example 14

An arc that is $\frac{7\pi}{4}$ inches long is intercepted by a central angle measuring $\frac{2\pi}{3}$ inches. How long is the radius of this circle?

Solution: $S = r\theta$

$$\frac{7\pi}{4} = r\cdot\frac{2\pi}{3}$$

$$r = \frac{\frac{7\pi}{4}}{\frac{2\pi}{3}} = \frac{7\not{\pi}}{4}\cdot\frac{3}{2\not{\pi}}$$

$$= \frac{21}{8} \text{ inches or } 2\frac{5}{8}.$$

Example 15

A cart has a 4-inch wheel and is moved along a path 78 inches long. How many rotations has the wheel made (express the answer to the nearest tenth of an inch)?

Solution: Each rotation has a length of 2π inches. Since $S = r\theta$, then $78 = 2\theta$ and $\theta = \frac{78}{2} = 39$. The measure of the central angle is 39. Therefore,

$\frac{39}{2\pi}$ represents the number of rotations made by the wheel. A calculator can approximate the value to the nearest tenth of an inch.

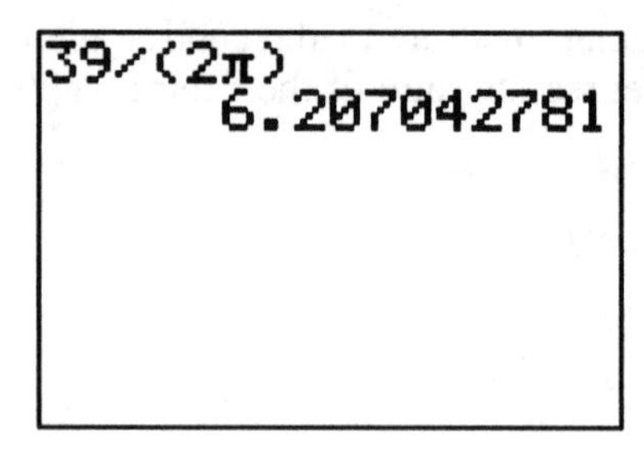

So the answer is approximately 6.2 rotations. Notice that it is necessary to put parentheses around the 2π to calculate this answer.

Check Your Understanding of Section 8.1

1. How many radians are equivalent to 120°?

(1) $\frac{3\pi}{2}$ (2) $\frac{2\pi}{3}$ (3) $\frac{5\pi}{6}$ (4) $\frac{7\pi}{12}$

2. Convert $\frac{6\pi}{5}$ to the equivalent degree measure.

(1) 216° (2) 144° (3) 108° (4) 36°

3. How many radians are equivalent to a 2/3 rotation on a circle?

(1) $\frac{4\pi}{3}$ (2) $\frac{2\pi}{3}$ (3) $\frac{5\pi}{6}$ (4) $\frac{3\pi}{2}$

4. How long does the minute hand on a clock take to move through $\frac{5\pi}{6}$ radians?

(1) 5 minutes (2) 12 minutes (3) 20 minutes (4) 25 minutes

5. An arc is intercepted by a central angle of 200° in a circle of radius 4 cm. How many centimeters long is this arc?

(1) $\frac{5\pi}{18}$ (2) $\frac{20\pi}{9}$ (3) $\frac{40\pi}{9}$ (4) $\frac{80\pi}{9}$

6. An arc measuring 12 inches on a circle whose diameter is 14 inches is intercepted by a central angle whose radian measure is:

(1) $\frac{12}{7}$ (2) $\frac{6}{7}$ (3) $\frac{12}{7\pi}$ (4) $\frac{12\pi}{7}$

7. A central angle of a circle measuring $\frac{7\pi}{6}$ radians intercepts an arc 14π cm long. How long is the radius of this circle?

(1) 6 cm (2) 12 cm (3) 4π cm (4) 24 cm

8. A Ferris wheel has a diameter of 20 feet. Determine the length of the arc traveled by one of its cars (as measured by the point where the car is attached to the wheel) when the wheel rotates 300°.

9. Which has the greater radian measure: a 240° angle, a $\frac{3}{5}$ rotation of a point on a circle, or the length of an arc on a circle whose radius is 2 subtended by a central angle of $\frac{7\pi}{12}$? Explain your answer.

10. A wall clock has a pendulum 6 inches long. If the pendulum swings through an arc whose length is 4 radians, what is the degree measure of the angle formed by the pendulum arm during its swing?

8.2 TRIGONOMETRIC RATIOS IN A RIGHT TRIANGLE

The ratios of the lengths of a pair of the sides of a right triangle can be defined in terms of trigonometric functions of the acute angles of that triangle.

Definitions of the Six Trigonometric Functions

The following diagram illustrates right triangle *ABC* with a right angle at vertex *C*.

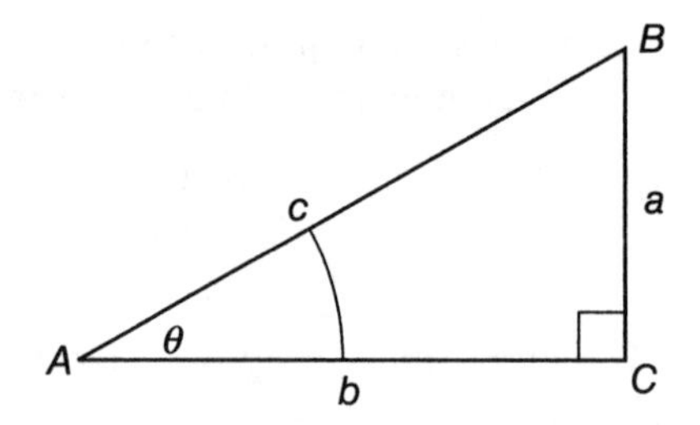

The lengths of the sides of the triangle are indicated in lowercase letters corresponding to the vertex of the opposite angle. The three basic trigonometric functions are defined in the following table.

Definition	Ratio as a Function of θ From Previous Diagram
Sine of an acute angle $= \dfrac{\text{length of opposite side}}{\text{length of hypotenuse}}$	$\sin\theta = \dfrac{a}{c}$
Cosine of an acute angle $= \dfrac{\text{length of adjacent side}}{\text{length of hypotenuse}}$	$\cos\theta = \dfrac{b}{c}$
Tangent of an acute angle $= \dfrac{\text{length of opposite side}}{\text{length of adjacent side}}$	$\tan\theta = \dfrac{a}{b}$

In addition to these three basic trigonometric functions, there are three more trigonometric ratios that are the reciprocals. These functions are the cotangent(cot), secant(sec), and cosecant(csc). These reciprocal relations are defined as follows:

$$\csc\theta = \frac{1}{\sin\theta}$$

$$\sec\theta = \frac{1}{\cos\theta}$$

$$\cot\theta = \frac{1}{\tan\theta}$$

Notice also that $\tan\theta = \dfrac{\sin\theta}{\cos\theta}$, so $\cot\theta = \dfrac{\cos\theta}{\sin\theta}$. Using the previous triangle, the reciprocal functions become $\cot\theta = \dfrac{b}{a}$, $\sec\theta = \dfrac{c}{b}$, and $\csc\theta = \dfrac{c}{a}$.

Example 1

In the following diagram, determine the values of all six trigonometric functions of θ.

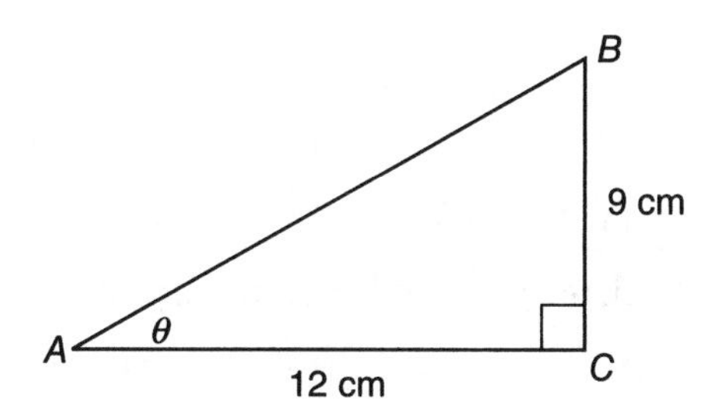

Solution: First use the Pythagorean Theorem to solve for the length of the hypotenuse of the triangle: $c^2 = a^2 + b^2 = 9^2 + 12^2 = 81 + 144 = 225$. So $c = 15$. As an alternative, notice that this is a multiple of the 3-4-5 triangle. Each side of the triangle is multiplied by 3. Now that the length of the hypotenuse is known, each of the trigonometric ratios can be determined.

$$\sin\theta = \frac{9}{15} = \frac{3}{5}$$

$$\cos\theta = \frac{12}{15} = \frac{4}{5}$$

$$\tan\theta = \frac{9}{12} = \frac{3}{4}$$

$$\cot\theta = \frac{12}{9} = \frac{4}{3}$$

$$\sec\theta = \frac{15}{12} = \frac{5}{4}$$

$$\csc\theta = \frac{15}{9} = \frac{5}{3}$$

A calculator can be used to evaluate a trigonometric function for any angle. Before evaluating a trigonometric function, *it is important to recognize whether the calculator is set in radian or degree mode*. A degree is subdivided into minutes. Similar to units of time, 1 degree = 60 minutes and

1 minute = 60 seconds. In order to use a calculator to determine the value of a trigonometric function of an angle measured in degrees and minutes, the minutes and degrees must be converted to a number of degrees in decimal form.

Example 2

Express 27°15′ to a number of degrees in decimal form.

Solution: $15' = \dfrac{15}{60} = 0.25°$. Therefore, 27°15′ = 27.25°.

Example 3

Use a calculator to evaluate cos 52°.

Solution: First check the mode to make sure that the calculator is working in degrees. Press **MODE** to see if the calculator is set correctly for this example. The display should look like this.

Press **2ND** **MODE** to get back to the original screen.

Now press **5** **2** **)** **ENTER**. The display is illustrated below.

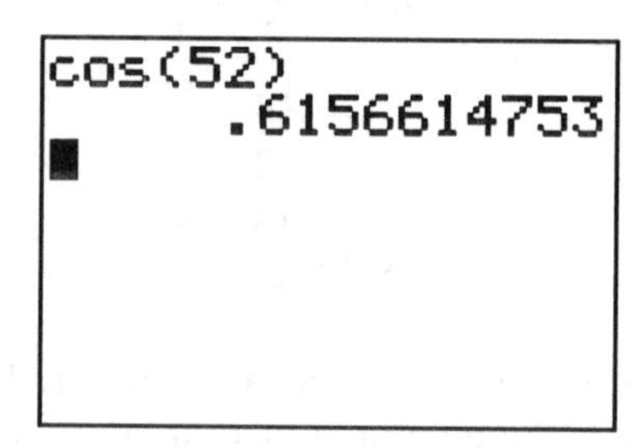

Example 4

Use a calculator to evaluate $\tan\frac{5\pi}{6}$.

Solution: First check the mode to make sure that the calculator is working in radians. Then press **TAN** **5** **2ND** **∧** **÷** **6** **)** **ENTER**. The answer is approximately –.5774.

Example 5

Use a calculator to evaluate cos 42°45′.

Solution: cos 42°45′ = cos 42.75°. Now, check the mode to make sure that the calculator is working in degrees. Then press **COS** **4** **2** **.** **7** **5** **)** **ENTER**. The answer is approximately 0.7343.

To utilize the calculator to evaluate an angle in the cotangent, cosecant, or secant functions requires recalling their reciprocal relationships. This is illustrated in the following example.

Example 6

Use a calculator to evaluate sec 1.36.

Solution: Notice that there is no degree symbol next to the 1.36. That means that the angle is measured in radians. Now check the mode to make sure that the calculator is working with radians. Then press **1** **÷** **COS** **1** **.** **3** **6** **)** **ENTER**. The answer is approximately 4.7792.

Notice that the six trigonometric functions come in pairs: sine and cosine, tangent and cotangent, secant and cosecant. These pairs are called cofunctions. A special rule relates the cofunctions of angles.

MATH FACTS

The Cofunction Rule: A trigonometric function of an angle is equal to the cofunction of the complement of that angle.

Example 7

If $\sec\theta = 1.3145$, what is the value of $\csc(90 - \theta)$?

Solution: $\csc(90 - \theta) = \sec\theta = 1.3145$

In previous courses, applications of trigonometric functions were made in right triangles, similar to the following two examples.

Example 8

If the side opposite a 34° angle of a right triangle measures 16 cm, how long (to the nearest thousandth of a cm) is the length of the leg of the right triangle adjacent to that angle?

Solution: $\tan 34° = \frac{16}{x}$ or $\cot 34° = \frac{x}{16}$

$.6745 \approx \frac{16}{x}$ $\qquad 1.4826 \approx \frac{x}{16}$

$.6745x \approx 16$ $\qquad 1.1826(16) \approx x$

$x = \frac{16}{.6745} \approx 23.721 \text{ cm}$ $\qquad x \approx 23.721 \text{ cm}$

Example 9

A 15-foot ladder is placed at a height of 12 feet on a wall. How many radians are in the measure of the angle formed between the ladder and the ground?

Solution: $\sin\theta = \frac{12}{15} = \frac{4}{5} = 0.8$. Using a graphing calculator, make sure the calculator is in radian mode and then press **2ND** **SIN** **.** **8** **)** **ENTER**. The result is approximately .9273 radians.

Check Your Understanding of Section 8.2

1. In the following diagram, determine the values of all six trigonometric functions.

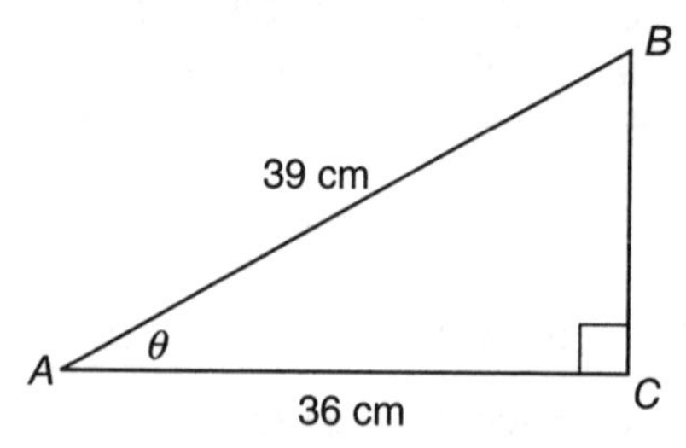

2. In the following diagram, determine the values of all six trigonometric functions of θ.

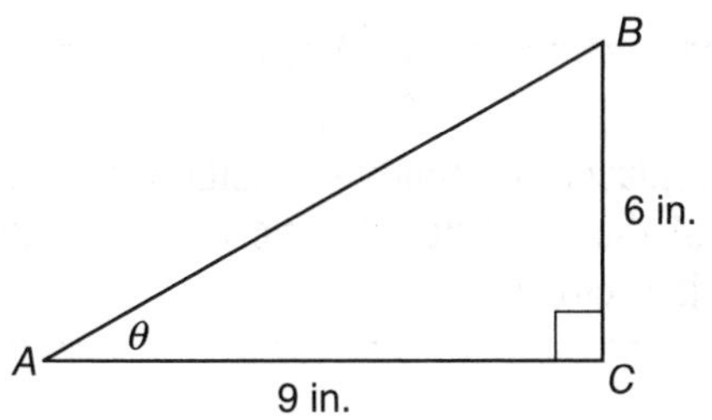

3. Use a calculator to evaluate each of the following (round your answers to the nearest ten-thousandth):

a. $\tan 23°$ b. $\sec\frac{7\pi}{12}$ c. $\cos\frac{3\pi}{4}$ d. $\cot\frac{\pi}{6}$

e. $\csc 173°$ f. $\sin 85°$

4. In right triangle ABC with the right angle at vertex C, if $\cot B = 0.75$, determine each of the following:

a. $\sec B$ b. $\cos A$ c. $\tan A$ d. $\sin B$

e. $\cot A$ f. $\csc B$ g. $\cos B$ h. $\csc A$

5. If $\tan 68° = 2.4751$, evaluate $\cot 22°$.

6. If $\cos\frac{\pi}{12} = 0.9659$, then the sine of what angle (as measured in radians) is also equal to 0.9569?

7. A surveyor stands 10 feet away from one end of a pond and measures 52° toward the other end of the pond. To the nearest tenth of a foot, determine the length of the pond.

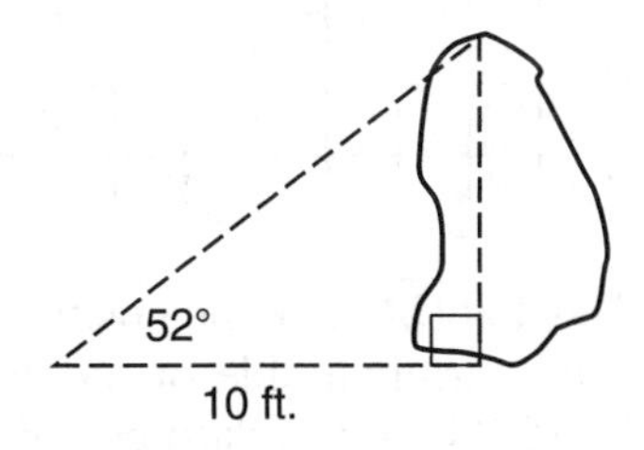

8.3 ANGLES IN STANDARD POSITION

KEY IDEAS

An angle can be placed in standard position on a circle centered at the origin so that the six trigonometric functions can be associated with a point on that circle.

Standard Position

To begin to understand what a radian was in Section 8.1, an angle was placed so that its initial side is a ray positioned on the positive half of the x-axis with its endpoint at the origin. A rotation was applied to this ray in a counterclockwise direction to find the terminal side of the angle. Let's examine this again.

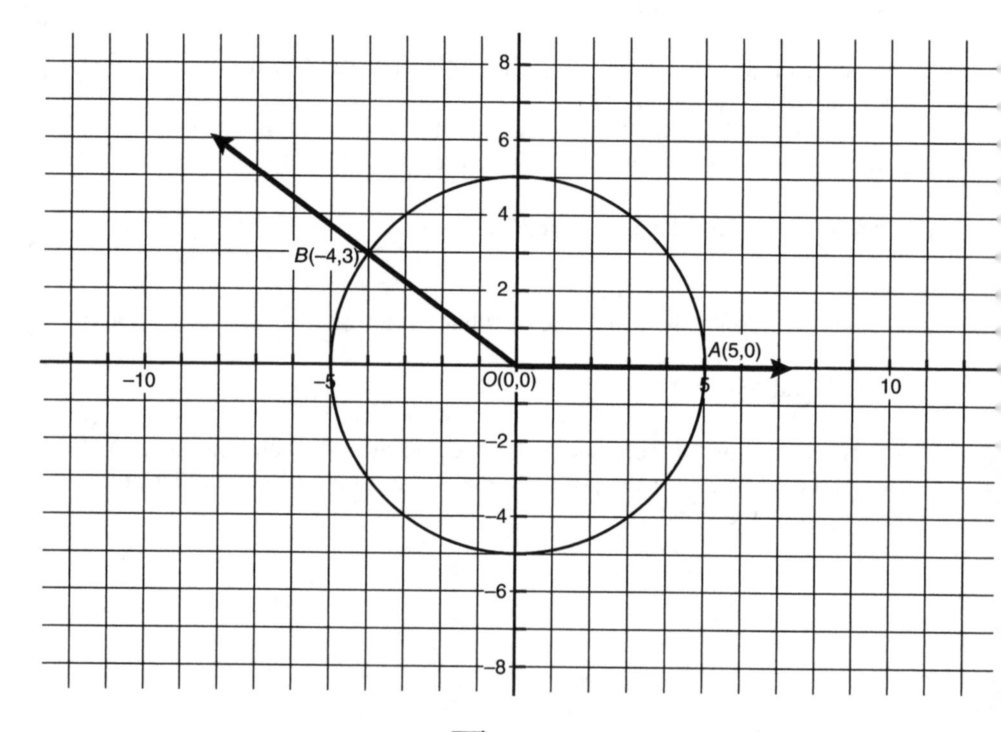

Angle AOB is positioned so that $\overrightarrow{OA}$ is on the positive x-axis and is the initial side of the angle. $\overrightarrow{OB}$ is the terminal side of the angle. Point B intersects circle O at (–4, 3). Angle AOB is in standard position.

Trigonometric Functions of Angles in Standard Position

For an angle, θ, in standard position, the six trigonometric functions are defined in terms of the radius r of the circle and the x and y values of the ordered pair at the intersection of the terminal side of the angle and the circle as follows:

$$\sin\theta = \frac{y}{r}$$

$$\cos\theta = \frac{x}{r}$$

$$\tan\theta = \frac{y}{x}$$

$$\cot\theta = \frac{x}{y}$$

$$\sec\theta = \frac{r}{x}$$

$$\csc\theta = \frac{r}{y}$$

In the diagram on page 230, $\angle AOB$ intersects a circle with radius 5 at the ordered pair (–4, 3). Therefore, $\sin\angle AOB = \frac{3}{5}$, $\cos\angle AOB = -\frac{4}{5}$, $\tan\angle AOB = -\frac{3}{4}$, $\cot\angle AOB = -\frac{4}{3}$, $\sec\angle AOB = -\frac{5}{4}$, and $\csc\angle AOB = \frac{5}{3}$.

Example 1

$\overrightarrow{OA}$ is the initial side and $\overrightarrow{OB}$ is the terminal side of $\angle AOB$. The coordinates of point B are (4, –3).

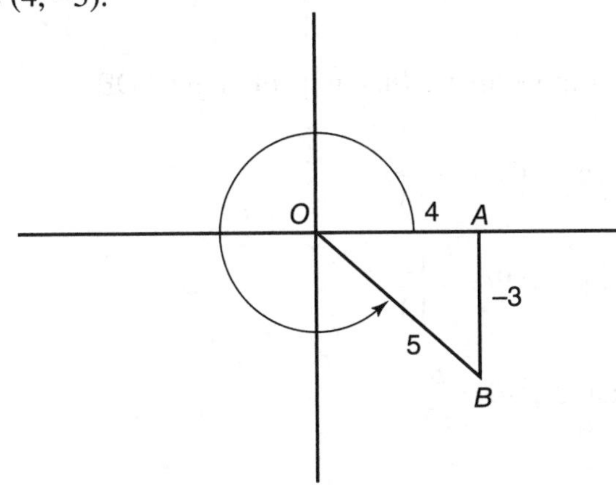

Determine the numerical values of all six trigonometric functions for $\angle AOB$.

Solution: $\sin\angle AOB = -\frac{3}{5}$

$$\cos\angle AOB = \frac{4}{5}$$

$$\tan\angle AOB = -\frac{3}{4}$$

$$\cot\angle AOB = -\frac{4}{3}$$

$$\sec\angle AOB = \frac{5}{4}$$

$$\csc\angle AOB = -\frac{5}{3}$$

Example 2

$\overrightarrow{OA}$ is the initial side and $\overrightarrow{OB}$ is the terminal side of $\angle AOB$. The coordinates of point B are (–12, –5).

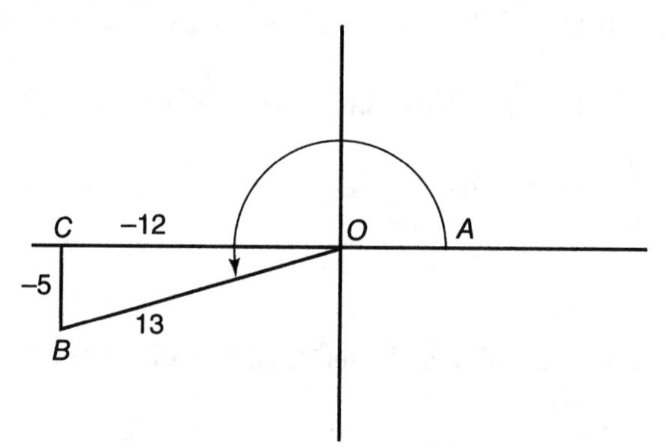

Evaluate all six trigonometric functions of angle AOB.

Solutions: $\sin\angle AOB = -\frac{5}{13}$

$$\cos\angle AOB = -\frac{12}{13}$$

$$\tan\angle AOB = \frac{5}{12}$$

$$\cot \angle AOB = \frac{12}{5}$$

$$\sec \angle AOB = -\frac{13}{12}$$

$$\csc \angle AOB = -\frac{13}{5}$$

The radius of the circle is always positive. However, the x-values and y-values of the ordered pairs on the circle vary in signs according to the quadrant.

Quadrant	Sign of			Positive Trigonometric Functions
	x	y	r	
I	+	+	+	All six functions
II	−	+	+	sin and csc
III	−	−	+	tan and cot
IV	+	−	+	cos and sec

MATH FACTS

An easy way to remember which trigonometric functions are positive in which quadrants is with the statement "All Students Take Calculus." Each word in the phrase is in the order of the quadrant it refers to. "A" stands for all six trigonometric functions, "S" stands for sin and its reciprocal csc, "T" stands for tan and its reciprocal cot, and "C" stands for cos and its reciprocal sec.

Example 3

If $\sin\theta = -\frac{4}{7}$ and $\cot\theta = \frac{\sqrt{33}}{4}$, in what quadrant is the terminal side of θ? Give the numerical value of sec θ.

Solution: Since sin θ is negative and cot θ is positive, θ is in quadrant III. $\sec\theta = -\frac{7\sqrt{33}}{33}$

Check Your Understanding of Section 8.3

1. If $\tan B < 0$ and $\sec B > 0$, in what quadrant does the terminal side of angle B lie?

(1) I (2) II (3) III (4) IV

2. If $\cot\theta = \frac{2}{3}$ and $\csc\theta < 0$, what is the numerical value of $\sec\theta$?

(1) $\sqrt{13}$ (2) -2 (3) $-\frac{\sqrt{13}}{2}$ (4) $-\frac{2\sqrt{13}}{13}$

3. If $\cos A = -\frac{6}{13}$ and $\tan A < 0$, evaluate $\sin A$.

(1) $\frac{\sqrt{133}}{13}$ (2) $-\frac{\sqrt{133}}{13}$ (3) $\frac{6\sqrt{133}}{133}$ (4) $-\frac{6\sqrt{133}}{133}$

4. If $\sin\beta = -\frac{5}{8}$ and $\sec\beta < 0$, evaluate $\cot\beta$.

(1) $-\frac{\sqrt{39}}{5}$ (2) $-\frac{5\sqrt{39}}{39}$ (3) $\frac{\sqrt{39}}{5}$ (4) $\frac{\sqrt{89}}{5}$

5. If the point $(-7, 3)$ is on the terminal side of an angle, what is the cosine of that angle?

(1) $\frac{\sqrt{58}}{7}$ (2) $\frac{3\sqrt{58}}{58}$ (3) $-\frac{3\sqrt{10}}{20}$ (4) $-\frac{7\sqrt{58}}{58}$

6. If $\tan C = -\frac{5}{4}$ and angle C is not in the second quadrant, what is the exact value of $\sin C$?

(1) $\frac{4\sqrt{41}}{41}$ (2) $-\frac{5\sqrt{41}}{41}$ (3) $-\frac{5}{3}$ (4) $-\frac{3}{5}$

7. Based on the diagram below where the terminal side of angle AOB includes the ordered pair (15, –8), evaluate each of the following:

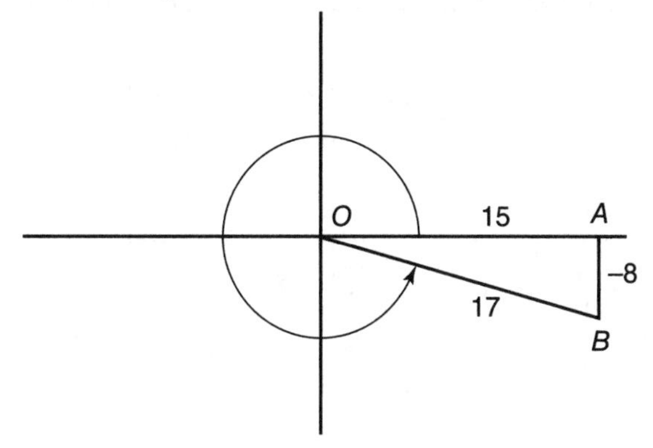

a. $\sec AOB$ b. $\cot AOB$ c. $\sin AOB$
d. $(\tan AOB)(\cot AOB)$ e. $(\sin AOB)(\cos AOB)$
f. $(\sin AOB)^2 + (\cos AOB)^2$

8. If $\sin A = \frac{5}{13}$ when $\tan A < 0$ and if $\csc B = -\frac{17}{8}$ when $\cot B < 0$, determine the numerical value of $\tan A \cdot \tan B$?

9. If angle A is in the third quadrant and $\sin A = -\frac{4}{x}$, express $\sec A$ in terms of x.

10. If θ is measured in radians and $\sec\theta = \frac{9}{7}$ when $\tan\theta < 0$, evaluate $\sin\left(\frac{\pi}{2} - \theta\right)$.

8.4 THE UNIT CIRCLE

A circle whose radius is equal to 1 unit is called the unit circle. In this section, the trigonometric functions will be redefined as angles in standard position whose terminal side endpoints are on the unit circle.

Angles in Standard Position and the Unit Circle

Pictured below is an angle in standard position and a unit circle.

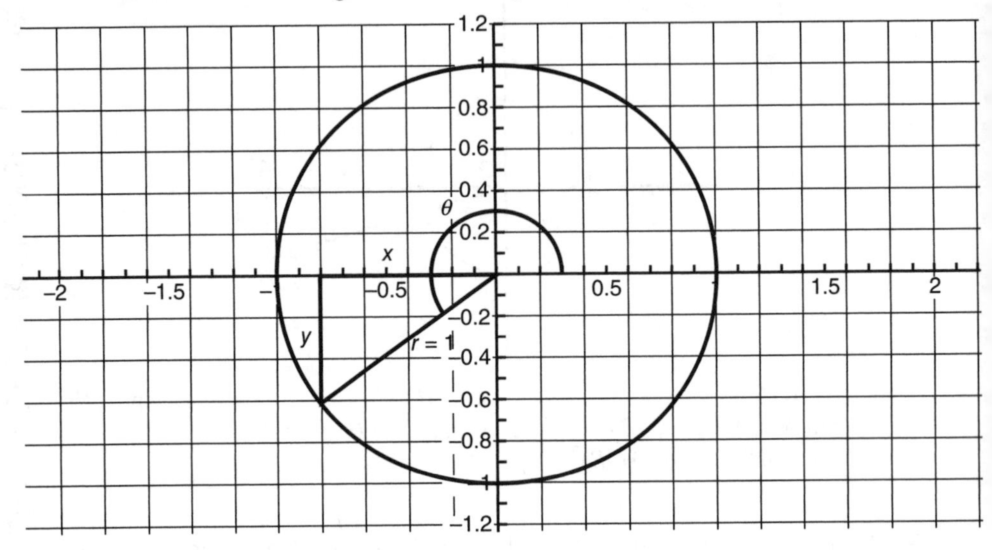

Now that $r = 1$, the definitions of the six trigonometric functions can be revised accordingly.

Trigonometry Definitions for Angles in Standard Position	Revised Trigonometry Definitions for Angles in Standard Position on the Unit Circle
$\sin\theta = \frac{y}{r}$	$\sin\theta = y$
$\cos\theta = \frac{x}{r}$	$\cos\theta = x$
$\tan\theta = \frac{y}{x}$	$\tan\theta = \frac{y}{x}$
$\cot\theta = \frac{x}{y}$	$\cot\theta = \frac{x}{y}$
$\sec\theta = \frac{r}{x}$	$\sec\theta = \frac{1}{x}$
$\csc\theta = \frac{r}{y}$	$\csc\theta = \frac{1}{y}$

Example 1

Determine the exact value of all of the six trigonometric functions of an angle, θ, whose terminal side intersects the unit circle at the ordered pair $\left(-\frac{12}{13}, \frac{5}{13}\right)$.

Solution: $\sin\theta = \frac{5}{13}$

$$\cos\theta = -\frac{12}{13}$$

$$\tan\theta = \frac{\frac{5}{13}}{-\frac{12}{13}} = -\frac{5}{12}$$

$$\cot\theta = \frac{-\frac{12}{13}}{\frac{5}{13}} = -\frac{12}{5}$$

$$\sec\theta = \frac{1}{-\frac{12}{13}} = -\frac{13}{12}$$

$$\csc\theta = \frac{1}{\frac{5}{13}} = \frac{13}{5}$$

Consider an angle of $\frac{\pi}{6}$ radians or 30°. Recall that the 30-60-90 right triangle has sides in the ratio $1:\sqrt{3}:2$ correspondingly opposite these angles. Since the hypotenuse of any right triangle drawn in the unit circle is the length of the radius of the circle or 1, then the ratio of sides can be set at $\frac{1}{2}:\frac{\sqrt{3}}{2}:1$. This can be used to determine the six trigonometric functions of an angle of $\frac{\pi}{6}$ radians or 30°.

Example 2

Determine the exact numerical values of the trigonometric functions of an angle measuring $\frac{\pi}{6}$ radians.

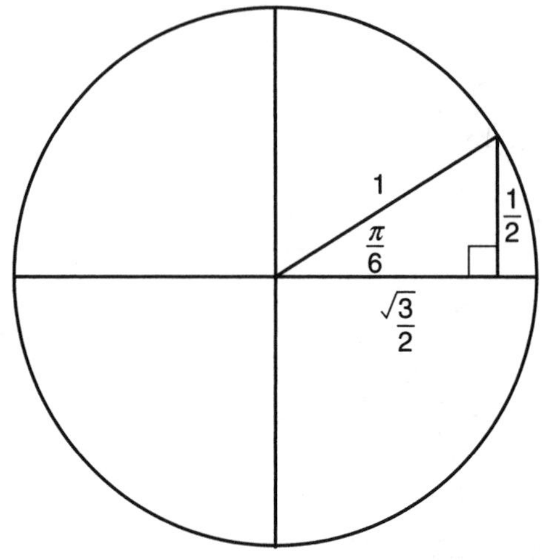

$$\sin\frac{\pi}{6}=\frac{1}{2}$$

$$\cos\frac{\pi}{6}=\frac{\sqrt{3}}{2}$$

$$\tan\frac{\pi}{6}=\frac{\frac{1}{2}}{\frac{\sqrt{3}}{2}}=\frac{1}{\sqrt{3}}=\frac{\sqrt{3}}{3}$$

$$\cot\frac{\pi}{6}=\frac{\frac{\sqrt{3}}{2}}{\frac{1}{2}}=\sqrt{3}$$

$$\sec\frac{\pi}{6}=\frac{1}{\frac{\sqrt{3}}{2}}=\frac{2}{\sqrt{3}}=\frac{2\sqrt{3}}{3}$$

$$\csc\frac{\pi}{6}=\frac{1}{\frac{1}{2}}=2$$

The same approach can be used to solve for any trigonometric functions for any angle that is a multiple of 30° or 60° or of $\frac{\pi}{6}$ or $\frac{\pi}{3}$ radians. The following example illustrates this fact.

Example 3

Determine the exact numerical values of the trigonometric functions of an angle measuring 300°.

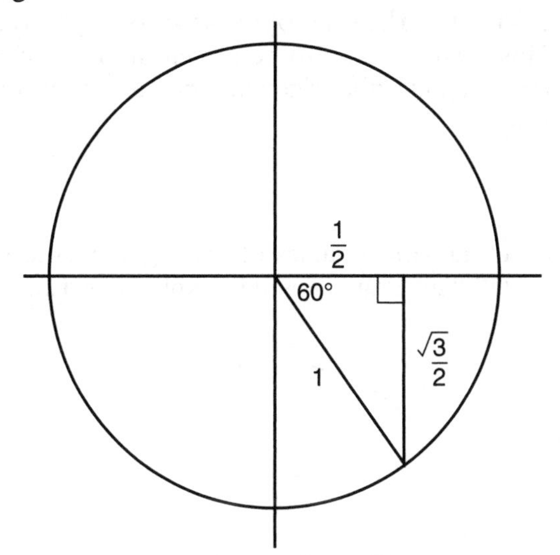

Solution: $300° = -\frac{\sqrt{3}}{2}$

$$\cos 300° = \frac{1}{2}$$

$$\tan 300° = \frac{-\frac{\sqrt{3}}{2}}{\frac{1}{2}} = -\frac{\sqrt{3}}{1} = -\sqrt{3}$$

$$\cot 300° = \frac{\frac{1}{2}}{-\frac{\sqrt{3}}{2}} = -\frac{1}{\sqrt{3}} = -\frac{\sqrt{3}}{3}$$

$$\sec 300° = \frac{1}{\frac{1}{2}} = \frac{2}{1} = 2$$

$$\csc 300° = \frac{1}{-\frac{\sqrt{3}}{2}} = -\frac{2}{\sqrt{3}} = -\frac{2\sqrt{3}}{3}$$

However, the 30-60-90 right triangle is not the only right triangle whose sides are in a known ratio. The sides of the 45-45-90 right triangle are in the ratio $1:1:\sqrt{2}$. This triangle can also be placed in different quadrants. The following example examines the trigonometric values for a triangle placed in the third quadrant.

Example 4

Determine the exact numerical values of the trigonometric functions of an angle measuring 225°. This example will be explored in a circle whose radius is $\sqrt{2}$.

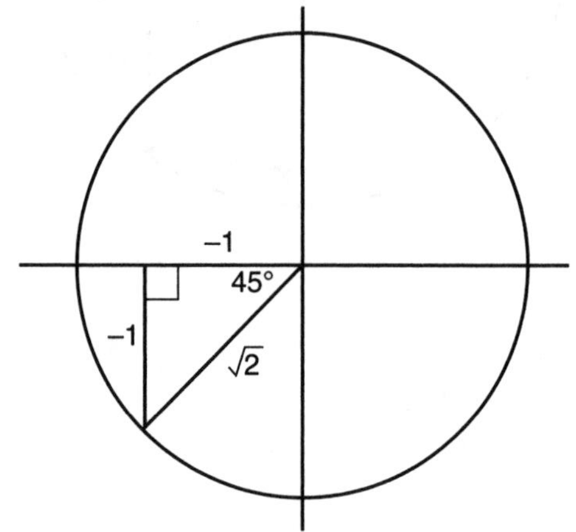

Solution: $\sin 225° = -\frac{1}{\sqrt{2}} = -\frac{\sqrt{2}}{2}$

$\cos 225° = -\frac{1}{\sqrt{2}} = -\frac{\sqrt{2}}{2}$

$\tan 225° = 1$

$\cot 225° = 1$

$$\sec 225° = \frac{\sqrt{2}}{-1} = -\sqrt{2}$$

$$\csc 225° = \frac{\sqrt{2}}{-1} = -\sqrt{2}$$

Notice that an angle whose measure is more than 360° or 2π can also be placed in a triangle. The following example explores how to determine the cotangent of an angle whose measure is $\frac{11\pi}{4}$ radians or 495°.

Example 5

Determine the exact numerical values of $\cot\frac{11\pi}{4}$.

Solution: $\frac{11\pi}{4} = \frac{8\pi}{4} + \frac{3\pi}{4} = 2\pi + \frac{3\pi}{4}$. That places the vertex of the right triangle in the second quadrant as illustrated below:

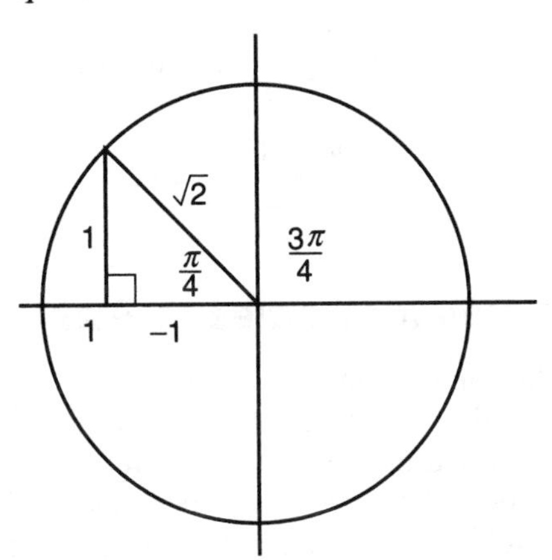

$$\cot\frac{11\pi}{4} = \cot\frac{3\pi}{4} = \frac{-1}{1} = -1$$

Quadrantal Angles

Angles whose terminal sides lie on one of the axes are called quadrantal angles. One example is an angle of 270° or $\frac{3\pi}{2}$ radians. All the trigonometric ratios for this angle can be determined by using the coordinates of the ordered pair (0, −1).

Example 6

Determine the numerical values of all six trigonometric functions for $\frac{3\pi}{2}$.

Solution: Since the terminal side of an angle measuring $\frac{3\pi}{2}$ radians intersects the unit circle at (0, –1),

$$\sin\frac{3\pi}{2} = -1$$

$$\cos\frac{3\pi}{2} = 0$$

$$\tan\frac{3\pi}{2} = \frac{-1}{0} \text{ or undefined}$$

$$\cot\frac{3\pi}{2} = \frac{0}{-1} = 0$$

$$\sec\frac{3\pi}{2} = \frac{1}{0} \text{ or undefined}$$

$$\csc\frac{3\pi}{2} = \frac{1}{-1} = -1$$

Reference Angles for Angles in Standard Position

When an angle is in standard position, its measure from the positive or negative x-axis can be used to determine the value of a trigonometric function of an angle in the first quadrant. For instance, an angle measuring 257° is in the third quadrant and is 77° from the x-axis ($180° + 77° = 257°$). An angle of 77° is called the reference angle for 257°. Recall that only the tangent and cotangent are positive in quadrant III. The following example illustrates how this can be used.

Example 7

Express the sec 257° as a trigonometric function of a positive acute angle less than 45°.

Solution: $\sec 257° = -\sec 77° = -\csc 13°$. This is because the secant function is negative in the third quadrant and the reference angle for 257° is 77°. Using the cofunction rule resolves this to an angle measuring less than 45°.

Note that if this example had asked for the numerical value of sec 257°, a graphing calculator could be used. The same value would be found for sec 257° as for –csc 13°. To do this, first remember to check the mode to see if the calculator is in degrees. Now, press 1 ÷ COS 2 5 7) ENTER. The result displayed is –4.445411483. Similarly, the result obtained by pressing – 1 ÷ SIN) 1 3 ENTER also yields –4.445411483.

Trigonometric Functions of Negative Angles

An angle whose terminal side is the result of a clockwise rotation from the initial side has a negative measure. The sine, cosine, and tangent functions of these angles are clearly related to the same functions of a positive angle measure. The list below shows that $y_1 = \sin(x)$ and $y_2 = \sin(-x)$ for angles in degree measure in multiples of 15.

X	Y1	Y2
0	0	0
15	.25882	-.2588
30	.5	-.5
45	.70711	-.7071
60	.86603	-.866
75	.96593	-.9659
90	1	-1

X=0

MATH FACTS

$\sin(-\theta) = -\sin\theta$ $\quad\cos(-\theta) = \cos\theta$ $\quad\tan(-\theta) = -\tan\theta$

$\csc(-\theta) = -\csc\theta$ $\quad\sec(-\theta) = \sec\theta$ $\quad\cot(-\theta) = -\cot\theta$

Example 8

Express cot (–163°) as a trigonometric function of a positive acute angle less than 45°.

Solution: $\cot(-163°) = -\cot(163°) = -\cot(180° - 17°) = \cot 17°$

Check Your Understanding of Section 8.4

1. Determine the numerical value of each of the following:

a. $\cot\dfrac{5\pi}{3}$ b. sec 225° c. cos 420° d. $\sin\dfrac{7\pi}{6}$

e. $\tan\dfrac{3\pi}{2}$ f. $\sec\dfrac{-7\pi}{3}$ g. csc 135° h. sec 870°

2. Express each of the following as a trigonometric function of a positive acute angle less than 45° or $\dfrac{\pi}{4}$ radians:

a. cos 242° b. $\tan\dfrac{4\pi}{5}$ c. csc 139° d. sin 589°

e. $\sec\dfrac{18\pi}{7}$ f. cot (–344°) g. tan 492° h. sin 673°

3. Which of the following is not equivalent to cot 498°?

(1) cot 138°
(2) –cot 42°
(3) tan 48°
(4) cot 858°

4. If $\theta = \pi$, then which of the following functions are undefined?

(1) sin θ and cos θ
(2) cot θ and csc θ
(3) cos θ and sec θ
(4) tan θ and sec θ

Chapter Nine

TRIGONOMETRIC FUNCTIONS AND THEIR GRAPHS

9.1 GRAPHING THE SINE AND COSINE FUNCTIONS

The domain of the sine and cosine functions is $\Re$ while their range is $\{y: -1 \leq y \leq 1\}$. However, notice that the members of the range are mapped to by many members of the domain. For instance, $\sin\frac{\pi}{6} = \sin\frac{13\pi}{6} = \sin\frac{25\pi}{6} = \frac{1}{2}$. In fact, the range elements are reached in a cyclic manner. These functions are called periodic.

What Is a Periodic Function?

A function $f(x)$ is periodic if and only if for all x in the domain of f, $f(x) = f(x + p)$ for some positive number p. The smallest number p for which this statement is true is called the period of the function. Examine the diagram that follows Example 1 to recognize how the sine function maps the y-values of ordered pairs on the circle to the graph.

Example 1

Explain why $f(x) = \sin x$ is a periodic function.

Solution: The curve pictured on the next page is that of a portion of the sine curve. The dotted line indicates places where the y-values of ordered pairs are equal to 0.5. These values are $\frac{\pi}{6} + 2k\pi$ and $\frac{5\pi}{6} + 2k\pi$ or $30° \pm k \cdot 360°$ and $150° \pm k \cdot 360°$, where k is an integer. Notice that the y-value of 0.5 occurs in intervals of 2π or 360°. Also note that any horizontal line drawn through the unit circle will generate similar results, with repetition of the same y-values occurring in intervals of 2π or 360°. Therefore, the sine function is periodic, and the period of the sine function is 2π.

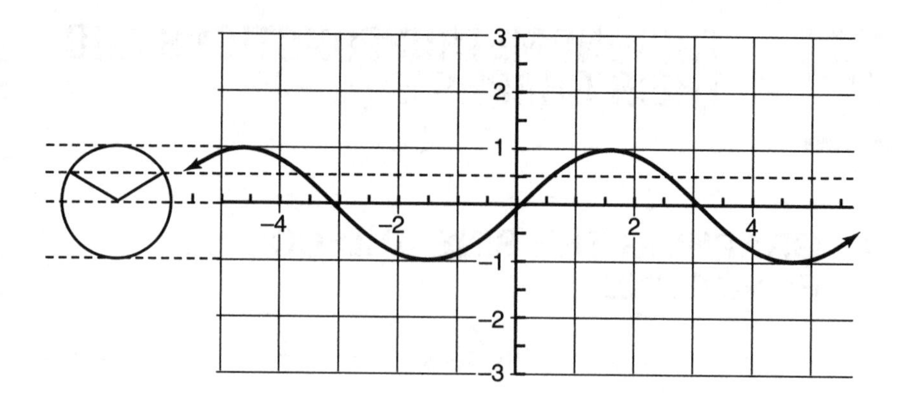

The Sine Function

Here are some important features regarding the graph of the sine function.

Features of Sine	$f(x) = \sin x$
Amplitude (the distance above or below the x-axis)	1
Period (p such that $f(x) = f(x + p)$)	2π
Frequency (the number of full cycles of the curve between 0 and 2π)	1
y-intercept	0
x-intercept(s)	$0, \pm\pi, \pm 2\pi, \ldots$
Symmetry	Origin

The graph of the sine function can be generated on a graphing calculator. Press **Y=** **SIN** **x** **)** **ENTER** **ZOOM** **7**. Zoom 7, which is ZTrig, is a new zoom feature being introduced on graphing calculators. ZTrig zooms into a window as shown below.

```
WINDOW
 Xmin=-6.152285...
 Xmax=6.1522856...
 Xscl=1.5707963...
 Ymin=-4
 Ymax=4
 Yscl=1
 Xres=1
```

The graph of the sine function appears.

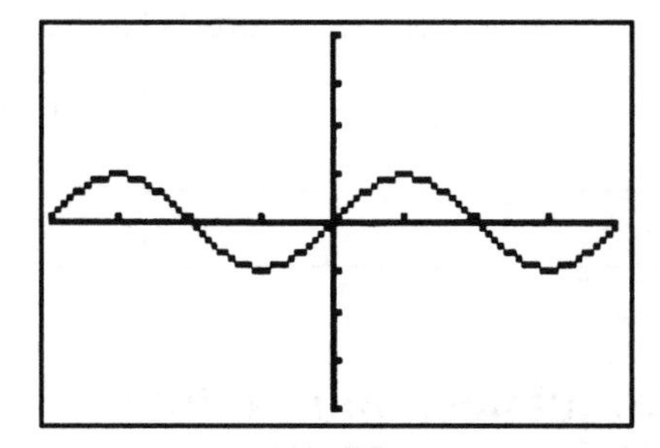

The Cosine Function

On a graphing calculator, the cosine function appears as shown.

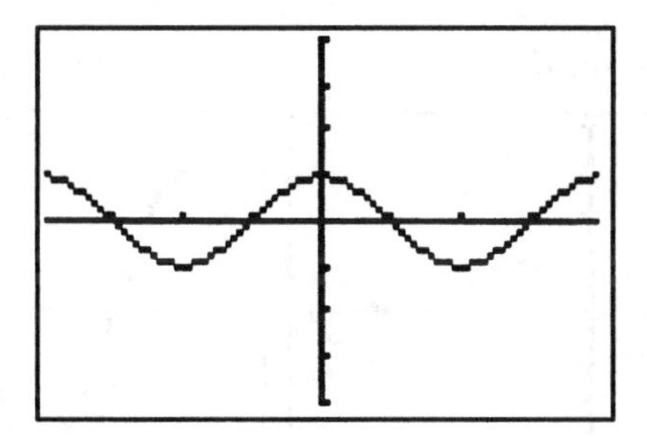

Notice how the graph of cos x is the same as the graph of sin x except that the graph is shifted over by $\frac{\pi}{2}$ units. This is because of the cofunction rule: $\cos\theta = \sin\left(\frac{\pi}{2} - \theta\right)$.

The table below summarizes the features of the cosine function:

Features of Cosine	$f(x) = \cos x$
Amplitude (the distance above or below the x-axis)	1
Period (p such that $f(x) = f(x + p)$)	2π
Frequency (the number of full cycles of the curve between 0 and 2π)	1
y-intercept	1
x-intercept(s)	$0, \pm\frac{\pi}{2}, \pm\frac{3\pi}{2}, \ldots$
Symmetry	y-axis

Amplitude, Frequency, and Period: $y = a \sin bx$ and $y = a \cos bx$

What is the effect of multiplying the function by a constant or multiplying the angle by a constant? For $y = a \sin bx$ and $y = a \cos bx$, the a stands for the amplitude of the function. Compare the graphs of $f(\theta) = \sin \theta$ and $g(\theta) = 3\sin \theta$ as shown below.

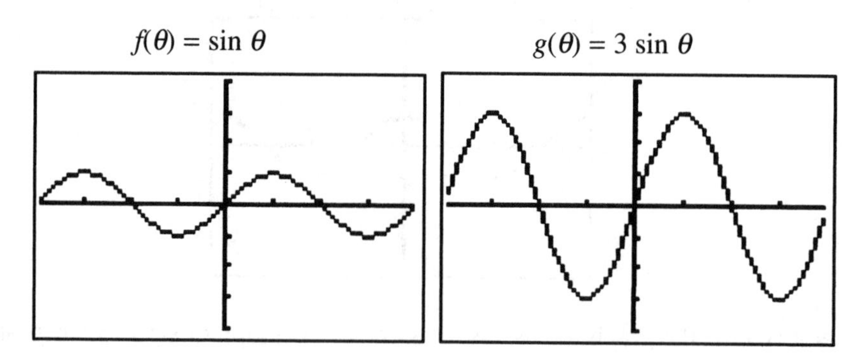

These graphs are the same basic curve, but the amplitudes (the distance above or below the x-axis) are different; $f(\theta) = \sin \theta$ has an amplitude of 1 and $g(\theta) = 3\sin \theta$ has an amplitude of 3.

Now compare the graphs of $f(\theta) = \sin \theta$ and $h(\theta) = \sin (2\theta)$.

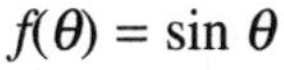

$h(\theta) = \sin(2\theta)$

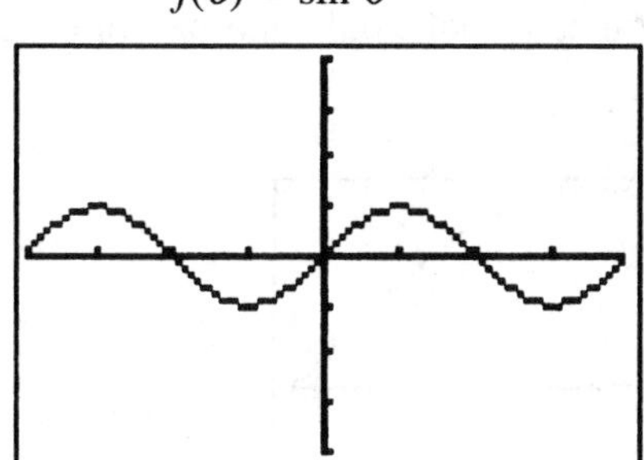

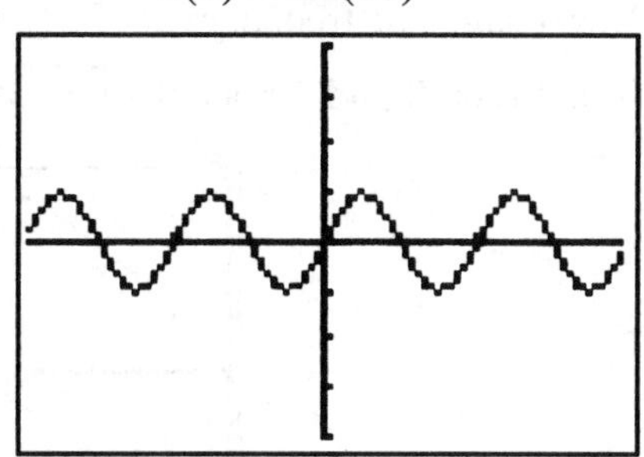

The difference between these curves is the number full cycles of the curve between 0 and 2π, that is, the frequency. So b is the frequency of the function. The frequency for $f(\theta) = \sin \theta$ is 1, while the frequency for $h(\theta) = \sin(2\theta)$ is 2.

The frequency of a trigonometric function also affects its period. The period of a trigonometric function is $\frac{2\pi}{b}$. The period for $f(\theta) = \sin \theta$ is 2π, while the period for $h(\theta) = \sin(2\theta)$ is π.

Example 2

Sketch a graph of $y = 2\sin\left(\frac{1}{2}x\right)$ on $[0, 2\pi]$.

Solution: Start with one full cycle of the basic sine curve as it appears below from $[0, 2\pi]$.

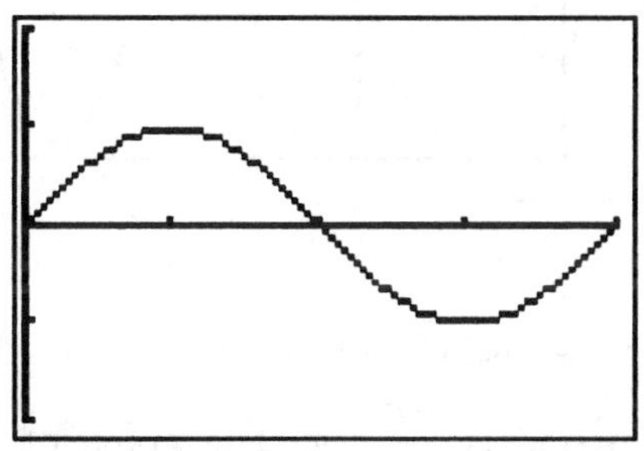

Clearly, the amplitude is 1 and the frequency is 1 for $y = \sin x$. Now take this and stretch the amplitude to 2.

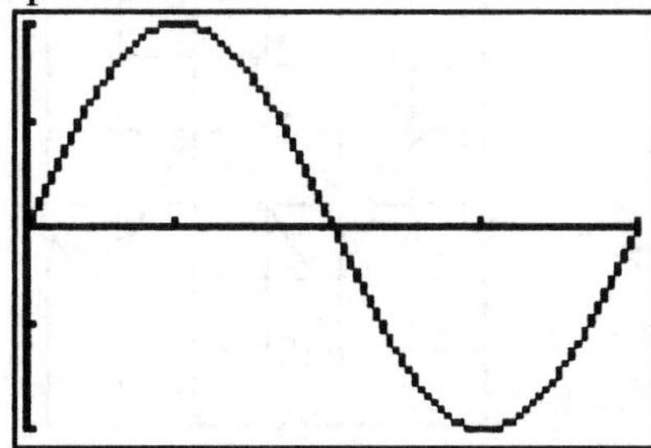

Now change its frequency to $\frac{1}{2}$. That means make only one-half of the curve fit between 0 and 2π, as pictured below.

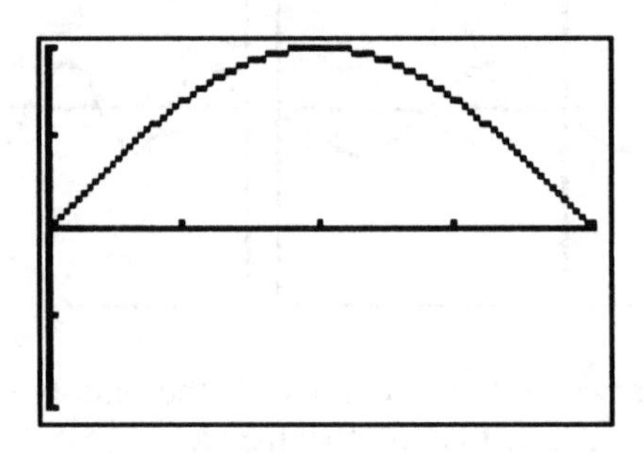

Example 3

Sketch a graph of $y = 2\sin\left(\frac{1}{2}x\right)$ on $[-2\pi, 4\pi]$.

Solution: Take the last graph and continue the curve into the region from -2π to 0 and from 2π to 4π. Here is what it looks like now:

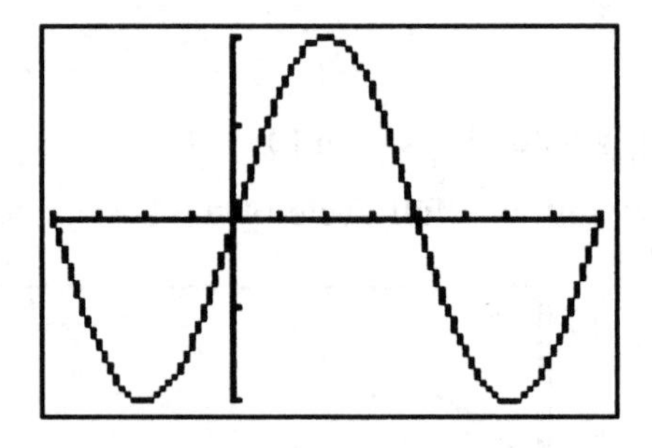

Example 4

Sketch a graph of $y = -\sin(2x)$ on $[-\pi, \pi]$.

Solution: The period is $\frac{2\pi}{2} = \pi$. The graph of $y = \sin(2x)$ is shown below.

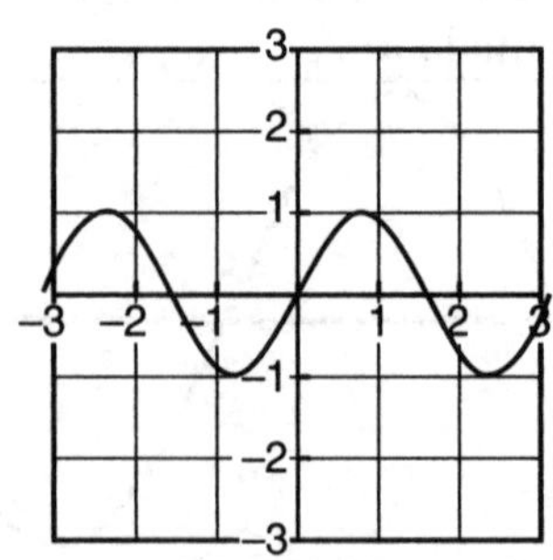

The function $y = -\sin(2x)$ is a reflection of $y = \sin(2x)$ over the x-axis. The graph follows.

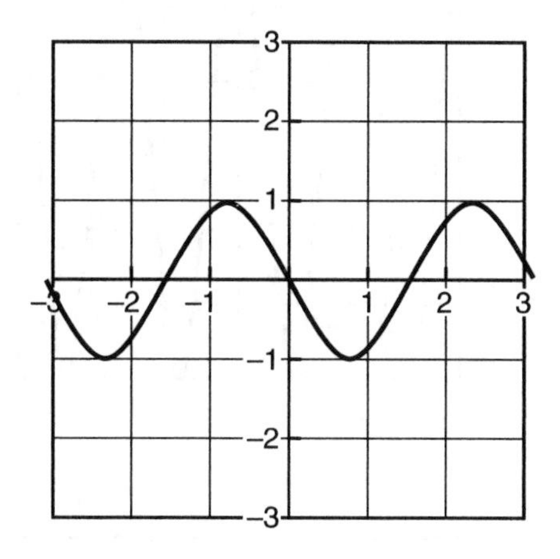

Example 5

Sketch a graph of $y = 3\cos(4x) + 2$ on $[-\pi, \pi]$.

Solution: The function $y = 3\cos(4x)$ has an amplitude of 3 and period of $\frac{2\pi}{4} = \frac{\pi}{2}$. Its graph is illustrated.

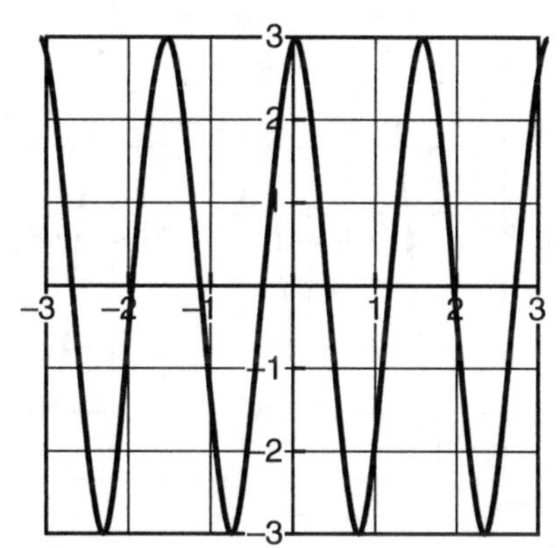

The function $y = 3\cos(4x) + 2$ is a vertical shift of $y = 3\cos(4x)$ up 2 units. Therefore, the midline of the cosine function in the solution is at 2 rather than at 0. The graph is displayed.

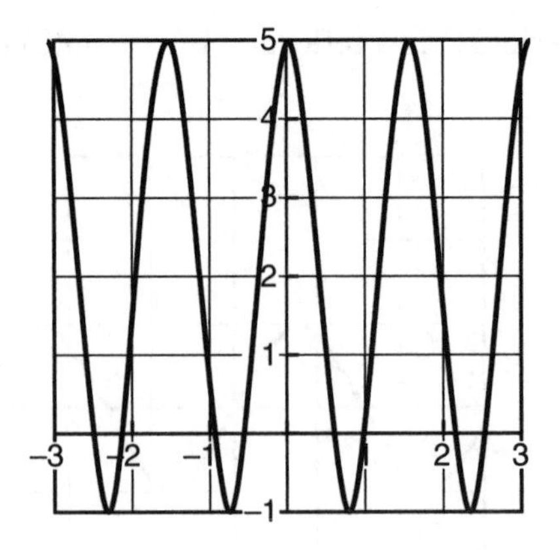

Check Your Understanding of Section 9.1

1. Which equation is represented in the graph below?

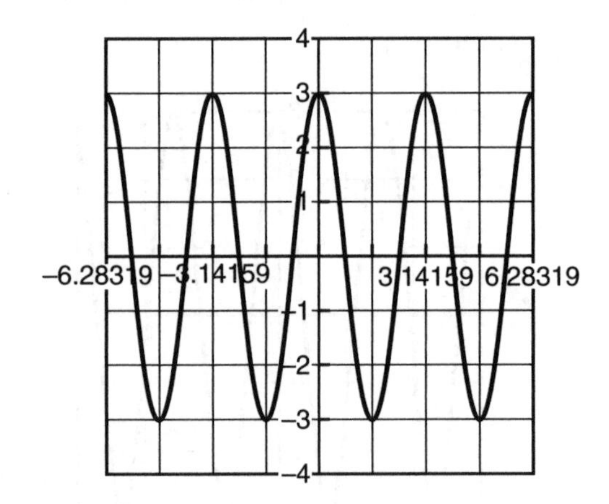

(1) $y = 2\sin(3\theta)$
(2) $y = 3\cos(2\theta)$
(3) $y = 3\sin\left(\frac{1}{2}\theta\right)$
(4) $y = 2\cos(3\theta)$

2. Which equation is represented in the graph below?

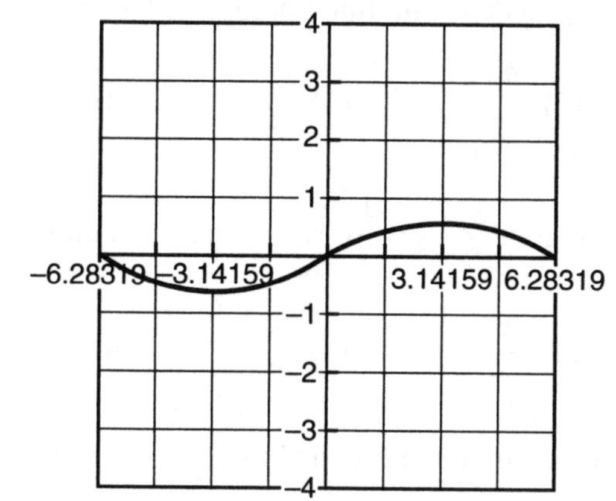

(1) $y=\frac{1}{2}\cos(2\theta)$

(2) $y=\frac{1}{2}\cos\left(\frac{1}{2}\theta\right)$

(3) $y=\frac{1}{2}\sin(2\theta)$

(4) $y=\frac{1}{2}\sin\left(\frac{1}{2}\theta\right)$

3. The graph of which function has an amplitude of 2 and a period of 3π?

(1) $y = 3\cos(2\theta)$

(2) $y = 2\cos(3\theta)$

(3) $y=2\sin\left(\frac{2}{3}\theta\right)$

(4) $y=\frac{2}{3}\sin(2\theta)$

4. A wave displayed on an oscilloscope is represented by the function $y = 2\sin(3x)$. What is the period of this function?

(1) 3 (2) 2 (3) π (4) $\frac{2\pi}{3}$

5. The graph of $y = 4\sin(2\theta)$ has a line of symmetry at

(1) $\frac{\pi}{4}$ (2) $\frac{\pi}{2}$ (3) π (4) 2π

6. What is the amplitude of the function $y = \frac{2}{3}\cos\left(\frac{3\theta}{4}\right)$?

(1) $\frac{3}{4}$ (2) $\frac{8\pi}{3}$ (3) $\frac{2}{3}$ (4) 3π

7. Where does the graph of $y = \frac{1}{2}\cos(2\theta)$ reach its minimum value on $[0, \pi]$?

(1) 0 (2) $\frac{\pi}{4}$ (3) $\frac{\pi}{2}$ (4) π

8. The graph of which of the following functions passes through the ordered pair $(\frac{\pi}{2}, \sqrt{3})$?

(1) $y = \sin\left(\frac{3}{2}\theta\right)$

(2) $y = 2\cos\left(\frac{1}{3}\theta\right)$

(3) $y = 2\sin\left(\frac{1}{3}\theta\right)$

(4) $y = \cos\left(\frac{3}{2}\theta\right)$

9. The graph of which function passes through (0, 3) and has a period of 4π?

(1) $y = 3\cos\left(\frac{1}{2}\theta\right)$

(2) $y = 3\sin\left(\frac{1}{2}\theta\right)$

(3) $y = 3\cos(4\theta)$

(4) $y = 4\cos(3\theta)$

10. What is the range of the function $y = \frac{1}{4}\sin\left(\frac{2}{3}\theta\right)$?

(1) $\left\{y: -\frac{2}{3} \le y \le \frac{2}{3}\right\}$

(2) all real numbers

(3) $\left\{y: -\frac{1}{4} \le y \le \frac{1}{4}\right\}$

(4) $\{y: -3\pi \le y \le 3\pi\}$

11. Sketch a graph of $y = 2\cos\left(\frac{1}{2}\theta\right)$ on $[-\pi, \pi]$.

12. Sketch a graph of $y = \frac{3}{4}\sin(4\theta)$ on $[-2\pi, \pi]$.

13. On the same set of axes, sketch the graphs of $y = 4\sin(2\theta)$ and $y = \frac{1}{2}\cos\theta$ on $[-2\pi, 2\pi]$. In how many points do these two functions intersect on $[-2\pi, 2\pi]$?

14. On the same set of axes, sketch the graphs of $y = 2\sin\left(\frac{1}{4}\theta\right)$ and $y = 4\cos(2\theta)$ on $[0, 4\pi]$. In how many points do these two functions intersect on $[0, 4\pi]$?

15. Sketch a graph of $y = \frac{2}{3}\sin(2\theta) - 1$. Specify the amplitude, frequency, period, and equation of the midline of this function.

16. A model of a shot over the net in volleyball is pictured below.

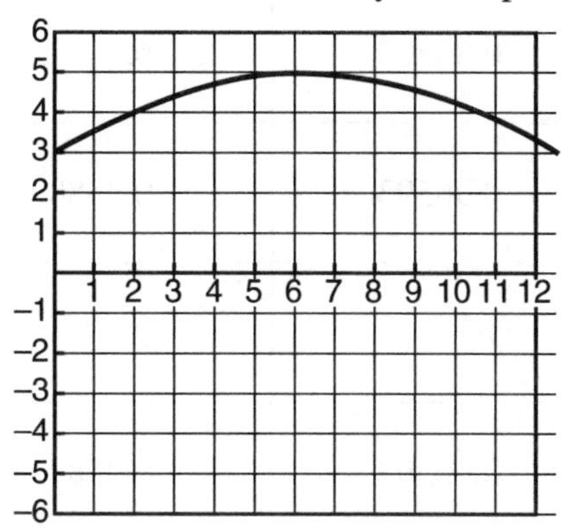

If the function $y = A\sin(Bx) + C$ models this shot, determine the values of A, B, and C.

9.2 GRAPHING THE TANGENT, COTANGENT, SECANT, AND COSECANT FUNCTIONS

The graphs of the tangent and cotangent functions can be generated by looking at the ratios between the sine and cosine of a series of angles in radian form. Graphs of the secant and cosecant functions can be determined using their definitions in terms of sine and cosine.

Graphs of the Tangent and Cotangent Functions

Since the $\tan\theta = \dfrac{\sin\theta}{\cos\theta}$, examine the table below.

θ	$\sin\theta$	$\cos\theta$	$\tan\theta$
$\frac{\pi}{12}$	0.25882	0.96593	0.26795
$\frac{\pi}{6}$	0.5	0.86603	0.57735
$\frac{\pi}{4}$	0.70711	0.70711	1
$\frac{\pi}{3}$	0.86603	0.5	1.7321
$\frac{5\pi}{12}$	0.96593	0.25882	3.7321
$\frac{\pi}{2}$	1	0	Undefined
$\frac{7\pi}{12}$	0.96593	–0.2588	–3.732
$\frac{2\pi}{3}$	0.86603	–0.5	–1.732
$\frac{3\pi}{4}$	0.70711	–0.70711	–1
$\frac{5\pi}{6}$	0.5	–0.866	–0.5774
$\frac{11\pi}{12}$	0.25882	–0.96593	–0.26795
π	0	–1	0

From this table, it becomes obvious that the tangent function has a different range than both the sine and cosine functions. Here is a graph of the tangent function.

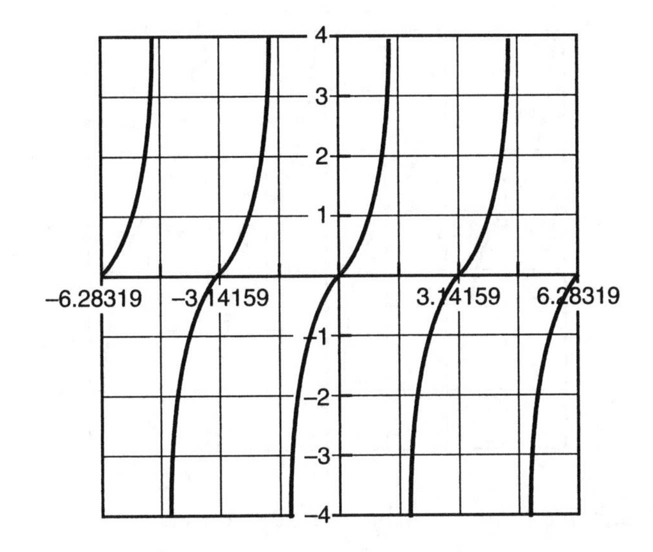

From the graph, the following table is easily verified.

Features of Tangent	$f(x) = \tan x$
Domain	$\left\{x : x \neq \frac{\pi}{2} + k\pi\right\}$
Range	$\{y: -\infty \leq y \leq \infty\}$
Period (p such that $f(x) = f(x + p)$)	π
y-intercept	0
x-intercept(s)	$0, \pm\pi, \pm 2\pi, \ldots$
Symmetry	Origin

The cotangent function has a similar type of graph, as shown.

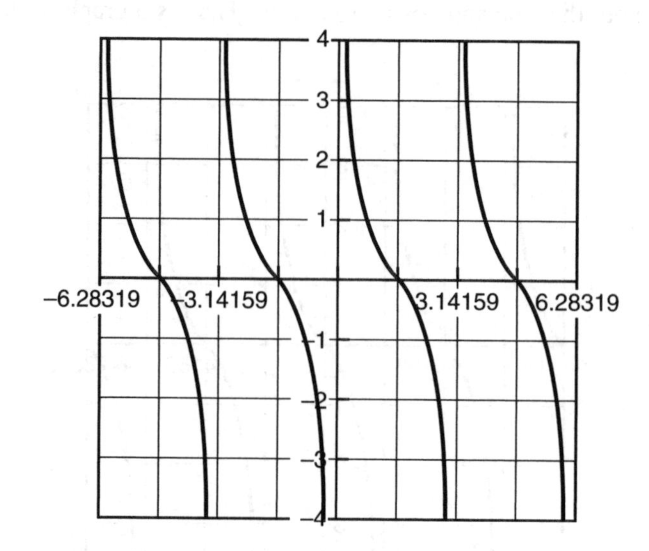

From the graph, the following table is easily verified

Features of Cotangent	***f*(*x*) = cot *x***
Domain	$\{x: x \neq \pi + k\pi\}$
Range	$\{y: -\infty \leq y \leq \infty\}$
Period (p such that $f(x) = f(x + p)$)	π
y-intercept	None
x-intercept(s)	$0, \pm\frac{\pi}{2}, \pm\frac{3\pi}{2}, \ldots$
Symmetry	Origin

Since the period of the tangent and cotangent functions is π instead of 2π, the frequency is measured in terms of how many full cycles are found between 0 and π. The period for tangent and cotangent equals $\frac{\pi}{b}$.

Example 1

Sketch a graph of $y = \cot\left(\frac{1}{2}\theta\right)$ on $[-2\pi, 2\pi]$.

Solution: The period $\frac{\pi}{\frac{1}{2}} = 2\pi$. This means the graph of the basic cotangent function is stretched so that one full cycle goes from 0 to 2π, as shown below.

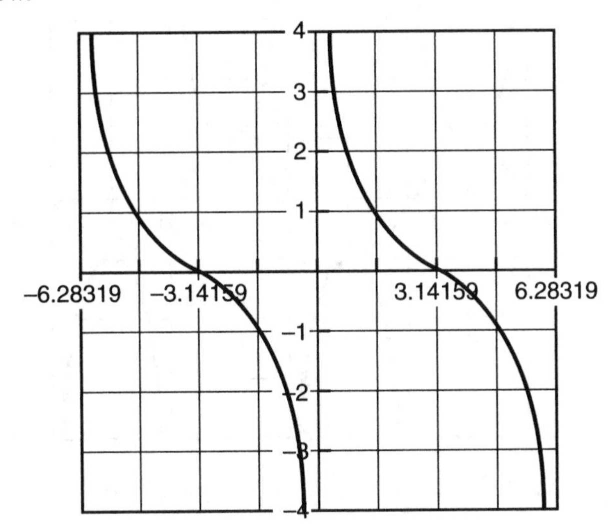

Graphs of the Secant and Cosecant Functions

Since $\sec\theta = \frac{1}{\cos\theta}$ and the range of the cosine function is $\{y: -1 \le y \le 1\}$, the range of the secant function is $\{y: y \le -1 \text{ or } y \ge 1\}$. The graph of the secant function is shown along with its reciprocal, the cosine function (dotted curve in the graph).

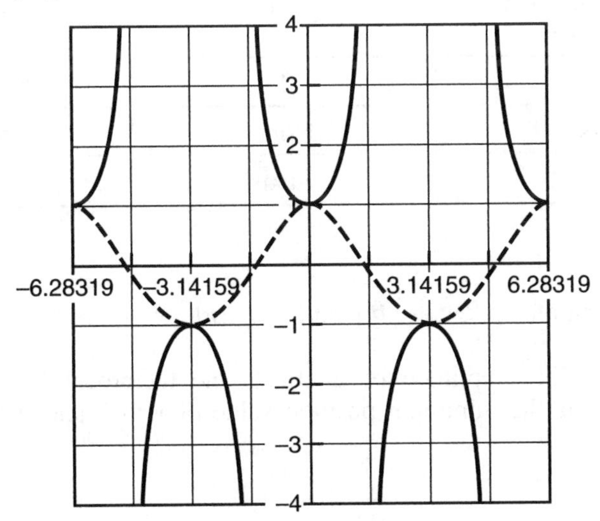

Similarly, the graph of the cosecant is shown below with its reciprocal sine (as a dotted curve).

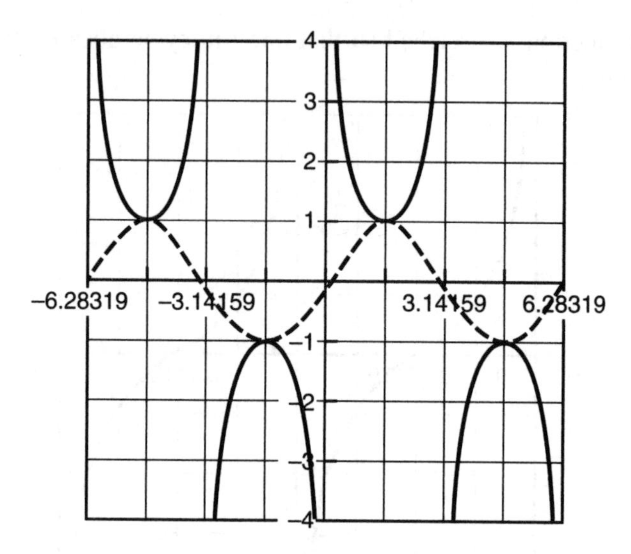

From these graphs, the following table is easily verified.

Features of Secant and Cosecant	$f(x) = \sec x$	$f(x) = \csc x$
Domain	$\left\{x : x \neq \frac{\pi}{2} + k\pi\right\}$	$\{x: x \neq \pi + k\pi\}$
Range	$\{y: y \leq -1 \text{ or } y \geq 1\}$	$\{y: y \leq -1 \text{ or } y \geq 1\}$
Period (p such that $f(x) = f(x + p)$)	2π	2π
y-intercept	1	None
x-intercept(s)	None	None
Symmetry	y-axis	Origin

Example 2

Sketch a graph of $y = 2\sec\left(\frac{1}{2}\theta\right)$ on $[-2\pi, 2\pi]$.

Solution: The amplitude of 2 affects the reciprocal function $\cos\theta$. It therefore sets the minimum positive value of y to 2 and the maximum negative value to –2. The period is $\frac{2\pi}{\frac{1}{2}} = 4\pi$ so that only half of the graph is shown from 0 to 2π. The graph is displayed below.

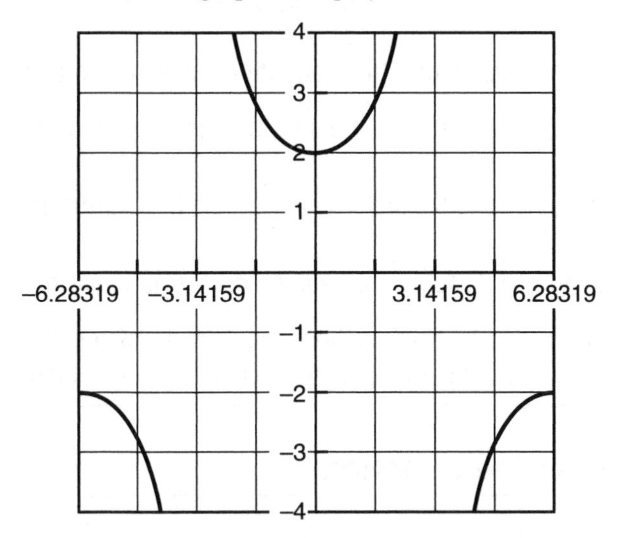

Check Your Understanding of Section 9.2

1. Which of the following functions has a domain of $\left\{x : x \neq \frac{\pi}{2} + 2k\pi\right\}$ and a range of all real numbers?
(1) $y = \sin\theta$ (2) $y = \cot\theta$ (3) $y = \sec\theta$ (4) $y = \tan\theta$

2. The graph of which of the following functions has a θ-intercept at $\frac{\pi}{2}$ and a y-intercept at 1?
(1) $y = \sin\theta$ (2) $y = \sec\theta$ (3) $y = \cos\theta$ (4) $y = \tan\theta$

3. Sketch the graphs of $y = 2\sin\left(\frac{1}{2}\theta\right)$ and $y = \sec 2\theta$ on $[-2\pi, 2\pi]$.

4. Sketch the graphs of $y = 2\sec\theta$ and $y = \cot 2\theta$ on $[-2\pi, 2\pi]$.

5. Use a sketch of the graphs of $y = \sec\theta$ and $y = 2\sin\theta$ to determine where $\sec\theta = 2\sin\theta$ on $[0, \frac{\pi}{2}]$.

9.3 INVERSE TRIGONOMETRIC FUNCTIONS

Recall that the inverse of a function interchanges the x- and y-values of the ordered pairs that make up the two functions. Since $\sin\frac{\pi}{6} = 0.5$, the inverse sine of 0.5 must equal $\frac{\pi}{6}$. In other words, the inverse sine function of 0.5 asks what angle has a sine equal to $\frac{\pi}{6}$. Since $\sin\theta$ is not a one-to-one function, the domain of the inverse sine function must be limited to a single cycle of the sine function.

The Inverse Sine Function

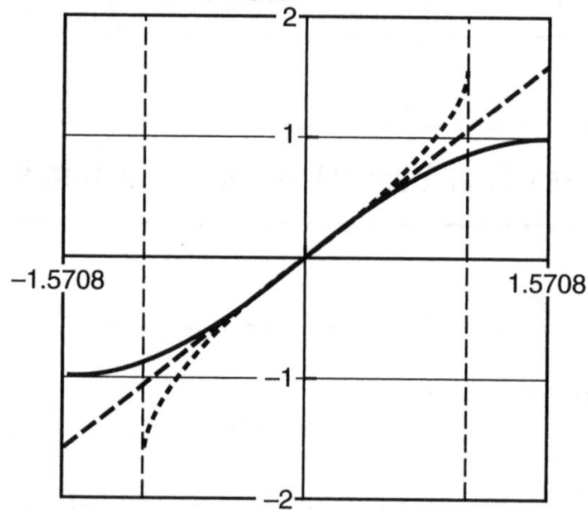

Since the sine function is not one-to-one, reflect one period of the sine curve over the line $y = x$.

Since the range associated with the restricted domain $\left\{x: -\frac{\pi}{2} \leq x \leq \frac{\pi}{2}\right\}$ of the sine function is $\{y: -1 \leq y \leq 1\}$, the inverse sine function has a domain of $\{x: -1 \leq x \leq 1\}$ and a range of $\left\{y: -\frac{\pi}{2} \leq y \leq \frac{\pi}{2}\right\}$.

The notation for inverse sine is $\sin^{-1}$ or arcsin. The principle values of $y = \sin^{-1}x$ or $\arcsin x$ are $\left\{y: -\frac{\pi}{2} \leq y \leq \frac{\pi}{2}\right\}$.

The Inverse Cosine Function

Restricting the cosine function to a domain in which it can be one-to-one is easiest from 0 to π as shown below.

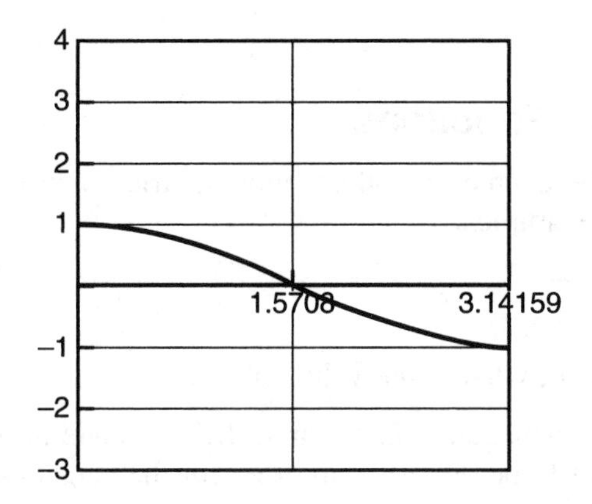

So the range of the inverse cosine function is $\{y: 0 \le y \le \pi\}$. The notation for inverse cosine is $\cos^{-1}$ or arccos.

Example 1

Evaluate: $\sin^{-1}(0.5)$

Solution: The value for $\sin^{-1}(0.5)$ is $\frac{\pi}{6}$ because $\sin\frac{\pi}{6} = \frac{1}{2} = 0.5$.

Example 2

Evaluate: $\text{Arc}\cos\left(\frac{-\sqrt{3}}{2}\right)$

Solution: The cosine function is negative in the second quadrant. So draw a diagram of a right triangle that fits this situation.

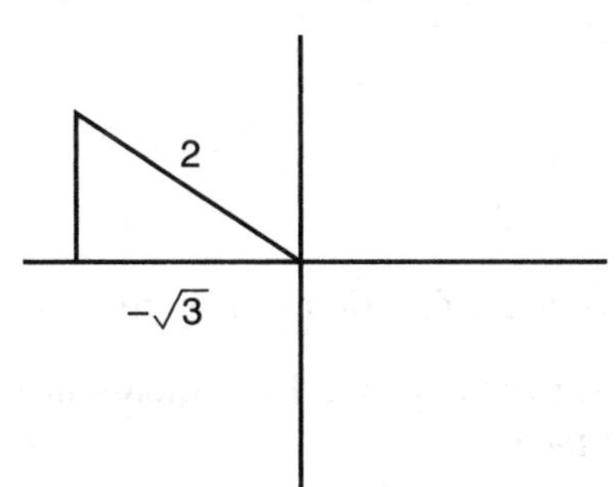

The reference angle is $\frac{\pi}{6}$ in the second quadrant, so the angle is $\pi - \frac{\pi}{6} = \frac{6\pi}{6} - \frac{\pi}{6} = \frac{5\pi}{6}$.

Other Inverse Functions

The inverse can be taken of the other trigonometric functions. Some are used in the following examples.

Example 3

If $y = \cos(\tan^{-1}(-1))$, what is the value of y?

Solution: $\tan^{-1}$ is negative in quadrant II. A triangle in quadrant II needs the x- and y-values to be opposites in order for the tangent to equal -1. This is pictured below.

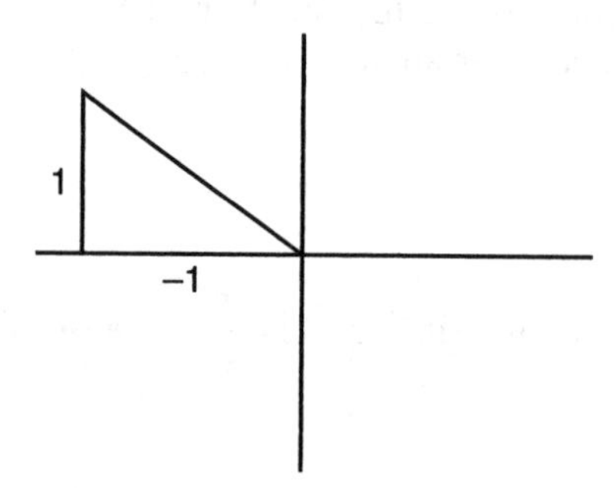

The reference angle is $\frac{\pi}{4}$. In the second quadrant, that yields an angle measure of $\frac{3\pi}{4}$. So, $\cos\frac{3\pi}{4} = \frac{-1}{\sqrt{2}} = -\frac{\sqrt{2}}{2}$.

Example 4

Evaluate: $\cos\left(\csc^{-1}\frac{-13}{12}\right)$.

Solution: The cosecant function is negative on $\left[\frac{-\pi}{2}, \frac{\pi}{2}\right]$. That places the angle in the fourth (or negative first) quadrant in a 5-12-13 triangle as pictured on the next page.

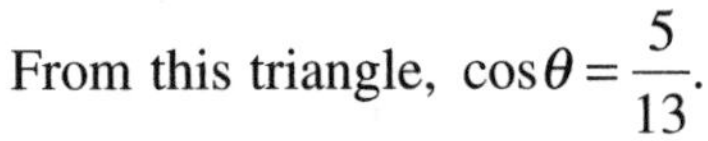

From this triangle, $\cos\theta = \frac{5}{13}$.

Example 5

Does $\cos(\cos^{-1}(1.3)) = 1.3$?

Solution: No, 1.3 is not in the domain of $\cos^{-1}\theta$.

Check Your Understanding of Section 9.3

1. Evaluate: $\tan\left(\sin^{-1}\frac{3}{5}\right)$.

(1) $\frac{3}{4}$ (2) $\frac{-4}{3}$ (3) $\frac{3\sqrt{7}}{7}$ (4) $\frac{-3}{4}$

2. If $\theta = \sec^{-1}\left(-\sqrt{2}\right)$, determine θ.

(1) $\frac{2\pi}{3}$ (2) $\frac{\pi}{4}$ (3) $\frac{3\pi}{4}$ (4) $-\frac{\pi}{4}$

3. Determine $\sin\theta$ if $\theta = \tan^{-1}\frac{-\sqrt{3}}{3}$.

(1) $-\frac{1}{3}$ (2) -1 (3) 1 (4) $-\frac{1}{2}$

4. Determine y if $y = \tan^{-1}(-\sqrt{3}) + \cos^{-1}\left(\frac{\sqrt{3}}{2}\right)$.

(1) $\frac{\pi}{2}$ (2) $-\frac{\pi}{6}$ (3) $\frac{\pi}{6}$ (4) $-\frac{\pi}{2}$

5. Evaluate: $\sin\left(\arccos\frac{-8}{17}\right)$.

(1) $\frac{17}{15}$ (2) $\frac{8}{17}$ (3) $-\frac{15}{17}$ (4) $\frac{15}{17}$

6. If $f(x) = \tan x$ and $g(x) = \csc^{-1}x$, determine $(f \circ g)\left(\frac{-5}{3}\right)$.

(1) $\frac{-3}{4}$ (2) $\frac{-4}{3}$ (3) $\frac{3}{5}$ (4) $\frac{4}{3}$

7. Evaluate: $\left(\cos^{-1}\left(\frac{-\sqrt{2}}{2}\right)\right)^2$.

(1) $\frac{-\pi}{4}$ (2) $\frac{\pi^2}{16}$ (3) $\frac{\pi}{4}$ (4) $\frac{3}{4}$

8. Determine the numerical value of $\tan\left(\csc^{-1}\left(\frac{-\sqrt{13}}{2}\right)\right)$.

(1) $\frac{-2}{3}$ (2) $\frac{\sqrt{13}}{3}$ (3) $\frac{-3}{2}$ (4) $\frac{-3\sqrt{13}}{13}$

9. Evaluate: $\tan^{-1}\left(\tan\left(\frac{5\pi}{6}\right)\right)$.

(1) $\frac{5\pi}{6}$ (2) $\frac{-\pi}{6}$ (3) $\frac{\pi}{6}$ (4) $\frac{\pi}{3}$

10. Evaluate: $\sec\left(\cot^{-1}0\left(\frac{x}{y}\right)\right)$

11. Does $\csc(\csc^{-1}(-0.6)) = -0.6$? Explain your answer.

TRIGONOMETRIC IDENTITIES AND EQUATIONS

10.1 PYTHAGOREAN IDENTITIES

If an ordered pair (x, y) is on the unit circle, by the Pythagorean Theorem $x^2 + y^2 = 1$. Therefore, $\sin^2\theta + \cos^2\theta = 1$.

The Pythagorean Theorem and the Sine and Cosine Functions

Recall the definitions of the sine and cosine functions on the unit circle.

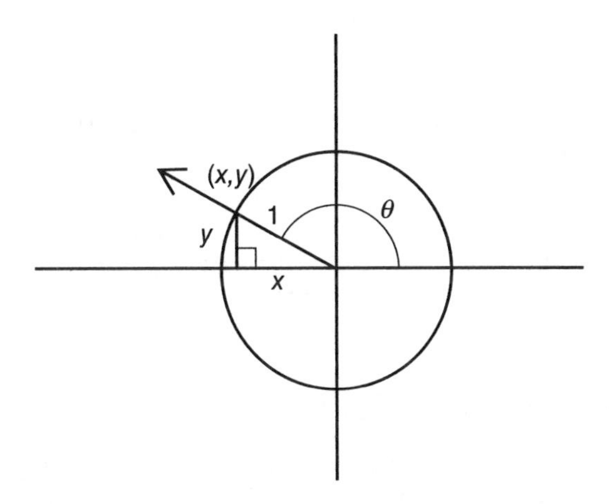

Since $y = \sin\theta$ and $x = \cos\theta$, the ordered pair (x, y) is equivalent to $(\cos\theta, \sin\theta)$. By applying the Pythagorean Theorem to the right triangle in the diagram, $x^2 + y^2 = 1$. By substitution, $\cos^2\theta + \sin^2\theta = 1$. By the commutative law, $\sin^2\theta + \cos^2\theta = 1$. Note that this is not simply an equation because it is true for all values of θ.

Identities

An equation that is true for all values of a variable is called an identity. This means that $\sin^2\theta + \cos^2\theta = 1$ is an identity.

The reciprocal relationships discussed in Section 8.2 are actually identities.

MATH FACTS

The reciprocal identities are:

$$\csc\theta = \frac{1}{\sin\theta} \text{ and } \sin\theta = \frac{1}{\csc\theta}$$

$$\sec\theta = \frac{1}{\cos\theta} \text{ and } \cos\theta = \frac{1}{\sec\theta}$$

$$\cot\theta = \frac{1}{\tan\theta} \text{ and } \tan\theta = \frac{1}{\cot\theta}$$

$$\tan\theta = \frac{\sin\theta}{\cos\theta} \text{ and } \cot\theta = \frac{\cos\theta}{\sin\theta}$$

Example 1

Using $\sin\beta$, $\cos\beta$, or both, write $\csc\beta + \cot\beta$ as an equivalent expression as a single fraction in lowest terms.

Solution: $\csc\beta + \cot\beta = \dfrac{1}{\sin\beta} + \dfrac{\cos\beta}{\sin\beta} = \dfrac{1+\cos\beta}{\sin\beta}$

The Pythagorean Identities

The identity $\sin^2\theta + \cos^2\theta = 1$ can be modified by subtracting either $\sin^2\theta$ or $\cos^2\theta$ from both sides of the identity. If every term of the identity is divided by $\sin^2\theta$, this results in a new identity, $\cot^2\theta + 1 = \csc^2\theta$. Dividing instead by $\cos^2\theta$ develops the identity $1 + \tan^2\theta = \sec^2\theta$.

MATH FACTS

The three basic Pythagorean identities are:

$$\sin^2\theta + \cos^2\theta = 1$$
$$1 + \tan^2\theta = \sec^2\theta$$
$$\cot^2\theta + 1 = \csc^2\theta$$

Each of these can be transposed to get the following identities:

$$\sin^2\theta = 1 - \cos^2\theta$$
$$\cos^2\theta = 1 - \sin^2\theta$$
$$\tan^2\theta = \sec^2\theta - 1$$
$$\sec^2\theta - \tan^2\theta = 1$$
$$\cot^2\theta = \csc^2\theta - 1$$
$$\csc^2\theta - \cot^2\theta = 1$$

Example 2

If $\sin\theta = \frac{-\sqrt{3}}{4}$ and θ is an angle in the fourth quadrant, determine the numerical value of $\cos\theta$.

Solution: Since $\cos^2\theta = 1 - \sin^2\theta$, $\cos^2\theta = 1 - \left(\frac{-\sqrt{3}}{4}\right)^2 = 1 - \frac{3}{16} = \frac{13}{16}$ and $\cos\theta = \pm\frac{\sqrt{13}}{4}$. In quadrant IV, $\cos\theta > 0$. Therefore, $\cos\theta = \frac{\sqrt{13}}{4}$.

Example 3

Write a simple expression using only one trigonometric function equivalent to $\sin^2 A + \cos^2 A + \tan^2 A$.

Solution: Since $\sin^2 A + \cos^2 A = 1$, then $\sin^2 A + \cos^2 A + \tan^2 A$ becomes $1 + \tan^2 A$, which is equivalent to $\sec^2 A$.

Example 4

Rewrite $\tan x \cdot \sec x$ as a single fraction in terms of only $\sin x$.

Solution: Use the fact that $\tan x = \frac{\sin x}{\cos x}$ and $\sec x = \frac{1}{\cos x}$. Substitute these into the original expression $\tan x \cdot \sec x$.

$$\tan x \cdot \sec x = \frac{\sin x}{\cos x} \cdot \frac{1}{\cos x} = \frac{\sin x}{\cos^2 x} = \frac{\sin x}{1 - \sin^2 x}$$

Example 5

Express $\cot^2\theta - \cos^2\theta$ as a single fraction in terms of $\sin\theta$ and $\cos\theta$.

Solution: $\cot^2\theta - \cos^2\theta = \dfrac{\cos^2\theta}{\sin^2\theta} - \cos^2\theta$

$$= \frac{\cos^2\theta}{\sin^2\theta} - \frac{\cos^2\theta\sin^2\theta}{\sin^2\theta} = \frac{\cos^2\theta - \cos^2\theta\cdot\sin^2\theta}{\sin^2\theta}$$

Check Your Understanding of Section 10.1

1. What is the expression $\tan y(\cot y + \tan y)$ is equivalent to?
(1) $\cos^2 y$ (2) $\sec^2 y$ (3) $\csc^2 y$ (4) $\sin^2 y$

2. The expression $\dfrac{\sin^2 A}{\cos^2 A}$ is equivalent to?
(1) $\tan^2 A$ (2) $\csc^2 A$ (3) $\cot^2 A$ (4) $\sec^2 A$

3. What is $\sec^4\alpha - \sec^2\alpha\cdot\tan^2\alpha$ equivalent to?
(1) $\sec^2\alpha$ (2) $\sec\alpha\tan\alpha$ (3) $\cot^2\alpha$ (4) $\cos\alpha\cot\alpha$

4. Express $\cot x + \tan x$ as a single fraction in simplest form.
(1) $\dfrac{\cos x + \sin x}{\sin x\cos x}$ (3) $\dfrac{1}{\sin x\cos x}$
(2) $\dfrac{\sec x + 1}{\cos x}$ (4) $\dfrac{1}{\sin x + \cos x}$

5. Expressed in terms of $\sin\theta$ and $\cos\theta$, $\csc\theta\cot\theta =$
(1) $\dfrac{\sin\theta}{\cos^2\theta}$ (2) $\sin\theta\cos\theta$ (3) $\dfrac{\cos^2\theta}{\sin\theta}$ (4) $\dfrac{\cos\theta}{\sin^2\theta}$

6. $(\tan B + 1)(1 - \tan B) =$
(1) $\sec^2 B$ (2) $1 - \sec^2 B$ (3) $\sec^2 B - 2$ (4) $2 - \sec^2 B$

7. If $\sin x = a$, then $\cos x =$
(1) $90 - a$ (2) $\pm\sqrt{1-a^2}$ (3) $1 - a$ (4) $\pm\sqrt{a^2-1}$

8. Rewrite $\sec A(\cot A - 1) + \csc A + \sec A$ as a single fraction in terms of $\sin\theta$ and $\cos\theta$.

9. Rewrite $\csc^2 x - 1$ as a single fraction in terms of $\sin\theta$.

10. Express the sum of $\dfrac{1}{\tan^2 A} + \dfrac{1}{\sec A - 1}$ as a single fraction in terms of only $\sec A$.

11. Rewrite the expression $\cot\theta(\tan\theta + \cos\theta\csc\theta)$ by first changing the entire expression in terms of $\sin\theta$ and $\cos\theta$ and then simplifying the expression to a single fraction in lowest terms. Show all work.

12. Prove that $\cot^2\theta + 1 = \csc^2\theta$ by starting with $\sin^2\theta + \cos^2\theta = 1$.

13. An ordered pair (x, y) on the unit circle can be expressed in terms of $\sin\theta$ and $\cos\theta$. Use this fact and the Pythagorean Theorem for a right triangle constructed by connecting any ordered pair (x, y) on the unit circle to the origin and drawing a perpendicular line from (x, y) to the x-axis.

10.2 SUM AND DIFFERENCE IDENTITIES

Sometimes it is helpful to determine a trigonometric function of the sum or difference of two angles. This is not as simple as adding the function values. For instance, $\cos 20^\circ \approx 0.9397$ and $\cos 12^\circ \approx 0.9781$. However, $\cos 32^\circ \approx 0.8480$ and clearly not 1.9178 since $-1 \le \cos x \le 1$.

Sine and Cosine of the Sum of Two Angles

As demonstrated above, $\cos(A + B) \neq \cos A + \cos B$. Similarly, $\sin(A + B) \neq \sin A + \sin B$.

MATH FACTS

$$\sin(A + B) = \sin A\cos B + \sin B\cos A$$

and

$$\cos(A + B) = \cos A\cos B - \sin A\sin B$$

Example 1

Evaluate: $\sin 75^\circ$

Solution: $\sin 75^\circ = \sin(30^\circ + 45^\circ) = \sin 30^\circ \cos 45^\circ + \sin 45^\circ \cos 30^\circ$

$$\frac{1}{2}\cdot\frac{\sqrt{2}}{2}+\frac{\sqrt{2}}{2}\cdot\frac{\sqrt{3}}{2}=\frac{\sqrt{2}}{4}+\frac{\sqrt{6}}{4}=\frac{\sqrt{2}+\sqrt{6}}{4}$$

This result can be checked on a calculator. First make sure that the calculator mode is in radians. See the display below.

```
sin(75)
          .9659258263
(√(2)+√(6))/4
          .9659258263
█
```

Example 2

If α is an angle in the second quadrant, $\sin\alpha = \frac{5}{13}$, and β is an angle in the third quadrant such that $\sin\beta = -\frac{4}{5}$, determine $\cos(\alpha + \beta)$.

Solution: $\cos(\alpha + \beta) = \cos\alpha\cos\beta - \sin\alpha\sin\beta$.

If $\sin\alpha = \frac{5}{13}$, then $\cos\alpha = -\frac{12}{13}$ (a 5-12-13 triangle in the second quadrant).

If $\sin\beta = -\frac{4}{5}$, then $\cos\beta = -\frac{3}{5}$ (a 3-4-5 triangle in the third quadrant).

$$\cos(\alpha+\beta)=\frac{-12}{13}\cdot\frac{-3}{5}-\frac{5}{13}\cdot\frac{-4}{5}=\frac{36}{65}+\frac{20}{65}=\frac{56}{65}$$

Example 3

Express $\sin 22 \cos 37 + \sin 37 \cos 22$ as a sine function of a single angle.

Solution: $\sin 22 \cos 37 + \sin 37 \cos 22 = \sin(22 + 37) = \sin 59$.

Example 4

Use the formulas for sin(A + B) and cos(A + B) to derive a formula for tan(A + B).

Solution: $\tan(A+B)=\dfrac{\sin(A+B)}{\cos(A+B)}=\dfrac{\sin A\cos B+\sin B\cos A}{\cos A\cos B-\sin A\sin B}$

It would be better if this formula could be expressed in terms of the tangent function. To do this, divide each term of the expression on the right side of the equation by $\cos A\cos B$.

$$\begin{aligned}\tan(A+B)&=\frac{\sin A\cos B+\sin B\cos A}{\cos A\cos B-\sin A\sin B}\\&=\frac{\dfrac{\sin A\cos B}{\cos A\cos B}+\dfrac{\sin B\cos A}{\cos A\cos B}}{\dfrac{\cos A\cos B}{\cos A\cos B}-\dfrac{\sin A\sin B}{\cos A\cos B}}\\&=\frac{\tan A\cdot 1+\tan B\cdot 1}{1-\tan A\tan B}\\&=\frac{\tan A+\tan B}{1-\tan A\tan B}\end{aligned}$$

Tangent of the Sum of Two Angles

MATH FACTS

$$\tan(A+B)=\frac{\tan A+\tan B}{1-\tan A\tan B}$$

Example 5

Determine the exact numerical value of tan 195°.

Solution: $\tan 195°=\tan 150°+\tan 45°=\dfrac{\tan 150°+\tan 45°}{1-\tan 150°\tan 45°}$

$\tan 45° = 1$

and $\tan 150°=-\tan(180°-30°)=-\tan 30°=\dfrac{-\sqrt{3}}{3}$

$$\tan 195° = \frac{1+\frac{-\sqrt{3}}{3}}{1-1\cdot\frac{-\sqrt{3}}{3}} = \frac{1-\frac{\sqrt{3}}{3}}{1+\frac{\sqrt{3}}{3}} = \frac{\frac{3-\sqrt{3}}{3}}{\frac{3+\sqrt{3}}{3}}$$

$$= \frac{3-\sqrt{3}}{3}\cdot\frac{\cancel{3}}{\cancel{3}+\sqrt{3}} = \frac{\left(3-\sqrt{3}\right)}{\left(3+\sqrt{3}\right)}\cdot\frac{\left(3-\sqrt{3}\right)}{\left(3-\sqrt{3}\right)}$$

$$= \frac{9-6\sqrt{3}+3}{9-3} = \frac{12-6\sqrt{3}}{6} = 2-\sqrt{3}$$

Difference Formulas

MATH FACTS

$$\sin(A-B) = \sin A\cos B - \sin B\cos A$$
$$\cos(A-B) = \cos A\cos B + \sin A\sin B$$
$$\tan(A-B) = \frac{\tan A - \tan B}{1+\tan A\tan B}$$

Example 6

Evaluate cos 105°

Solution: $\cos 105° = \cos(150° - 45°) = \cos 150° \cos 45° + \sin 150° \sin 450°$

$$= -\cos 30° \cos 45° + \sin 30° \sin 45°$$

$$= -\frac{\sqrt{3}}{2}\cdot\frac{\sqrt{2}}{2} + \frac{1}{2}\cdot\frac{\sqrt{2}}{2}$$

$$= -\frac{\sqrt{6}}{4} + \frac{\sqrt{2}}{4}$$

$$= \frac{-\sqrt{6}+\sqrt{2}}{4}$$

Example 7

Express $\frac{\tan 31 - \tan 12}{1 + \tan 31 \cdot \tan 12}$ as a tangent function of a single angle.

Solution: $\frac{\tan 31 - \tan 12}{1 + \tan 31 \cdot \tan 12} = \tan(31 - 12) = \tan 19$

Check Your Understanding of Section 10.2

1. The expression $\sin 74 \cos 33 - \sin 33 \cos 74$ is equivalent to?
(1) $\sin 107$ (2) $\sin 17$ (3) $\cos 107$ (4) $\sin 41$

2. What is the expression $\cos\frac{\pi}{3}\cos\frac{3\pi}{4} - \sin\frac{3\pi}{4}\cos\frac{\pi}{3}$ is equivalent to?
(1) $\cos\frac{13\pi}{12}$ (2) $\sin\frac{5\pi}{12}$ (3) $\cos\frac{5\pi}{12}$ (4) $\cos\frac{\pi}{12}$

3. The expression $\tan\left(x - \frac{\pi}{3}\right)$ is equivalent to?
(1) $\sqrt{3} + \tan x$
(2) $\dfrac{\tan x - \sqrt{3}}{1 + \sqrt{3}\tan x}$
(3) $\dfrac{\tan x + \sqrt{3}}{1 - \sqrt{3}\tan x}$
(4) $\dfrac{3\tan x - \sqrt{3}}{3 + \sqrt{3}\tan x}$

4. Evaluate: $\sin\left(\tan^{-1}\left(-\frac{1}{3}\right) - \csc^{-1}\left(\frac{7}{3}\right)\right)$
(1) $\frac{1}{7}$ (2) $\frac{20 - 9\sqrt{10}}{70}$ (3) $\frac{-20 - 9\sqrt{10}}{70}$ (4) $\frac{60 - 3\sqrt{10}}{70}$

5. Evaluate: $\cos\left(\sin^{-1}\left(\frac{1}{2}\right) + \tan^{-1}(1)\right)$
(1) $\frac{\sqrt{3}}{2}$ (2) $\frac{15\pi}{36}$ (3) $\frac{\sqrt{6} - \sqrt{2}}{4}$ (4) $\frac{\sqrt{6} + \sqrt{2}}{4}$

6. If $\sin A = \frac{8}{17}$ where A is in the second quadrant and $\cos B = \frac{-5}{13}$ where B is in the third quadrant, evaluate $\tan(A - B)$.
(1) $\frac{133}{220}$ (2) $\frac{220}{21}$ (3) $\frac{-285}{84}$ (4) $\frac{-171}{140}$

7. If $\cos\theta = x$ and $\sin\theta = y$, what is the expression $\cos(45 + \theta)$ equivalent to?
(1) $\frac{x - y}{\sqrt{2}}$ (2) $\frac{x + y}{\sqrt{2}}$ (3) $x\sqrt{2} - y\sqrt{2}$ (4) $x\sqrt{2} + y\sqrt{2}$

8. What is the expression $\sin(\pi - a)$ equivalent to?
(1) $\cos a$ (2) $-\cos a$ (3) $\sin a$ (4) $-\sin a$

9. If $\tan A = \frac{3}{4}$ and $\cot B = 2$ and A and B are acute angles, what is the value of $\sin(A - B)$?
(1) $\frac{11\sqrt{5}}{5}$ (2) $\frac{-11\sqrt{5}}{5}$ (3) $\frac{2\sqrt{5}}{5}$ (4) $\frac{-2\sqrt{5}}{5}$

10. If $x = \sec^{-1}\frac{-2\sqrt{3}}{3}$ and $y = \csc^{-1}2$, then $\cos(x + y) =$
(1) -1 (2) 1 (3) $\frac{\sqrt{3}}{2}$ (4) $\frac{-\sqrt{3}}{2}$

11. If α is in the third quadrant, $\tan\alpha = \frac{12}{5}$, β is in the fourth quadrant, $\csc\beta = -\frac{17}{8}$, determine the exact numerical value of each of the following:
a. $\sin(\alpha + \beta)$ b. $\cos(\alpha + \beta)$ c. $\tan(\alpha + \beta)$
d. $\cot(\alpha + \beta)$ [**Hint**: Use the results of parts a and b or the result of part c to determine this value.]

12. Multiply $\cos(x + y)$ by $\cos(x - y)$ and express the answer in simplest form in terms of only $\cos x$ and $\sin y$. [**Hint**: To get this to the final form, it is necessary to use the fact that $\cos^2 y = 1 - \sin^2 y$.]

10.3 DOUBLE-ANGLE IDENTITIES

Clearly, $\sin 2\theta \neq 2\sin\theta$. This can be verified by noticing that the 2 in the function on the left side of this statement represents frequency while the 2 in the function on the right side represents amplitude. Therefore, $\sin 2\theta$ and $2\sin\theta$ do not have the same graph. Also notice that $2\sin\frac{\pi}{6} = (2)\left(\frac{1}{2}\right) = 1$ and $\sin\frac{\pi}{3} = \frac{\sqrt{3}}{2} \neq 1$.

Developing the Double-Angle Identities

The double-angle identities can be easily derived from the identities for the sum of two angles. For instance, $\sin 2x = \sin(x + x) = \sin x \cos x + \sin x \cos x = 2\sin x \cos x$.

MATH FACTS

Double-Angle Identities

$\sin 2\theta = 2\sin\theta\cos\theta$

$\cos 2\theta = \cos^2\theta - \sin^2\theta$
$\cos 2\theta = 1 - 2\sin^2\theta$
$\cos 2\theta = 2\cos^2\theta - 1$

$\tan 2\theta = \dfrac{2\tan\theta}{1-\tan^2\theta}$

Notice that there are three different forms of the identity for $\cos 2\theta$. This is possible because the Pythagorean Identities can be used to create forms of the identity that involve only one trigonometric function.

Example 1

If A is an angle in the first quadrant and $\cos A = \frac{1}{3}$, determine $\sin 2A$.

Solution: If $\cos A = \frac{1}{3}$, then by the Pythagorean Theorem, the third side of a triangle placed in the first quadrant in a circle whose radius is 3 is $2\sqrt{2}$ and $\sin A = \frac{2\sqrt{2}}{3}$. Therefore, $\sin 2A = 2\sin A\cos A = 2\cdot\frac{2\sqrt{2}}{3}\cdot\frac{1}{3} = \frac{4\sqrt{2}}{9}$.

Example 2

If $\sin\alpha = \frac{2}{5}$, determine $\cos 2\alpha$.

Solution: Since the numerical value of $\sin\alpha$ is given, use

$$\cos 2\alpha = 1 - 2\sin^2\alpha = 1 - 2\left(\frac{2}{5}\right)^2 = 1 - 2\left(\frac{4}{25}\right) = 1 - \frac{8}{25} = \frac{17}{25}.$$

Example 3

If $\tan\theta = 0.2$, determine $\tan 2\theta$.

Solution: Since $\tan 2\theta = \dfrac{2\tan\theta}{1-\tan^2\theta}$

$$\tan 2\theta = \frac{2(0.2)}{1-(0.2)^2} = \frac{0.4}{1-0.04} = \frac{0.4}{0.96} = 0.41666\ldots$$

Example 4

Evaluate: $\cos\left(2\sin^{-1}\frac{-\sqrt{3}}{2}\right)$

Solution: If $\theta = \sin^{-1}\frac{-\sqrt{3}}{2}$ and $\cos 2\theta = 1 - 2\sin^2\theta$, then

$$\begin{aligned}\cos\left(2\sin^{-1}\frac{-\sqrt{3}}{2}\right) &= 1-2\sin^2\left(\sin^{-1}\frac{-\sqrt{3}}{2}\right)\\ &= 1-2\left(\frac{-\sqrt{3}}{2}\right)^2\\ &= 1-2\left(\frac{3}{4}\right)\\ &= 1-\frac{3}{2}\\ &= -\frac{1}{2}.\end{aligned}$$

Check Your Understanding of Section 10.3

1. If $\sin A = \frac{1}{8}$, evaluate $\cos 2A$.

(1) $-\frac{31}{32}$ (2) $\frac{31}{32}$ (3) $\frac{7}{8}$ (4) $-\frac{7}{8}$

2. If $\tan\alpha = \frac{5}{4}$, determine the numerical value of $\tan 2\alpha$.

(1) $\frac{10}{9}$ (2) -10 (3) $-\frac{40}{9}$ (4) $-\frac{40}{11}$

3. Determine the numerical value of $\sin\left(2\cos^{-1}\frac{-3}{5}\right)$.

(1) $\frac{24}{25}$ (2) $\frac{-24}{25}$ (3) $\frac{16}{25}$ (4) $\frac{-16}{25}$

4. If θ is in the second quadrant and $\csc\theta = \frac{7}{3}$, evaluate $\sin 2\theta - \cos 2\theta$.

(1) $-\frac{12\sqrt{10}+31}{49}$ (3) $\frac{-12\sqrt{10}+31}{49}$

(2) $-\frac{12\sqrt{10}-31}{49}$ (4) $-\frac{12\sqrt{10}+40}{49}$

5. What is $\cos 2A + 2\sin^2 A + \sin 2A$ equivalent to?

(1) $(\sin A - \cos A)^2$ (3) $1 + 2\sin 2A$

(2) $1 - 2\sin 2A$ (4) $(\sin A + \cos A)^2$

6. $\cos 2\theta + \sin^2\theta$ is equivalent to?

(1) 1 (2) $\sin^2\theta$ (3) $\cos^2\theta$ (4) $1 - 3\sin^2\theta$

7. Use the formulas for $\sin(A + B)$ and $\sin 2A$ to develop a formula for $\sin 3A$.

8. If θ is an angle in the second quadrant with $\cos 2\theta = \frac{-5}{13}$ and $\cos\theta = \frac{2\sqrt{13}}{13}$, determine the numerical value of $\sin\theta$.

9. Start with the identity $\cos 2\theta = 1 - 2\sin^2\theta$ and solve for $\sin\theta$.

10.4 HALF-ANGLE IDENTITIES

Since $\sin 2\theta \neq 2\sin\theta$, it seems reasonable to assume that $\sin\frac{1}{2}\theta \neq \frac{1}{2}\sin\theta$. This can also be easily verified by noticing that the $\frac{1}{2}$ in the function on the left side represents frequency while the $\frac{1}{2}$ in the function on the right side represents amplitude.

Using the Double-Angle Identities to Develop the Half-Angle Identities

Exercise 9 in "Check Your Understanding" for Section 10.3 gives a good hint about how to use the Double-Angle Identities to develop the Half-Angle Identities.

Start with the identity $\cos 2\theta = 2\cos^2\theta - 1$. So $2\cos^2\theta = 1 + \cos 2\theta$ and $\cos^2\theta = \dfrac{1+\cos 2\theta}{2}$. Let $A = 2\theta$, and substitute this into the last step to yield $\cos^2\dfrac{A}{2} = \dfrac{1+\cos A}{2}$ and $\cos\dfrac{A}{2} = \pm\sqrt{\dfrac{1+\cos A}{2}}$.

The Half-Angle Identities

MATH FACTS

$$\sin\frac{1}{2}\theta = \pm\sqrt{\frac{1-\cos\theta}{2}} \qquad \cos\frac{1}{2}\theta = \pm\sqrt{\frac{1+\cos\theta}{2}} \qquad \tan\frac{1}{2}\theta = \pm\sqrt{\frac{1-\cos\theta}{1+\cos\theta}}$$

Example 1

If $\cos\alpha = \dfrac{1}{3}$, determine $\cos\dfrac{\alpha}{2}$.

Solution: $\cos\dfrac{\alpha}{2} = \pm\sqrt{\dfrac{1+\cos\alpha}{2}} = \pm\sqrt{\dfrac{1+\frac{1}{3}}{2}} = \pm\sqrt{\dfrac{\frac{4}{3}}{2}} = \pm\sqrt{\dfrac{2}{3}} = \pm\dfrac{\sqrt{6}}{3}$

Example 2

Determine $\tan\dfrac{1}{2}\left(\sec^{-1}\left(\dfrac{-13}{5}\right)\right)$.

Solution: $\tan\dfrac{1}{2}\theta = \pm\sqrt{\dfrac{1-\cos\theta}{1+\cos\theta}}$

$$\tan\frac{1}{2}\left(\sec^{-1}\left(\frac{-13}{5}\right)\right) = \pm\sqrt{\frac{1-\cos\left(\sec^{-1}\left(\frac{-13}{5}\right)\right)}{1+\cos\left(\sec^{-1}\left(\frac{-13}{5}\right)\right)}} = \pm\sqrt{\frac{1-\frac{-5}{13}}{1+\frac{-5}{13}}}$$

$$= \pm\sqrt{\frac{\frac{18}{13}}{\frac{8}{13}}} = \pm\sqrt{\frac{18}{8}} = \pm\sqrt{\frac{9}{2}} = \pm\frac{3\sqrt{2}}{2}$$

Check Your Understanding of Section 10.4

1. If θ is in the third quadrant and $\tan\theta = 1$, determine the exact numerical value of $\sin\frac{\theta}{2}$.

(1) $-\frac{\sqrt{2-\sqrt{2}}}{2}$ (2) $\frac{\sqrt{2-\sqrt{2}}}{2}$ (3) $-\frac{\sqrt{2+\sqrt{2}}}{2}$ (4) $\frac{\sqrt{2+\sqrt{2}}}{2}$

2. If $\cos\alpha = 0.32$ and $90 < \alpha < 180$, determine $\tan\frac{\alpha}{2}$ to the nearest hundredth.

(1) −1.39 (2) −0.72 (3) 0.72 (4) 1.39

3. Determine $\tan\left(\frac{1}{2}\cot^{-1}\left(\frac{3}{4}\right)\right)$.

(1) $\frac{\sqrt{2}}{2}$ (2) $\frac{1}{2}$ (3) $\frac{\sqrt{3}}{2}$ (4) $\frac{\sqrt{6}}{4}$

4. If $\sin\frac{\alpha}{2} = \frac{3}{5}$, determine $\cos\alpha$.

(1) $\frac{7}{25}$ (2) $-\frac{7}{25}$ (3) $-\frac{4}{5}$ (4) $\frac{16}{25}$

5. What is the expression $2\cos^2\left(\frac{1}{2}x\right)$ is equivalent to?

(1) $\sin x + \cos x$
(2) $1 + \cos x$
(3) $1 - \cos x$
(4) $\cos x - 1$

6. If $\cos A - 1 = -0.37$ and A is an acute angle, what is the value of $\tan\frac{A}{2}$ to the nearest hundredth?

(1) 0.47 (2) 2.10 (3) 0.48 (4) −0.47

7. Use the half-angle formulas to determine each of the following:

a. $\tan 22.5^\circ$ b. $\cos\frac{7\pi}{8}$ c. $\sin\left(\frac{-\pi}{12}\right)$ d. $\cos 105^\circ$

10.5 IDENTITY PROOFS

Note: This is an optional section that is not tested on the New York State Algebra 2/Trigonometry Regents Examination.

An identity is an equation that is true for all values of the variable.

Proving Identities

Proving that an identity is true is similar to doing a geometric proof. The first basic rule is to convert all terms to sine and cosine. Generally start on the more complicated side of the identity and then simplify using algebraic skills until the expression becomes the other side of the identity.

Example 1

Prove: $\tan x \cdot \sec x = \dfrac{\sin x}{1-\sin^2 x}$

Solution: In this example, it is possible to work on either side of the identity, as shown in the following table.

Work on the Left Side of the Identity	Work on the Right Side of the Identity
$\tan x \cdot \sec x = \dfrac{\sin x}{1-\sin^2 x}$	$\tan x \cdot \sec x = \dfrac{\sin x}{1-\sin^2 x}$
$\dfrac{\sin x}{\cos x} \cdot \dfrac{1}{\cos x} = \dfrac{\sin x}{1-\sin^2 x}$	$\tan x \cdot \sec x = \dfrac{\sin x}{\cos^2 x}$
$\dfrac{\sin x}{\cos^2 x} = \dfrac{\sin x}{1-\sin^2 x}$	$\tan x \cdot \sec x = \dfrac{\sin x}{\cos x} \cdot \dfrac{1}{\cos x}$
$\dfrac{\sin x}{1-\sin^2 x} = \dfrac{\sin x}{1-\sin^2 x}$	$\tan x \cdot \sec x = \tan x \sec x$

Example 2

Prove: $\cot^2\theta - \cos^2\theta = \cos^2\theta \cdot \cot^2\theta$

Solution: $\cot^2\theta - \cos^2\theta = \cos^2\theta \cdot \cot^2\theta$

$$\frac{\cos^2\theta}{\sin^2\theta} - \cos^2\theta = \cos^2\theta \cdot \cot^2\theta$$

$$\frac{\cos^2\theta}{\sin^2\theta} - \frac{\cos^2\theta \cdot \sin^2\theta}{\sin^2\theta} = \cos^2\theta \cdot \cot^2\theta$$

$$\frac{\cos^2\theta - \cos^2\theta \cdot \sin^2\theta}{\sin^2\theta} = \cos^2\theta \cdot \cot^2\theta$$

$$\frac{\cos^2\theta(1 - \sin^2\theta)}{\sin^2\theta} = \cos^2\theta \cdot \cot^2\theta$$

$$\frac{\cos^2\theta \cdot \cos^2\theta}{\sin^2\theta} = \cos^2\theta \cdot \cot^2\theta$$

$$\cos^2\theta \cdot \frac{\cos^2\theta}{\sin^2\theta} = \cos^2\theta \cdot \cot^2\theta$$

$$\cos^2\theta \cdot \cot^2\theta = \cos^2\theta \cdot \cot^2\theta$$

Example 3

Prove: $\cos 2A + \sin 2A + 2\sin^2 A = (\sin A + \cos A)^2$

Solution: $\cos 2A + \sin 2A + 2\sin^2 A = (\sin A + \cos A)^2$

$$\cos^2 A - \sin^2 A + 2\sin A\cos A + 2\sin^2 A = (\sin A + \cos A)^2$$

$$\cos^2 A + 2\sin A\cos A + \sin^2 A = (\sin A + \cos A)^2$$

$$(\sin A + \cos A)^2 = (\sin A + \cos A)^2$$

Example 4

Prove that $\dfrac{2\sin^2\dfrac{\alpha}{2}}{\cos\alpha} = \sec\alpha - 1$.

Solution: $$\frac{2\sin^2\dfrac{\alpha}{2}}{\cos\alpha} = \sec\alpha - 1$$

$$\frac{2\left(\pm\sqrt{\dfrac{1-\cos\alpha}{2}}\right)^2}{\cos\alpha} = \sec\alpha - 1$$

$$\frac{2\left(\dfrac{1-\cos\alpha}{2}\right)}{\cos\alpha} = \sec\alpha - 1$$

$$\frac{1-\cos\alpha}{\cos\alpha} = \sec\alpha - 1$$

$$\frac{1}{\cos\alpha} - \frac{\cos\alpha}{\cos\alpha} = \sec\alpha - 1$$

$$\sec\alpha - 1 = \sec\alpha - 1$$

Check Your Understanding of Section 10.5

1. Prove: $\sec A(\cot A + 1) = \csc A + \sec A$

2. Prove: $\csc^4 x - 1 = \dfrac{\cos^2 x(1+\sin^2 x)}{\sin^4 x}$

3. Prove: $(\sin\theta + \cos\theta)^2 = 1 + \sin 2\theta$

4. Prove: $\dfrac{\tan x + \cot x}{2} = \csc 2x$

5. Prove: $\dfrac{\cos(\alpha - \beta)}{\sin\alpha\cos\beta} = \cot\alpha + \tan\beta$

6. Prove: $\dfrac{\sin 2x}{\cos 2x} = \dfrac{2\tan x}{1-\tan^2 x}$

10.6 TRIGONOMETRIC EQUATIONS

An equation that contains trigonometric functions of one angle can be solved for the measure of that angle.

Linear Trigonometric Equations

To solve a trigonometric equation, first solve for the trigonometric function and then use that to solve for the angle.

The following chart shows the solution of a linear equation compared to a very similar trigonometric equation.

Linear Equation	Linear Trigonometric Equation
Solve for x: $2x - 1 = 0$	Solve for θ if $0 \leq \theta \leq 2\pi$: $2\sin\theta - 1 = 0$
$2x - 1 = 0$ $2x = 1$ $x = \frac{1}{2}$	$2\sin\theta - 1 = 0$ $2\sin\theta = 1$ $\sin\theta = \frac{1}{2}$ Remember that $\sin\theta > 0$ in quadrants I and II $\theta = \frac{\pi}{6}$ and $\frac{5\pi}{6}$

Tip

Note that it is sometimes easier to solve a trigonometric equation by substituting a single variable for the trigonometric function, solving for that variable, and finally solving for the inverse trigonometric function.

Example 1

Solve for A if $0° \leq A \leq 360°$ and $4\tan A + 5 = 1$.

Solution: If $4\tan A + 5 = 1$, let $a = \tan A$ and solve the resulting equation, $4a + 5 = 1$. So $4a = -4$ and $a = -1$. Now find $A = \tan^{-1}(-1)$. Since the tangent function is negative in the second and fourth quadrants and the reference angle for $\tan^{-1}(-1)$ is 45°, the solution is $A = 135°$ and $A = 315°$.

Quadratic Trigonometric Equations

If the quadratic equation is factorable, the Zero Product Property can be used to solve it. If not, then the quadratic equation can be solved by either completing the square or using the quadratic formula.

Example 2

Solve for θ if $0 \le \theta \le 2\pi$ and $\cos^2\theta - 2\cos\theta - 3 = 0$.

Solution: If $a = \cos\theta$, $\cos^2\theta - 2\cos\theta - 3 = 0$ becomes $a^2 - 2a - 3 = 0$.

This factors as $(a - 3)(a - 1) = 0$. By the Zero Product Property, $a - 3 = 0$ or $a + 1 = 0$. So $a = 3$ or $a = -1$, yielding $\cos\theta = 3$ or $\cos\theta = -1$. However, the range of the cosine function includes numbers from -1 to 1. So there is no solution for $\cos\theta = 3$. This means $\cos\theta = -1$ at $\theta = \pi$.

Example 3

Solve for θ if $0 \le \theta \le 2\pi$ and $4\sin^2\theta + \sin\theta - 2 = 0$.

Solution: Let $a = \sin\theta$ so that the equation becomes $4a^2 + a - 2 = 0$.

This is not factorable. Using the quadratic formula, $a = \dfrac{-1 \pm \sqrt{1^2 - 4 \cdot 4 \cdot -2}}{2 \cdot 4} = \dfrac{-1 \pm \sqrt{33}}{8}$. Now find arcsine of a. This calculation using a graphing calculator is shown below.

```
(-1+√(1²-4*4*(-2
)))/8
          .5930703308
sin⁻¹(Ans)
          .6348668711
```

```
(-1-√(1²-4*4*(-2
)))/8
         -.8430703308
sin⁻¹(Ans)
         -1.002966954
```

When $\sin\theta \approx 0.5931$, $\theta \approx 0.63$ radians. Since $\sin\theta$ is positive in the first and second quadrants, $\theta \approx 3.14 - 0.63 = 2.51$ also. When $\sin\theta \approx -0.8431$, $\theta \approx -1.00$ radians. From 0 to 2π, $\sin\theta$ is negative in the third and fourth quadrants, yielding $\theta \approx -1 + 6.28 = 5.28$ in the fourth quadrant and $\theta \approx 3.14 + 1 = 4.14$ in the third quadrant.

Solving Trigonometric Equations Using Identities

When an equation involves more than one trigonometric function, use an identity to create an equivalent equation with only one trigonometric function.

Example 4

Solve for x if $0° \leq x \leq 360°$ and $\sin x + 1 = 2\cos^2 x$.

Solution: Replace $\cos^2 x$ with $1 - \sin^2 x$ so that $\sin x + 1 = 2\cos^2 x$ becomes $\sin x + 1 = 2(1 - \sin^2 x)$. This becomes $\sin x + 1 = 2 - 2\sin^2 x$ or $2\sin^2 x + \sin x - 1 = 0$. Let $a = \sin x$ so that $2a^2 + a - 1 = 0$ or $(2a - 1)(a + 1) = 0$. This yields $a = \frac{1}{2}$ and $a = -1$. If $\sin x = \frac{1}{2}$, $x = 30°$ and $x = 150°$. If $\sin x = -1$, $x = 270°$. The solution set is $\{30°, 150°, 270°\}$.

Example 5

Solve for all values x (in degrees) if $\sin x + 1 = 2\cos^2 x$.

Solution: The only difference between this example and Example 4 is that the answers are not limited to $0° \leq x \leq 360°$. That means that any multiple of 360° can be added to each of the answers because of the periodic nature of the sine function. This can be shown by stating the solution set as $\{30° + 360°k, 150° + 360°k, 270° + 360°k\}$ where k is any integer.

Check Your Understanding of Section 10.6

1. Solve for θ on $[0, 2\pi]$: $8 - 3\csc\theta = 2$

(1) $\frac{\pi}{6}$ (2) $\frac{\pi}{3}$ (3) $\frac{\pi}{6}$ and $\frac{5\pi}{6}$ (4) $\frac{\pi}{3}$ and $\frac{2\pi}{3}$

2. Which angle is a solution of $9\tan^2 A - 3 = 0$?

(1) $\frac{\pi}{6}$ (2) $\frac{\pi}{3}$ (3) $\frac{2\pi}{3}$ (4) $\frac{5\pi}{4}$

3. If θ is in the second quadrant and $\sec^2\theta + 3\sec\theta = 0$, what is the value of θ to the nearest degree?

(1) 161° (2) 160° (3) 110° (4) 109°

4. Solve for α if $0 \le \alpha \le 2\pi$ and $\sin^2\alpha + 3\cos\alpha + 3 = 0$.

(1) $\frac{\pi}{2}$ (2) π (3) $\frac{3\pi}{2}$ (4) $\frac{\pi}{6}$

5. Solve the equation $2\cos x \cdot \tan x = \sqrt{3}$ on $\left[-\frac{\pi}{2}, \frac{\pi}{2}\right]$.

(1) $\frac{\pi}{2}$ (2) $\frac{\pi}{6}$ (3) $-\pi$ (4) $\frac{\pi}{3}$

6. A solution set for the equation $\tan^2 B - 1 = 0$ includes all multiples of

(1) $\frac{\pi}{2}$ (2) $\frac{\pi}{3}$ (3) $\frac{\pi}{4}$ (4) $\frac{\pi}{6}$

7. Solve for all values of x in the interval $0° \le x \le 360°$ that satisfy each of the following equations:

a. $\cos x = \sec x$
b. $\sin x = \cos^2 x + 1$
c. $\sec^2 x - 2 = 0$
d. $\sin 2x - \sin x = 0$
e. $\cos 2x = \sin x$

8. Determine to the nearest tenth of a degree all values of x between 0° and 360° such that:

a. $2\tan^2 x + 5\tan x = 12$
b. $13 - 7\cos x = 12\sin^2 x$

9. The velocity at which a baseball is batted is 96 feet per second. If the baseball leaves the bat at angle θ to the horizon, the distance traveled is given by the formula $d = \frac{1}{32}(v_0)^2 \sin 2\theta$ where $v_0 = 96$. Determine the degree measure of angle θ to the nearest degree if the ball travels 275 feet.

Chapter Eleven

APPLICATIONS OF TRIGONOMETRY

11.1 AREA OF A TRIANGLE

The area of a triangle is equal to one-half the measure of its base times the measure of its height. How can the area of a triangle be determined if the measure of the base is given along with the size of an adjacent angle?

The Area of a Triangle

Recall that in a triangle with base b and height h, the formula for the area is $A = \frac{1}{2}bh$. However, the height of a triangle can be expressed in terms of the adjacent side of that triangle as indicated below.

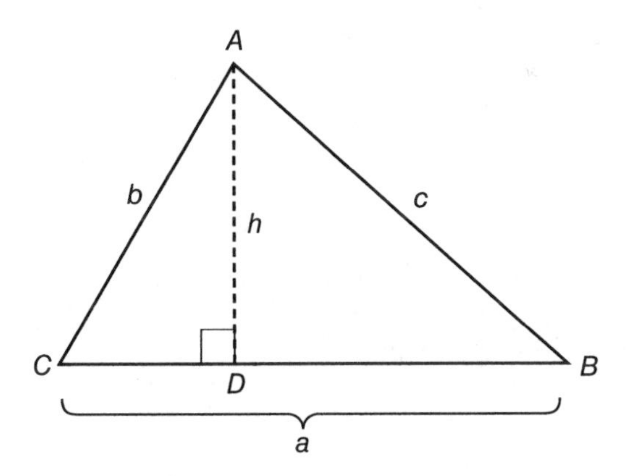

In triangle ABD, $\sin C = \frac{h}{b}$ or $h = b \cdot \sin C$. So the area of triangle ABC becomes $\frac{1}{2}ab\sin C$.

MATH FACTS

The formula for the area of a triangle is the product of two adjacent sides of the triangle times the sine of the included angle. So $A_{\Delta ABC} = \frac{1}{2}ab\sin C$ or $A_{\Delta ABC} = \frac{1}{2}bc\sin A$ or $A_{\Delta ABC} = \frac{1}{2}ac\sin B$.

Example 1

In ΔABC, if $a = 8$, $c = 5$, and $m\angle B = 72°$, determine the area of the triangle to the nearest thousandth.

Solution:

$$A_{\Delta ABC} = \frac{1}{2}ac\sin B = \frac{1}{2}(8)(5)(\sin 72) \approx 20(0.9511) = 19.022 \text{ square units}$$

Example 2

Determine the area (to the nearest thousandth) of the triangle pictured below.

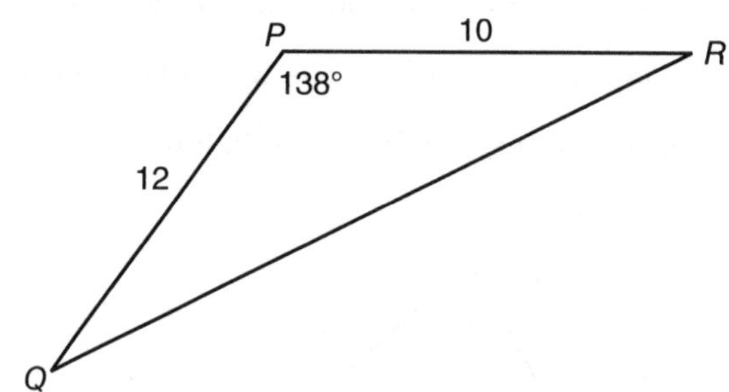

Solution: $A_{\Delta PQR} = \frac{1}{2}qr\sin P = \frac{1}{2}(12)(10)(\sin 138) \approx 60(0.6691)$
$= 40.148$ square units

Example 3

The adjacent sides of a parallelogram are 16 inches and 23 inches. If the angle between these two sides measures 60°, determine the *exact* area of this parallelogram.

Solution: Since a diagonal of a parallelogram separates the parallelogram into two congruent triangles, the area of the parallelogram is double the area of one of the congruent triangles. Therefore, the area of the parallelogram becomes $(16)(23)(\sin 60) = 368\sin 60 = 368 \cdot \frac{\sqrt{3}}{2} = 184\sqrt{3}$ square inches.

Check Your Understanding of Section 11.1

1. In ΔABC, if $a = 12$, $b = 10$, and $m\angle C = 45°$, determine the number of square units in the area of the triangle.
(1) $15\sqrt{2}$ (2) $30\sqrt{2}$ (3) 30 (4) $60\sqrt{2}$

2. The area of an obtuse triangle is 127.5 cm^2. If the lengths of the two shorter adjacent sides of the triangle are 17 cm and 30 cm, respectively, determine the measure of the angle included between these sides to the nearest degree.
(1) 30° (2) 60° (3) 120° (4) 150°

3. In ΔABC, if $a = 5$ inches, $b = 9$ inches, and $\cos C = \frac{-3}{5}$, determine the number of square inches in the area of the triangle.
(1) 13.5 (2) 18 (3) 27 (4) 36

4. The vertex angle of an isosceles triangle is 50° and the measure of each leg is 12 inches. Determine the area of the triangle to the nearest tenth of a square inch.
(1) 12 (2) 27.6 (3) 55.2 (4) 110.3

5. In triangle ABC, $m\angle A = 30°$, $m\angle C = 45°$, $AB = 12$ m, and $BC = 6\sqrt{2}$ m. Determine the area of the triangle to the nearest one-hundredth of a square meter.
(1) 31.18 (2) 16.39 (3) 43.46 (4) 49.18

6. Determine the exact area of each of the following triangles:

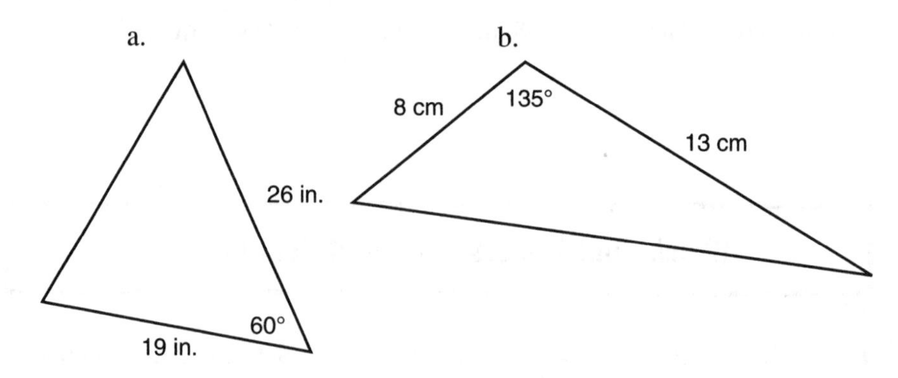

7. Hurricane Hal is approaching land. Its path is predicted to be limited to approximately a 28° range from its current position. Hurricane Hal is projected to travel somewhere through a triangular land region 120 miles long on one side of the 28° possible path and 140 miles long on the other side of the 28° possible path. What is the area of the region that may be affected by Hurricane Hal to the nearest tenth of a square mile?

8. A builder wants to invest in some property to be used to build new houses. The property is triangular. Two sides measure 1350 feet and 1210 feet. The angle between these two sides measures 137°. The builder estimates that he can subdivide the land into plots of approximately 5500 square feet. What is the maximum number of such plots the builder can create on this land?

9. Determine the area of a parallelogram to the nearest hundredth if it has sides of length 8 and 11 if the measure of an angle between these sides is 57°.

11.2 THE LAW OF SINES

The Law of Sines is a proportion involving two sides and their opposite angles that can be used to solve for one of these four measures.

The Area of a Triangle and the Law of Sines

The area of ΔABC has three different forms. Each determines the area of the same triangle, though. So $\frac{1}{2}ab\sin C = \frac{1}{2}bc\sin A$. Multiplying both sides of this equation by 2 and dividing by b yields $a\sin C = c\sin A$. If both sides are divided by $\sin A \cdot \sin C$, the resulting equation is the proportion $\frac{a}{\sin A} = \frac{c}{\sin C}$.

MATH FACTS

The Law of Sines:

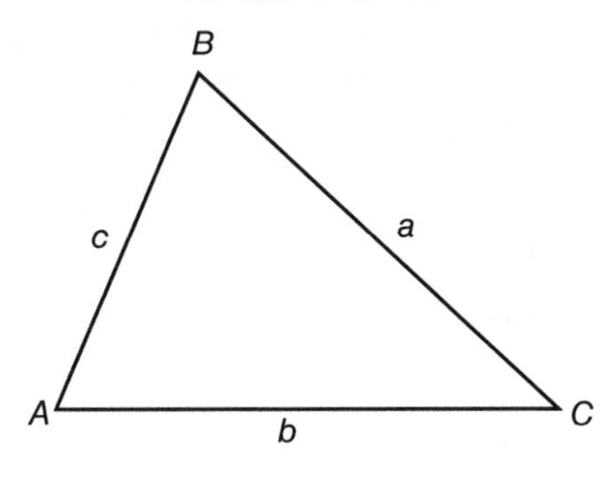

$$\text{In } \Delta ABC, \quad \frac{a}{\sin A} = \frac{b}{\sin B} = \frac{c}{\sin C}$$

Remember from geometry that there are several ways to prove triangles congruent. These can be summarized as SSS, SAS, ASA, and AAS. That means if three pairs of congruent parts are given, the other three parts of the triangle are fixed or determined. One might say that the triangle is rigid. The Law of Sines can be used to solve for the other three parts of a triangle when the measure of a side and its opposite angle are given along with either one other side or one other angle.

The Law of Sines in an AAS Situation

When the measures of two angles and a nonincluded side are given, the measures of the other three parts of the triangle can be determined by the Law of Sines.

Example 1

In ΔABC, $a = 6$, $\sin A = 0.23$, and $\sin C = 0.42$. Determine the length of side c to the nearest one-hundredth.

Solution: $\dfrac{a}{\sin A} = \dfrac{c}{\sin C}$

$$\frac{6}{0.23} = \frac{c}{0.42}$$

$$0.23c = 6(0.42)$$

$$0.23c = 2.52$$

$$c = 10.96$$

Example 2

In ΔABC, $b = 9$, $a = 7$, and $m\angle B = 58°$. Determine $m\angle A$ to the nearest degree. Then determine $m\angle C$ to the nearest degree and the length of side c to the nearest tenth. Finding these three measures is called solving the triangle.

Solution: $\dfrac{a}{\sin A} = \dfrac{b}{\sin B}$

$$\frac{7}{\sin A} = \frac{9}{\sin 58}$$

$$9 \sin A = 7 \sin 58$$

$$9 \sin A \approx 5.936336673$$

$$\sin A \approx 0.6595929637$$

$$A \approx 41.26883727 \approx 41°$$

Since $m\angle B = 58°$ and $m\angle A = 41°$, then $m\angle C = 180 - (58 + 41.26883727) = 80.73116273 \approx 81°$.

$$\frac{a}{\sin A} = \frac{c}{\sin C}$$

$$\frac{7}{\sin 41} = \frac{c}{\sin 80.73116273}$$

$$c \sin 41 = 7 \sin 80.73116273$$

$$c = \frac{7 \sin 80.73116273}{\sin 41} \approx 10.53046137 \approx 10.5$$

The Law of Sines in an ASA Situation

When the measures of two angles and the included side are given, the measure of the third angle of the triangle can be determined because the sum of the measures of the angles of a triangle is 180°. Once the measure of an angle and the length of its opposite side are known, the Law of Sines can be applied.

Example 3

In ΔABC, $b = 14$, $m\angle A = 56°$, and $m\angle C = 105°$. Solve the triangle.

Solution: $m\angle B = 180 - (56 + 105) = 19$

$$\frac{a}{\sin A} = \frac{b}{\sin B}$$

$$\frac{a}{\sin 56} = \frac{14}{\sin 19}$$

$$a \sin 19 = 14 \sin 56$$

$$a = \frac{14 \sin 56}{\sin 19} = 35.65006545 \approx 35.7$$

$$\frac{c}{\sin C} = \frac{b}{\sin B}$$

$$\frac{c}{\sin 105} = \frac{14}{\sin 19}$$

$$c = \frac{14 \sin 105}{\sin 19} = 41.53649976 \approx 41.5$$

The Law of Sines in an SSA Situation and the Ambiguous Case

Recall that SSA is not a valid reason for proving triangle congruence. However, when the lengths of two sides and the measure of a nonincluded angle are given, the Law of Sines can be used to solve for the other three parts of a triangle. This is because the measure of a side and its opposite angle are given along with one other side. Depending on whether the side opposite the given angle is less than, equal to, or greater than the length of the perpendicular segment from the vertex between the two given sides, either one, two, or no triangles will be formed. Therefore, this situation is called the ambiguous case.

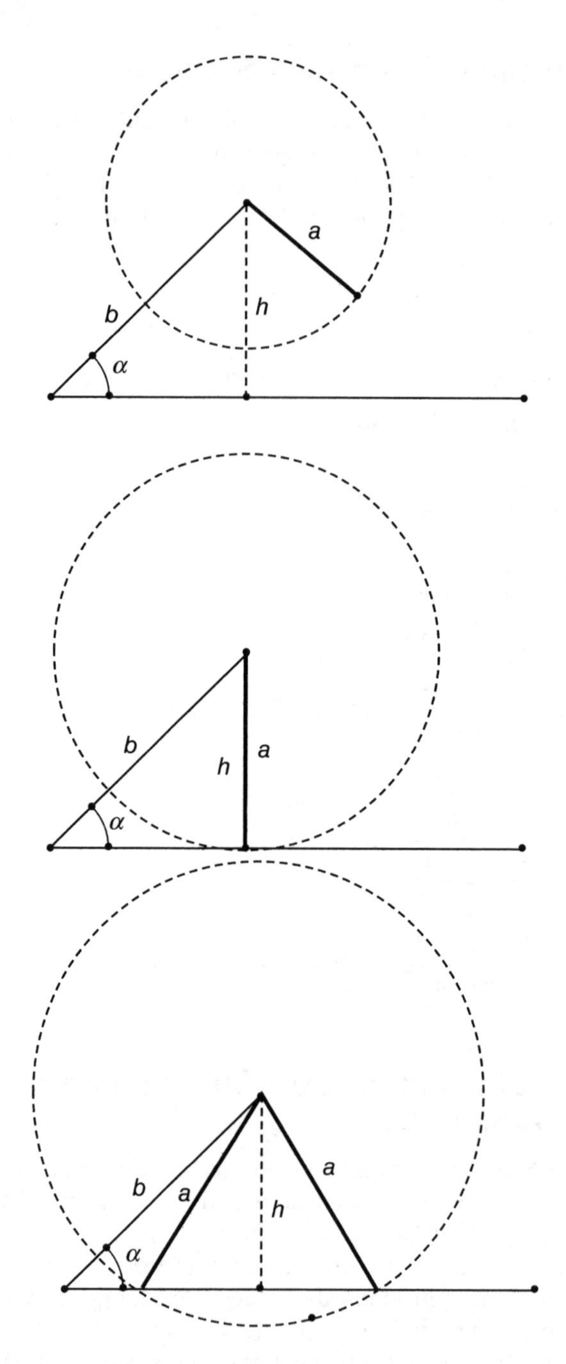

Each of the three previous diagrams shows a situation in which two sides and a nonincluded angle are given. In each of these, an arc of a circle is swung with the top vertex of the triangle as its center. This allows the length of the side opposite the angle α to be compared with the length of a perpendicular segment from that vertex, h. Notice that $h = b\sin\alpha$. In the first diagram, $a < h = b\sin\alpha$ illustrates why no triangle can be constructed. When $a = h = b\sin\alpha$, as in the second diagram, one right triangle can be formed. Finally, in the last diagram, when $a > h = b\sin\alpha$, the arc intersects the side opposite its center in two places. Therefore, two possible triangles can be constructed, one acute triangle and one obtuse triangle.

In each of the above situations, a is less than b. If $a \geq b$, then only one triangle can be formed.

MATH FACTS

The Ambiguous Case

In ΔABC, if the measures of two sides, a and b, and the measure of a nonincluded angle, $m\angle A$, are given, different numbers of possible triangles can be formed. This is summarized in the table below.

Sides a and b	Side a and $b\sin A$	Number of Triangles
$a < b$	$a < b\sin A = h$	No triangles
$a < b$	$a = b\sin A = h$	One right triangle
$a < b$	$a > b\sin A = h$	Two triangles
$a \geq b$		One triangle

Example 4

In ΔABC, $a = 6$, $b = 10$, $m\angle A = 42°$. How many triangles can be formed with these measures?

Solution: $h = b\sin A = 10\sin(42) \approx 6.69$. So $a < h$. Therefore, no triangle can be formed.

Example 5

In ΔABC, $c = 8$, $a = 16$, $m\angle C = 30°$. How many triangles can be formed with these measures? Solve the triangle.

Solution: $a\sin C = 16\sin(30) = 8$. Therefore, right triangle is formed. So $m\angle A = 90°$, $m\angle B = 60$, and $b = 8\sqrt{3}$.

Example 6

In ΔABC, $b = 26$, $a = 23$, $m\angle B = 72°$. How many triangles can be formed with these measures? Determine the measures of the other two angles to the nearest ten minutes and the length of the remaining side to the nearest hundredth.

Solution: Since $b > a$, one triangle can fit the given information. Now apply the Law of Sines.

$$\frac{a}{\sin A} = \frac{b}{\sin B}$$

$$\frac{23}{\sin A} = \frac{26}{\sin 72}$$

$$26 \sin A = 23 \sin 72$$

$$26 \sin A = 21.87429987$$

$$\sin A = 0.841319226$$

Now use the calculator to determine $m\angle A$.

```
23sin(72)
         21.87429987
Ans/26
          .841319226
sin⁻¹(Ans)
         57.27968993
█
```

Note that the question asks you to determine the measure of the angle to the nearest ten minutes. Since there are 60 minutes in each degree, subtract 57 from this last answer and multiply that by 60 to find the number of minutes past 57 for this angle. Alternatively, use **2ND** **APPS** **4** to convert the measure to degrees, minutes, and seconds.

```
          .841319226
sin⁻¹(Ans)
         57.27968993
Ans-57
          .279689926
Ans*60
         16.78139556
```

To the nearest ten minutes, $m\angle A = 57°20'$. In addition, $m\angle C = 180° - (72° + 57.27968993°) = 180° - 129.2796894° = 50.72031007° \approx 50°40'$. Now use the Law of Sines.

$$\frac{c}{\sin C} = \frac{b}{\sin B}$$

$$\frac{c}{\sin 50.72031007°} = \frac{26}{\sin 72}$$

$$c \sin 72 = 26 \sin 50.72031007°$$

$$c = \frac{26 \sin 50.72031007°}{\sin 72} = 21.16139404 \approx 21°20'$$

Example 7

In ΔABC, $c = 9$, $a = 11$, $m\angle C = 46°$. How many triangles can be formed with these measures? Determine the measures of the other two angles to the nearest ten minutes and the length of the remaining side to the nearest hundredth.

Solution: $c = 9 > a \sin C = 11 \sin 46 \approx 7.912737804$. Therefore, there are two solutions for this triangle.

$$\frac{a}{\sin A} = \frac{c}{\sin C}$$

$$\frac{9}{\sin 46} = \frac{11}{\sin C}$$

$$9 \sin C = 11 \sin 46$$

$$\sin C = \frac{11 \sin 46}{9} = 0.8791930893 \approx 0.8792$$

Using $\sin^{-1}(0.8791930893)$, $m\angle C = 61.54517895° \approx 61°30'$. In the second quadrant, $m\angle C = 118.4548211° \approx 118°30'$. Therefore, $m\angle B = 180° - (46° + 61.54517895°) = 72.45482105° \approx 72°30'$ and $m\angle B = 180° - (46° + 118.4548211°) = 15.54517895° \approx 15°30'$.

$$\frac{a}{\sin A} = \frac{b}{\sin B}$$
$$\frac{9}{\sin 46} = \frac{b}{\sin 72.45482105°}$$
$$b = \frac{9 \sin 72.45482105°}{\sin 46}$$
$$b = 11.92943289 \approx 11.93$$

$$\frac{a}{\sin A} = \frac{b}{\sin B}$$
$$\frac{9}{\sin 46} = \frac{b}{\sin 15.54517895°}$$
$$b = \frac{9 \sin 15.54517895°}{\sin 46}$$
$$b = 3.353051262 \approx 3.34$$

Multiple Triangle Problems

Sometimes it is necessary to solve for a side or an angle from one triangle to determine a part needed to solve a second triangle.

Example 8

Solve for MR to the nearest tenth if $LT = 9$, $m\angle T = 58°$, $m\angle R = 35°$, and $m\angle TML = 64°$.

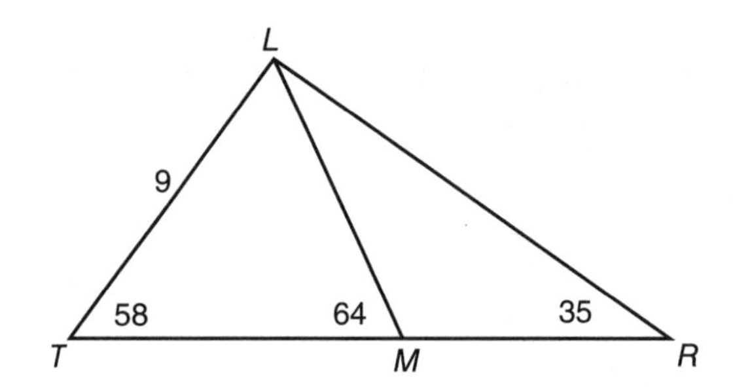

Solution: $$\frac{LT}{\sin \angle TML} = \frac{LM}{\sin \angle T}$$
$$\frac{9}{\sin 64} = \frac{LM}{\sin 58}$$
$$LM = \frac{9 \sin 58}{\sin 64} = 8.491859617 \approx 8.49$$

If $m\angle TML = 64°$, then $m\angle LMR = 116°$ because these two angles are supplementary. By triangle angle sum, $m\angle MLR = 29°$.

$$\frac{LM}{\sin \angle R} = \frac{MR}{\sin \angle MLR}$$

$$\frac{8.491859617}{\sin 35} = \frac{MR}{\sin 29}$$

$$LR = \frac{8.491859617 \sin 29}{\sin 35} = 7.177657545 \approx 7.2$$

Check Your Understanding of Section 11.2

1. In ΔABC, $c = 17$, $b = 19$, and $m\angle C = 53°$. Determine how many triangles can be constructed with these measurements.
(1) 0 (2) 1 (3) 2 (4) infinite

2. In ΔABC, $c = 16$, $a = 14$, and $m\angle C = 47°$. Determine $\sin A$ to the nearest ten-thousandth.
(1) 0.6399 (2) 0.6400 (3) 0.8358 (4) 39.7869

3. In ΔABC, $\sin A = \frac{3}{5}$ and $\sin C = \frac{2}{3}$. Determine the ratio $\frac{a}{c}$.
(1) $\frac{2}{5}$ (2) $\frac{5}{2}$ (3) $\frac{10}{9}$ (4) $\frac{9}{10}$

4. A tree is slanting at a 72° angle to the ground. When the sun is in back of the tree, the tree casts a shadow 18 feet long when the angle of depression from the sun is 62°. Determine, to the nearest tenth of a foot, the length of the tree.
(1) 23.8 feet (2) 19.4 feet (3) 22.1 feet (4) 14.7 feet

5. In ΔABC, $m\angle A = 51°$, $b = 10$, and $a = 8$. Determine $\sin \angle B$ to the nearest ten-thousandth.
(1) 0.6217 (2) 0.9714 (3) 0.7771 (4) 0.7627

6. Sara is working to solve for the measure of angle C in ΔABC. She knows that $m\angle A = 32°$, $a = 2$, and $c = 5$. She uses the Law of Sines and finds that $\sin \angle C = 2.6496$. How can that be since she knows that the sine function ranges from -1 to $+1$?

7. A huge mound of dirt is near a construction site. Andrew wants to find out how high the mound is. He measures the angle to the top of the mound to be 57°. Then he walks 8 feet away from the base of the mound and measures the angle to the top of the mound again to be 35°. To the nearest tenth of a foot, what is the height of the mound?

8. In ΔPQR, $p = 16$, $q = 13$, and $m\angle Q = 41°$. Solve for the other two angles of the triangle to the nearest ten minutes and the third side of the triangle to the nearest tenth.

9. A volleyball is hit by a player over the net toward a player on the opposing team. The player who hit the ball wants to estimate how high her ball went. She estimates the angle of elevation to the ball at its highest point to be 65°. The player it was hit toward estimates the angle of elevation from her perspective to be 45°. If the players were about 35 feet apart from each other at the time the ball reached its maximum height, determine the height of the ball to the nearest foot.

10. In parallelogram $HIJK$, side $HK = 12$, diagonal $HJ = 18$, and $m\angle HKJ = 125°$.
a. Determine $m\angle KJI$ to the nearest ten minutes.
b. Determine the length of side KJ to the nearest foot.

11. Two landmarks, a marina and a fishing pier, along the seashore are 20 miles apart. An observer is on an island that is across the shore at a point somewhere between the two landmarks but closer to the fishing pier. The observer measures the angle between the path of shortest distance from the island to the pier and the island to the marina to be 96°. His friend on the pier measures the angle between the path from the marina to the island and the path along the shoreline to the fishing pier to be 27°.
a. Determine the length of the path from the island to the marina to the nearest tenth of a mile.
b. Determine the length of the path from the island to the pier to the nearest tenth of a mile.

11.3 THE LAW OF COSINES

The Law of Cosines is another formula used to solve for missing sides or angles of a triangle. It is used in SAS and SSS situations. The formulas for the Law of Cosines begin in a similar format to the Pythagorean Theorem.

Three Versions of the Law of Cosines

MATH FACTS

The Law of Cosines

Similar to the Law of Sines, the Law of Cosines is used when three of the six parts of a triangle are given.

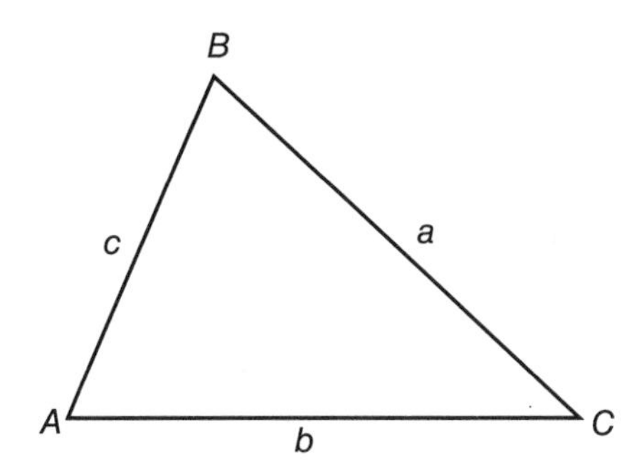

In ΔABC, any of the following three forms of the Law of Cosines are valid:

$$c^2 = a^2 + b^2 - 2ab\cos C$$
$$a^2 = b^2 + c^2 - 2bc\cos A$$
$$b^2 = a^2 + c^2 - 2ac\cos B$$

Notice that the first statement of the Law of Cosines starts with $c^2 = a^2 + b^2$, which is the Pythagorean Theorem. Since $\cos 90° = 0$, the Law of Cosines yields the Pythagorean Theorem when angle C measures 90°.

Using the Law of Cosines in an SAS Situation

Example 1

In ΔABC, $c = 5$, $a = 7$, and $m\angle B = 52°$. Determine the measures of the other two angles to the nearest tenth of a degree and the length of the remaining side to the nearest hundredth.

Solution: $b^2 = a^2 + c^2 - 2ac\cos B$

$$b^2 = 7^2 + 5^2 - 2\cdot 5\cdot 7\cos 52 = 49 + 25 - 70(0.6155514753)$$

$$= 30.90369673$$

$$b = 5.559109347 \approx 5.56$$

$$\frac{b}{\sin B} = \frac{a}{\sin A}$$

$$\frac{5.559109347}{\sin 52} = \frac{7}{\sin A}$$

$$5.559109347\sin A = 7\sin 52$$

$$\sin A = \frac{7\sin 52}{5.559109347} = 0.9922588191 \approx 0.99$$

$$A \approx 82.9°$$

$$C \approx 180° - (82.9° + 52°) = 45.1°$$

Using the Law of Cosines in an SSS Situation

Example 2

In ΔABC, $a = 13$, $b = 16$, and $c = 15$. Determine the measures of the three angles of the triangle to the nearest minute.

Solution: $c^2 = a^2 + b^2 - 2ab\cos C$

$$15^2 = 13^2 + 16^2 - 2\cdot 13\cdot 16\cdot \cos C$$

$$225 = 169 + 256 - 416\cos C$$

$$225 = 425 - 416\cos C$$

$$-200 = -416\cos C$$

$$\cos C \approx 0.4807692308$$

$$C = 61.26434626 \approx 61°16'$$

$$\frac{c}{\sin C} = \frac{a}{\sin A}$$

$$\frac{15}{\sin 61°16'} = \frac{13}{\sin A}$$

$$15\sin A = 13\sin 61°16'$$

$$15\sin A \approx 11.39901312$$

$$\sin A \approx 0.7599342077$$

$$A = 49.45839813 \approx 49°28'$$

$$\frac{c}{\sin C} = \frac{b}{\sin B}$$

$$\frac{15}{\sin 61°16'} = \frac{16}{\sin B}$$

$$15\sin B = 16\sin 61°16'$$

$$15\sin B \approx 14.0295546$$

$$\sin B \approx 0.9353036402$$

$$B = 69.27725562 \approx 69°17'$$

Check Your Understanding of Section 11.3

1. In ΔABC, $a = 4$, $b = 7$, and $c = 9$. Determine $\cos B$ to the nearest ten-thousandth.
(1) 0.7340 (2) 0.2857 (3) 0.6667 (4) −0.2857

2. In ΔRST, $r = 7$, $s = 8$, and $t = 10$. Determine the measure of $\angle S$ to the nearest degree.
(1) 35° (2) 52° (3) 53° (4) 85°

3. $PQRS$ is a parallelogram. If $PQ = 14$, $QR = 11$, and $m\angle PQR = 112°$, determine the length of diagonal $\overline{PR}$ to the nearest tenth.
(1) 8.5 (2) 20.8 (3) 14.2 (4) 11.8

4. The lengths of the sides of a triangle are 8, 12, and 15. What is the measure of the smallest angle of the triangle to the nearest degree?
(1) 32° (2) 53° (3) 55° (4) 85°

5. In ΔDEF, $d = 2$, $e = 5$, and $m\angle F = 48°$. Which of the following statements can be used to determine the numerical value of f?
(1) $5^2 = f^2 + 2^2 - 2 \cdot 2f \cos 48$ (3) $f^2 = 5^2 + 2^2 - 2 \cdot 5f \cos 48$
(2) $2^2 = f^2 + 5^2 - 2 \cdot 5f \cos 48$ (4) $f^2 = 5^2 + 2^2 - 2 \cdot 2 \cdot 5 \cos 48$

6. $ABCD$ is a parallelogram. $AB = 8$, $BC = 10$, and diagonal $AC = 13$.
a. Determine $m\angle ABC$ to the nearest minute.
b. Determine $m\angle BAD$ to the nearest minute.
c. Determine BD to the nearest tenth.

7. Dan wants to measure the width of the lake. He drew the picture below to illustrate how he walked away 70 feet from one end of the lake and measured an angle of 28° to the other end and walked to that end a distance of 90 feet. Show how he calculated the width of the lake. How wide is the lake to the nearest tenth of a foot?

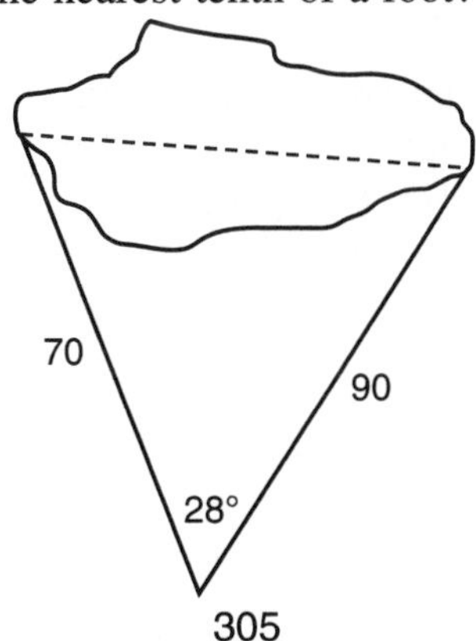

8. C and D are two points where C is on the initial side of θ, 4 units from the origin, and D lies on the terminal side of θ, 7 units from the origin. If $\theta = \frac{5\pi}{6}$, determine the length of $\overline{CD}$ to the nearest tenth of a unit.

9. Points A and B have coordinates $(2\cos\frac{\pi}{3},\ 2\sin\frac{\pi}{3})$ and $(5\cos\frac{2\pi}{3},\ 5\cos\frac{2\pi}{3})$ respectively. Point O is the origin.

a. Determine the length of $\overline{AB}$.

b. Determine the measure of the smallest angle of ΔAOB to the nearest hundredth.

10. In ΔABC, $c = 9$, $b = 6$, and $m\angle A = 63°$. Determine the measures of the other two angles to the nearest tenth of a degree and the length of the remaining side to the nearest hundredth.

11. Three friends, Mitch, Dwayne, and Jorge live in the same town. Dwayne lives 4.5 miles from Mitch. Mitch lives 6.6 miles from Jorge. If the angle between the path to Mitch and the path to Jorge is 74°, how far apart do Jorge and Mitch live from each other to the nearest tenth of a mile?

12. Dina is making a scale drawing of triangular garden in order to create a landscape design. Since the sides of the triangular garden are in ratio $7:10:12$, she is drawing sides of 7 inches and 10 inches. What size angle (to the nearest degree) should she draw between the these two sides?

11.4 APPLICATION TO FORCE PARALLELOGRAMS

Two forces acting on an object can be represented by vectors. The resultant force is the diagonal of a parallelogram whose sides are formed by and are parallel to these vectors.

Forces as Vectors

Example 1

One force acts on an object with a magnitude of 12 and a second force acts on the same object with a magnitude of 7. Represent these forces if the angle between them is 50°. Complete a parallelogram and draw in the resultant force.

Solution:

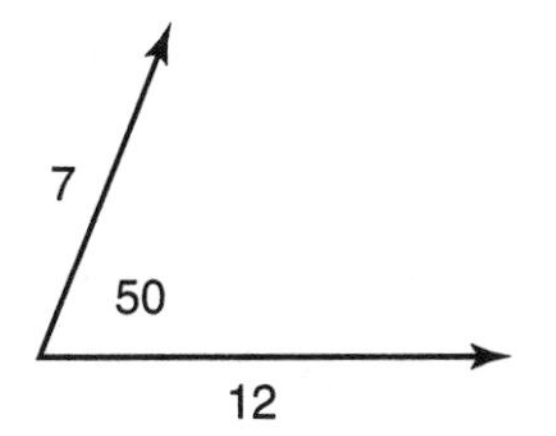

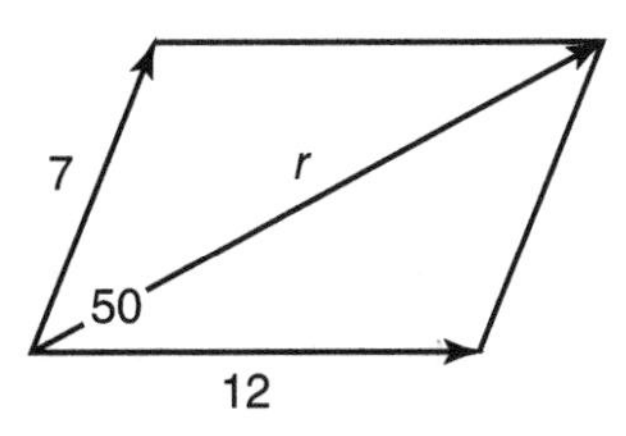

Example 2

Solve for the magnitude and direction of the resultant force in Example 1.

Solution: Since the consecutive angles of a parallelogram are supplementary and the opposite sides of a parallelogram are congruent, one of the triangles in the above diagram can be used to solve for the resultant force.

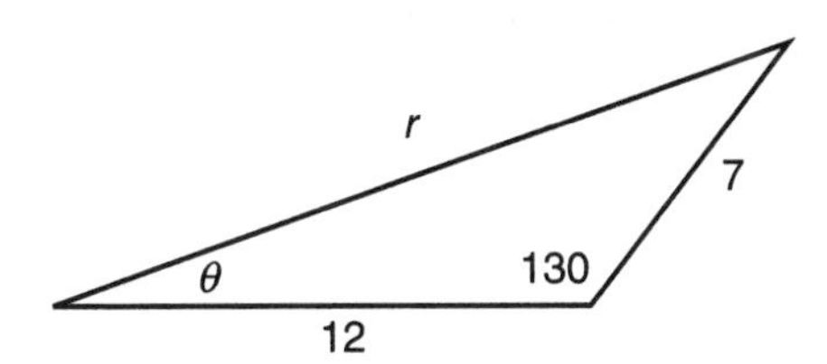

Since this is a SAS situation, use the Law of Cosines to solve for the magnitude of the resultant force as follows:

$$r^2 = 12^2 + 7^2 - 2 \cdot 12 \cdot 7 \cdot \cos 130 = 144 + 49 - 168 \cos 130 = 300.9883184$$
$$r \approx 17.35$$

To solve for the direction, solve for θ by using the Law of Sines as follows:

$$\frac{7}{\sin\theta} = \frac{17.35}{\sin 130}$$

$$17.35 \sin\theta = 7 \sin 130$$

$$\sin\theta = \frac{7\sin 130}{17.3490844713}$$

$$\sin\theta = 0.3090844713$$

$$\theta \approx 18°$$

Check Your Understanding of Section 11.4

1. Two forces acting on the same object have magnitudes of 32 and 41. If their resultant force has a magnitude of 49, determine the angle between these forces to the nearest degree.

2. A balloon is rising vertically at the rate of 9 feet per second. At the same time, the wind is blowing the balloon horizontally at 12 feet per second. Determine its actual velocity and the angle its path is making to the horizon to the nearest minute.

Chapter Twelve

PROBABILITY AND STATISTICS

12.1 THEORETICAL AND EMPIRICAL PROBABILITY

Theoretical probability is based on what is expected to happen in an experiment. In contrast, empirical probability is based on observed phenomena.

Theoretical Probability

To understand how to calculate a probability, three terms must first be defined.

- An outcome is one of the possible occurrences in a single trial in a probability experiment.
- The sample space is the set of all possible outcomes.
- An event is any subset of the sample space.

MATH FACTS

The theoretical probability of an event E equals:

$$\frac{\text{The number of outcomes that satisfy event } E}{\text{The total number of outcomes in the sample space}}$$

This can be written as:

$$P(E) = \frac{n(E)}{n(s)}$$

There are few extra rules that relate to the probability of an event.

MATH FACTS

1. The probability of an event ranges from 0 to 1 inclusive, i.e., $0 \leq P(E) \leq 1$.
2. The probability of an impossible event is 0.
3. The probability of an event that must happen is 1.
4. The sum of the probabilities of all the outcomes in a sample space is 1.
5. The complement of an event is the set of outcomes in the sample space that are not in the event. If an event is E, then its complement is $\bar{E}$ and $P(\bar{E}) = 1 - P(E)$.
6. $P(A \cup B) = P(A) + P(B) - P(A \cap B)$.
7. $P(A \cap B) = P(A) \cdot P(B \backslash A)$, where $P(B \backslash A)$ means the probability of B given A.

Example 1

If a coin is tossed 3 times, determine the probability that it lands on heads all three times.

Solution: The sample space is $S = \{TTT, TTH, THT, THH, HTT, HTH, HHT, HHH\}$. One of the 8 outcomes satisfies the condition that all 3 tosses come up heads. So $P(3 \text{ heads}) = \frac{1}{8}$.

Example 2

A single die is tossed once. Determine the probability that the die lands on an even number.

Solution: The sample space is $\{1, 2, 3, 4, 5, 6\}$. Three of these numbers are even. Therefore, $P(E) = \frac{n(E)}{n(S)} = \frac{3}{6} = \frac{1}{2}$.

Example 3

A card is selected from an ordinary deck. Determine the probability that the card selected at random is a king or a club.

Solution: The sample space contains 52 cards of which there are 4 kings and 13 clubs. However, one of the kings is also a club. $P(A \cup B) = P(A) + P(B) - P(A \cap B)$.

$$P(K \cup C) = P(K) + P(C) - P(K \cap C) = \frac{4}{52} + \frac{13}{52} - \frac{1}{52} = \frac{16}{52} = \frac{4}{13}$$

Example 4

A card is selected from an ordinary deck. Determine the probability that the card selected at random is the king of clubs.

Solution: The sample space contains 52 cards of which there are four kings and 13 clubs.

$$P(A \cap B) = P(A) \cdot P(B \backslash A)$$

$$P(\text{king of clubs}) = P(K) \cdot P(K \setminus C) = \frac{4}{52} \cdot \frac{1}{4} = \frac{1}{52}$$

Wasn't that obvious since there is only one card in the deck that is the king of clubs?

Example 5

The diagram shows a square whose side is 4. Inside the square is a triangle that shares the base of side 4 with the square and whose height is equal to a side of the square. What is the probability that a point selected inside the square is inside the triangle?

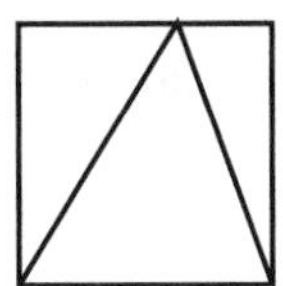

Solution: $P = \dfrac{\text{Area of the triangle}}{\text{Area of the square}} = \dfrac{\frac{1}{2} 4 \cdot 4}{4^2} = \dfrac{1}{2}$

MATH FACTS

The empirical probability of an event E is based on observed frequencies.

$$P(E) = \frac{\text{The frequency of the event } E}{\text{The total frequencies of all events}} = \frac{f}{n}$$

Example 6

A poll was taken in a high school to vote whether or not the legal driving age in New York State should be reduced to age 16. The results were 312 in favor of reducing the driving age to 16, 428 opposed to reducing the driving age to 16, and 33 with no preference either way. What is the probability that a student in this school was in favor of reducing the driving age to 16?

Solution: $P(\text{in favor}) = \dfrac{312}{773}$

Example 7

In a small high school, the adult staff were classified as follows:

Adult Staff			
	Male	**Female**	**Row Total**
Teachers	42	58	100
Civil service	16	24	40
Column total	58	82	140

Determine the probability that an adult staff member selected at random is:

a. A male
b. A teacher
c. A female civil service worker
d. A female or a teacher

Solution: a. The total number of males is 58 and there are 140 adult staff members.

$$P = \frac{58}{140} = \frac{29}{70}$$

b. $P = \dfrac{100}{140} = \dfrac{5}{7}$ c. $P = \dfrac{24}{140} = \dfrac{6}{35}$

d. $P(F \cup T) = P(F) + P(T) - P(F \cap T)$

$$= \frac{82}{140} + \frac{100}{140} - \frac{58}{140} = \frac{124}{140} = \frac{31}{35}$$

Check Your Understanding of Section 12.1

1. Ten cards are placed face down on a table. Each card has a different number written on its front from 1 to 10. If a card is selected at random and turned over, what is the probability that the number on the card is a multiple of three?

(1) $\frac{1}{10}$ (2) $\frac{1}{3}$ (3) $\frac{3}{10}$ (4) $\frac{1}{2}$

2. Marc has 7 unmatched socks in his draw—1 black sock, 2 white socks, and 4 blue socks. In the dark, he picks out one sock. What is the probability that that sock is white?

(1) $\frac{1}{7}$ (2) $\frac{2}{7}$ (3) $\frac{1}{3}$ (4) $\frac{2}{3}$

3. Triangle *ABC* is an equilateral triangle. Points *D*, *E*, *F*, *G*, *H*, and *I* are trisection points for the sides of the triangle.

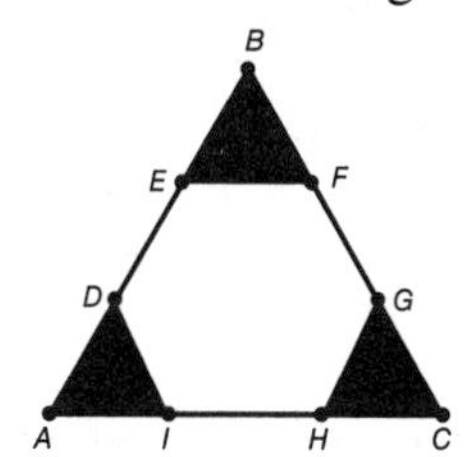

Determine the probability that a point inside triangle *ABC* is in one of the shaded regions.

(1) $\frac{3}{10}$ (2) $\frac{3}{5}$ (3) $\frac{1}{3}$ (4) $\frac{1}{4}$

4. The spinner is spun once. (Landing on a line does not count and the spinner is spun again.)

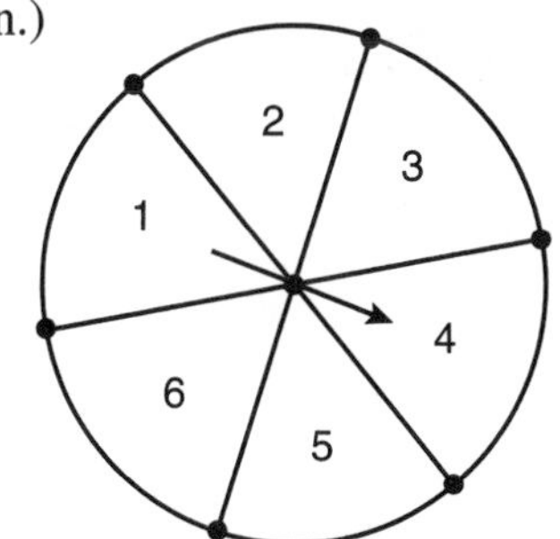

What is the probability that the spinner lands on a number that is both prime and less than 4?

5. The chart below represents data taken from a survey of people about their preferences about purchasing a new car.

Automobile Purchasing Preference				
	American	**European**	**Asian**	**Row Total**
Male	58	112	46	216
Female	27	76	133	236
Column total	85	188	179	452

Based on the data in the table, determine:

a. The probability that a person prefers an American-made automobile
b. The probability that the person taking the survey was female
c. The probability that a male prefers an Asian car
d. The probability that the person taking the survey was male given that he preferred a European automobile
e. The probability that the person taking the survey was either female or prefers an Asian automobile

6. Choose a month of the year at random.

a. What is the probability that the month selected begins with a "J"?
b. What is the probability that the month selected does not begin with a "J"?

7. If one card is drawn from a standard deck of cards, determine the probabilities of each of the following events:

a. The card selected is a jack.
b. The card selected is a heart.
c. The card selected is the jack of hearts.
d. The card selected is 5 or a spade.
e. The card selected is a black 7.

12.2 PERMUTATIONS

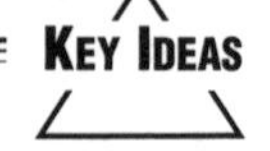

A permutation is used to determine the number of possible arrangements of n objects selected r at a time when the order is important.

The Fundamental Principle of Counting

If event A occurs in r number of ways and event B occurs in s number of ways, then the number of ways that both independent events A and B occur is $r \cdot s$.

Example 1

Maryanne has 3 blouses and 5 pairs of pants in her closet. How many outfits can she make with these clothes (assuming all blouses and pants will match with each other)?

Solution: The number of outfits will be $3 \cdot 5 = 15$.

Example 2

Shaun wants to make a sandwich. He has 3 types of bread to choose from, white, rye, and sourdough, and 2 types of deli meats, ham and turkey. He can spread either mustard or ketchup on the bread. How many different sandwiches can he make?

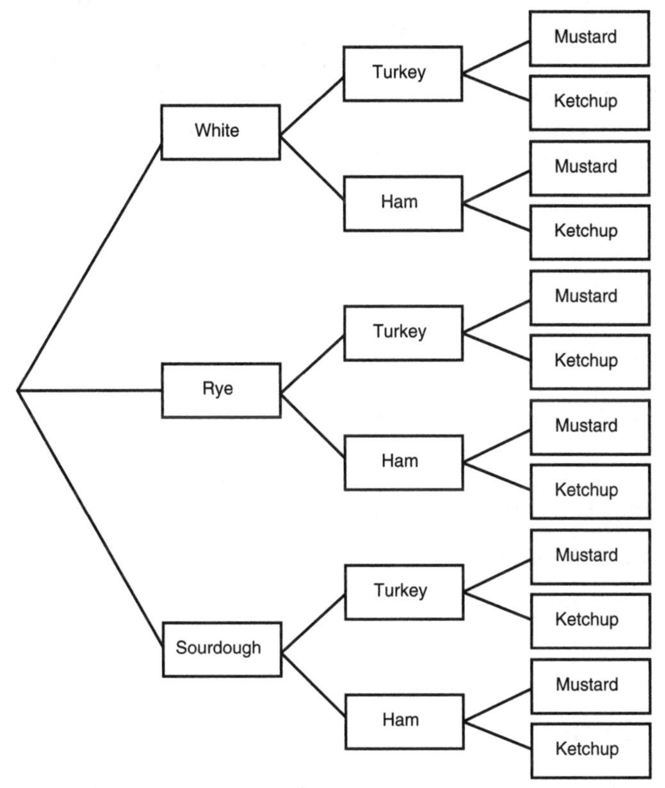

Shaun can make 12 different sandwiches.

Factorial Notation

An exclamation point used after a number indicates the factorial for that number.

MATH FACTS

Factorial Notation

For any counting number n:

$$n! = (n-1)(n-2)\ldots 2\cdot 1$$

$$0! = 1$$

$$1! = 1$$

Example 3

In how many ways can 4 books be arranged on a shelf?

Solution: Each book takes a different position on the bookshelf. There are 4 choices of books to take the first position on the shelf. After the first book is chosen, there are only 3 books to place in the other positions. After the second book is chosen, only 2 books are left for the last two positions on the shelf. After choosing the third book, there is only 1 book left for the last position on the shelf. Therefore, using the Fundamental Principle of Counting, there are $4 \cdot 3 \cdot 2 \cdot 1 = 24$ possible arrangements for the 4 books. Notice that $4! = 4 \cdot 3 \cdot 2 \cdot 1$ yields the same result.

Example 4

Using the integers 1, 2, 4, 5, 7, and 9,

a. How many 3-digit numbers can be formed?
b. How many 3-digit numbers can be formed if no repetition of digits is allowed?
c. How many 3-digit numbers greater than 500 can be formed if no repetition of digits is allowed?

Solution: a. $6 \cdot 6 \cdot 6 = 216$

b. $6 \cdot 5 \cdot 4 = 120$

c. $3 \cdot 5 \cdot 4 = 60$

Permutations

The number of ways of arranging n objects in order r at a time is called a permutation.

MATH FACTS

Permutation Notation

The symbol for the number of permutations of n objects taken r at a time is ${}_nP_r$.

$${}_nP_r = \frac{n!}{(n-r)!}$$

Example 5

Evaluate: a. ${}_7P_3$ b. ${}_5P_1$ c. ${}_{12}P_3$

Solution: a. $${}_7P_3 = \frac{7!}{(7-3)!} = \frac{7!}{4!} = \frac{7\cdot6\cdot5\cdot4\cdot3\cdot2\cdot1}{4\cdot3\cdot2\cdot1} = 7\cdot6\cdot5 = 210$$

b. $${}_5P_1 = \frac{5!}{(5-1)!} = \frac{5!}{4!} = \frac{5\cdot4\cdot3\cdot2\cdot1}{4\cdot3\cdot2\cdot1} = 5$$

c. $${}_{12}P_3 = \frac{12!}{(12-3)!} = \frac{12!}{9!} = \frac{12\cdot11\cdot10\cdot9\cdot8\cdot7\cdot6\cdot5\cdot4\cdot3\cdot2\cdot1}{9\cdot8\cdot7\cdot6\cdot5\cdot4\cdot3\cdot2\cdot1}$$
$$= 12\cdot11\cdot10 = 1320$$

There is a way to calculate ${}_{12}P_3$ on the calculator. First enter the number 12. Then press **MATH**. Scroll over to **PRB** and choose 2, **${}_nP_r$**, and then **ENTER** **3** **ENTER**. This is what it looks like on the calculator screen:

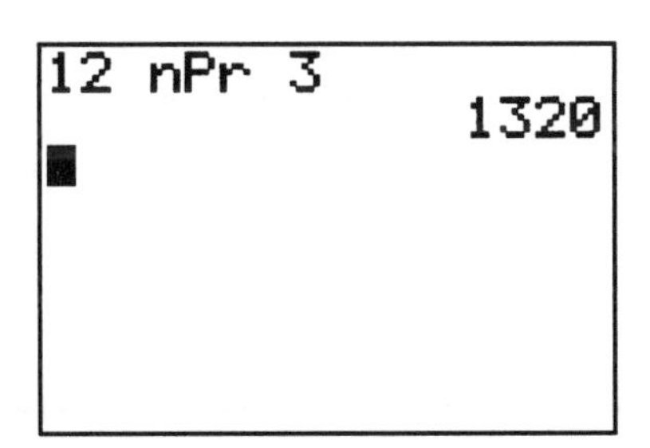

Example 6

The chess club decides to elect a slate of officers. First they will elect a president, then a vice president, and finally a treasurer for the club. If there are 10 members in the club, how many different groups of officers are possible?

Solution: The order in which these officers are elected is important because each elected office holds a different position. So, this is a permutation.

$$_{10}P_3 = \frac{10!}{7!} = 10 \cdot 9 \cdot 8 = 720$$

Example 7

How many 3-letter arrangements can be made using the letters in the word "SHELF"?

Solution: This is simply selecting 3 letters out of 5. $_5P_3 = \frac{5!}{2!} = 5 \cdot 4 \cdot 3 = 60$

Permutations with Repetition

Sometimes a permutation is used to determine the number of arrangements when some of the choices repeat or are indistinguishable. To do this, it is necessary to divide by the number of permutations of the repeated choices.

Example 8

How many 3-letter arrangements can be made using the letters in the word "EAGLE"?

Solution: This example is still selecting 3 out of 5 letters, but 2 of the 5 letters are the same letter, "E." One of the possible arrangements is EGA. It is not possible to distinguish between the EGA that uses the first E found in EAGLE and the EGA that uses the second E found in EAGLE. Therefore, it is necessary to divide by the number of arrangements of the two E's.

$$\frac{_5P_3}{_2P_2} = \frac{\frac{5!}{2!}}{\frac{2!}{0!}} = \frac{60}{2} = 30$$

Check Your Understanding of Section 12.2

1. Evaluate:

a. $_6P_4$ b. $_4P_1$ c. $_5P_4$ d. $_9P_6$

2. Using the integers 2, 3, 4, and 5,
   a. How many 3-digit numbers can be formed?
   b. How many 3-digit numbers can be formed if no repetition of digits is allowed?
   c. How many 3-digit numbers less than 300 can be formed if no repetition of digits is allowed?
   d. How many 3-digit odd numbers less than 300 can be formed if no repetition of digits is allowed?
   e. How many 3-digit numbers that are multiples of 5 can be formed if no repetition of digits is allowed?

3. Choose any 4 of the letters in the word "FORMULA." How many different 4-letter arrangements are possible if no letter is chosen more than once?
   (1) 42 (2) 210 (3) 840 (4) 2520

4. What is the value of $\frac{5!}{2!}$?
   (1) 60 (2) 20 (3) 6 (4) 3

5. Seven runners are racing against each other in a track meet. In how many ways can these runners be in first, second, and third place (assume that there cannot be any ties)?
   (1) 840 (2) 210 (3) 22 (4) 18

6. Three door prizes are being raffled off to the first 50 students who buy tickets to the senior play. The first prize is a free limousine to the Senior Prom, the second prize is a free ticket to the Senior Prom, and the third and last prize is $50 off the price of a gown or a tux for the Senior Prom. Each of the first 50 students is given one raffle ticket. How many possible arrangements are there for these raffle winners?

7. How many 5-letter arrangements can be made using the letters in the word "GEOLOGIST"?

8. A signal flag is made by putting 3 distinct colors on a flag of 3 vertical stripes. The flag maker has bolts of red, yellow, white, green, blue, brown, orange, and purple material that can be used for each stripe. How many different signals can the flag maker create with these colors?

9. A combination lock has 36 numbers on its dial. To open the lock, turn left to the first number, then turn right to a second different number, and then turn left again to a third number different from the first two numbers. How many possible sets of 3-number combinations are there for this lock?

10. A bingo card has 5 numbers in each column. The "B" column is filled with numbers ranging from 1 to 15. How many different orders of numbers from 1 to 15 can fill these 5 slots?

12.3 COMBINATIONS

A combination is used to determine the number of possible arrangements of *n* objects selected *r* at a time, when the order does not matter.

The Difference Between a Permutation and a Combination

When *n* objects are selected *r* at a time, the order in which they are selected may or may not matter. When the order matters, a permutation is used. When the order does not matter, a combination is used. For instance, there are 24 permutations when 3 letters are selected from the letters of the word "TIME." These 3-letter sequences are listed below:

TIM	ITM	MTI	ETI
TIE	ITE	MTE	ETM
TMI	IMT	MIT	EIT
TME	IME	MIE	EIM
TEM	IET	MET	EMT
TEI	IEM	MEI	EMI

Notice that TIM, TMI, ITM, IMT, MTI, and MIT consist of the same 3 letters. There are 6 ways to arrange 3 out of 3 items (in this case letters), ${}_3P_3 = \frac{3!}{0!} = \frac{3 \cdot 2 \cdot 1}{1} = 6.$

If the goal of selecting 3 out of the 4 letters was to determine how many combinations of 3 letters were selected, then the order in which they were selected would not matter. Therefore, the ${}_4P_3$ or 24 would have to be divided by the ${}_3P_3$ or by the number of ways the 3 letters selected can be arranged themselves. This would yield $\frac{24}{6} = 4$ sets of three letters, TIM, TIE, TME, and IME.

So a permutation is an arrangement of *n* items selected *r* at a time when the order of selection is important. If the order of selection is not important, then the arrangement is a combination. The chart that follows illustrates two classic examples of each type of arrangement.

Classic Example of a Permutation	Classic Example of a Combination
A group consisting of 40 members wants to select 4 of its members to be officers: president, vice president, secretary, and treasurer.	A group consisting of 40 members wants to select a committee of 4 of its members.
First elected will be president, second will be vice president, third will be secretary, and fourth will be treasurer.	All 4 committee members will have equal responsibility to the group and their order of selection makes no difference.

MATH FACTS

Combination Notation

The symbol for the number of combinations of n objects taken r at a time is ${}_nC_r$.

$${}_nC_r = \frac{n!}{r!(n-r)!}$$

Example 1

Evaluate: a. ${}_7C_3$ b. ${}_5C_1$ c. ${}_{12}C_3$ d. ${}_{12}C_9$

Solution: a. $${}_7C_3 = \frac{7!}{3!\cdot 4!} = \frac{7\cdot6\cdot5\cdot4\cdot3\cdot2\cdot1}{(3\cdot2\cdot1)\cdot(4\cdot3\cdot2\cdot1)} = \frac{7\cdot6\cdot5}{3\cdot2\cdot1} = 7\cdot5 = 35$$

Notice the cancellation of 4! in both the numerator and denominator of the fraction. In fact, either $r!$ or $(n - r)!$ always cancels, depending on which is larger.

b. $${}_5C_1 = \frac{5!}{1!\cdot 4!} = \frac{5}{1} = 5$$

Notice that this suggests that ${}_nC_1 = n$ for all values of n.

c. $${}_{12}C_3 = \frac{12!}{3!\cdot 9!} = \frac{12\cdot11\cdot10}{3\cdot2\cdot1} = 2\cdot11\cdot10 = 220$$

d. $${}_{12}C_9 = \frac{12!}{9!\cdot 3!} = \frac{12\cdot11\cdot10}{3\cdot2\cdot1} = 2\cdot11\cdot10 = 220$$

Notice that these last two examples suggest that ${}_nC_r = {}_nC_{n-r}$.

Combinations can be evaluated using the calculator. First enter the number 12. Then press MATH. Scroll over to PRB and choose 3, $_nC_r$, and then ENTER 3 ENTER. This is what it looks like on the calculator screen:

```
12 nCr 3
                220
```

MATH FACTS

Combination Computation Shortcuts

In order to calculate $_nC_r = \frac{n!}{r!(n-r)!}$, a few observable shortcuts make these calculations easy. For all positive integers, n and r where $r < n$:

$$_nC_n = 1$$
$$_nC_1 = n$$
$$_nC_r = {_nC_{n-r}}$$

Example 2

Calculate $_7C_3$ and use that to determine the value of $_7C_4$ and $_7P_4$.

Solution: $_7C_3 = \frac{7 \cdot 6 \cdot 5}{3 \cdot 2 \cdot 1} = 35$

$_7C_4 = 35$

$_7C_4 = \frac{_7P_4}{4!}$ so $_7P_4 = 4! \cdot {_7C_4} = 24 \cdot 35 = 840$

Example 3

A group consisting of 40 members wants to select a committee of 4 of its members from their group of 40 people.

$$\textit{Solution}: {}_{40}C_4 = \frac{\cancel{40}^{10} \cdot \cancel{39}^{13} \cdot \cancel{38}^{19} \cdot 37}{\cancel{4}_1 \cdot \cancel{3}_1 \cdot \cancel{2}_1 \cdot 1} = \frac{10 \cdot 13 \cdot 19 \cdot 37}{1} = 91{,}390.$$

Example 4

The astronomy club consists of 25 members. How many different sets of officers can be formed consisting of a president, vice president, secretary, and treasurer?

$$\textit{Solution}: {}_{25}P_4 = \frac{25!}{21!} = 25 \cdot 24 \cdot 23 \cdot 22 = 303{,}600$$

The order of selection is important since each officer has a different position.

Example 5

The astronomy club has 25 members, 11 girls and 14 boys.

The club wants to form a committee of 4 members to organize their annual Family Astronomy Night.

a. How many committees can be formed consisting of any of the 25 members?
b. How many committees can be formed consisting of only girls?
c. How many committees can be formed consisting of 2 girls and 2 boys?
d. What is the probability that a committee formed consists of 2 girls and 2 boys from the club?

Solution: a. The selection order is not important since each member of the committee has the same responsibility.

$${}_{25}C_4 = \frac{25!}{4! \cdot 21!} = \frac{25 \cdot \cancel{24}^{\cancel{6}^{2}} \cdot 23 \cdot \cancel{22}^{11}}{\cancel{4}_1 \cdot \cancel{3}_1 \cdot \cancel{2}_1 \cdot 1} = 25 \cdot 2 \cdot 23 \cdot 11 = 12{,}650$$

b. The selection order is not important since each member of the committee has the same responsibility. So use a combination selecting 4 out of the 11 girls and 0 out of the 14 boys.

$${}_{11}C_4 \cdot {}_{14}C_0 = \frac{11 \cdot 10 \cdot 9 \cdot 8}{4 \cdot 3 \cdot 2 \cdot 1} \cdot 1 = 11 \cdot 10 \cdot 3 = 330$$

c. Similarly

$${}_{11}C_2 \cdot {}_{14}C_2 = \frac{11 \cdot 10}{2 \cdot 1} \cdot \frac{14 \cdot 13}{2 \cdot 1} = 55 \cdot 7 \cdot 13 = 5{,}005$$

d. $P = \dfrac{{}_{11}C_2 \cdot {}_{14}C_2}{{}_{25}C_4} = \dfrac{5{,}005}{12{,}650} = \dfrac{91}{230}$

Example 6

In how many ways can 5 cards be selected from a standard deck of playing cards so that there are 3 kings and 2 queens?

Solution: Order is not important in this selection, so use combinations. Select 3 out of the 4 kings, 2 out of the 4 queens, and 0 out of the other 44 cards in the deck.

$$P = {}_4C_3 \cdot {}_4C_2 \cdot {}_{44}C_0 = 4 \cdot \frac{4 \cdot 3}{2 \cdot 1} \cdot 1 = 4 \cdot 6 = 24$$

Check Your Understanding Of Section 12.3

1. Evaluate:
a. ${}_7C_3$ b. ${}_6C_1$ c. ${}_{12}C_{10}$ d. ${}_9C_4$

2. There are 100 senators in the U.S. Congress. What formula can be used to determine how many ways a new committee of 8 members can be formed?
(1) ${}_{100}C_8$ (2) ${}_{100}P_8$ (3) ${}_{50}C_4 \cdot {}_{50}C_4$ (4) ${}_{50}P_4 \cdot {}_{50}P_4$

3. Your math teacher has a test bank of 20 multiple-choice questions for the topic on your next test. If she wants to use 10 multiple-choice questions from this test bank on the test, how many different sets of 10 multiple-choice questions can she select for the test?
(1) 15,504 (2) 3,628,800 (3) 184,756 (4) 10

4. There are 12 candidates for 4 of the same exact positions with the same exact salary for a new team in a company. How many different teams of 4 employees can be hired?
(1) 11,880 (2) 495 (3) 8 (4) 4

5. There are 20 possible jurors called in for jury selection. If the case requires 6 regular jurors, how many different possible juries can be selected?
(1) 125,970 (2) 924 (3) 2,790,200 (4) 38,760

6. A television station wants to create their new fall lineup of shows for Tuesday night. There are 7 half-hour shows and 5 full-hour shows that are possible for the Tuesday night lineup. From these 12 shows, it is decided that the station will use 2 half-hour and 2 full-hour shows for the Tuesday lineup. How many possible lineups can be created for Tuesday night?
(1) 495 (2) 210 (3) 21 (4) 10

7. From a standard deck of playing cards, a 5-card hand is selected.
   a. In how many ways can any 5 cards be selected from the deck?
   b. In how many ways can all 5 cards selected from the deck be hearts?
   c. In how many ways can 3 of the cards selected from the deck be hearts and 2 of the cards selected be diamonds?
   d. In how many ways can 3 of the cards selected from the deck be hearts and 2 of the cards selected not be hearts?
   e. What is the probability (rounded to the nearest hundredth) that 3 of cards selected from the deck be hearts and 2 of the cards selected not be hearts?

8. Your teacher randomly grades 5 problems from the homework assignment last night.
   a. If there were 14 problems in the assignment, how many possible sets of problems can be graded?
   b. Sonia completed only 8 of the assigned problems. If the teacher were grading only 5 problems from the 8 problems she completed, how many possible sets of problems can be graded?
   c. What is the probability that the set of 5 problems that the teacher uses to check this assignment include 5 problems that Sonia completed?

9. To calculate the number of a diagonals that can be drawn in a convex polygon, first determine the number of ways 2 vertices can be paired together and then subtract the number of sides of the polygon. Use this process to determine the number of diagonals in a convex polygon that has:
   a. Five sides
   b. Seven sides
   c. Ten sides

10. When expanding $(x + y)^8$, either an x or a y from each of the eight factors is used in the product. That means that x^3y^5 is going to occur several times and the number of times it occurs becomes the coefficient for x^3y^5. Based on this, what is the coefficient for x^3y^5?

12.4 BINOMIAL PROBABILITIES

When repeated independent trials of the same experiment are performed, the result in each trial can be thought of as a success or a failure. The probability of repeating the same outcome multiple times is called a binomial probability.

A Binomial Experiment

A binomial experiment has four categories that must be met.

- There must be a fixed number of trials for the experiment. This will be denoted by n.
- The n trials must be independent. That means the probability of an outcome in any trial remains the same.
- The outcomes in each trial can be classified as a success, S, or as a failure, F.
- For each individual trial, the probability of success $P(S) = p$ and the probability of failure $P(F) = q$. Additionally, $p + q = 1$.

For instance, Alan is practicing making foul shots in the gym. He takes 12 shots. Every time he shoots a basket, the probability of making the shot is 0.6. Therefore, there are 12 trials and on each trial $P(S) = 0.6$ and $P(F) = 0.4$. This is a binomial experiment.

MATH FACTS

Binomial Probability Formula

In a binomial experiment, if r is the number of successes out of n trials:

$$P(r) = {}_nC_r P^r q^{n-r}$$

Example 1

Alan is practicing making foul shots in the gym. He takes 12 shots. Every time he shoots a basket, the probability of making the shot is 0.6. What is the probability (to the nearest thousandth) that Alan makes 10 of the 12 foul shots?

Solution: $P(10) = {}_{12}C_{10}p^{10}q^2 = 66(0.6)^{10}(0.4)^2 \approx 0.064$

Example 2

Cara is a very good salesperson. The probability that she will make a sale whenever she sees a client is 0.7. Next week she has appointments to see 4 clients. Determine the probability that out of her 4 appointments, she makes

a. No sales
b. 1 sale
c. 2 sales
d. 3 sales
e. 4 sales

Solution: a. ${}_4C_0(0.7)^0(0.3)^4 = (0.0081) = 0.0081$

b. ${}_4C_1(0.7)^1(0.3)^3 = 4(0.7)(0.027) = 0.0756$

c. ${}_4C_2(0.7)^2(0.3)^2 = 6(0.49)(0.09) = 0.2646$

d. ${}_4C_3(0.7)^3(0.3)^1 = 4(0.343)(0.3) = 0.4116$

e. ${}_4C_4(0.7)^4(0.3)^0 = 1(0.2401)1 = 0.2401$

Notice that 0.0081 + 0.0756 + 0.2646 + 0.4116 + 0.2401 = 1.0. Also note that when using decimal approximations, the result may sometimes come out to a number very close to 1.

MATH FACTS

In a binomial experiment, the sum the probabilities from $r = 0$ to $r = n$, where r is the number of success out of n trials, is 1.

Example 3

A single die is tossed 4 times. The probability of tossing a 3 anytime the die is tossed is $\frac{1}{6}$. What is the probability (to the nearest thousandth) that the die lands on a 3 at least 3 times?

Solution: At least 3 tosses landing on a 3 out of 4 tosses means either 3 or 4 tosses result in a 3. The word "or" indicates addition. $P = {}_4C_3\left(\frac{1}{6}\right)^3\left(\frac{5}{6}\right)^1 + {}_4C_4\left(\frac{1}{6}\right)^4\left(\frac{5}{6}\right)^0 \approx 0.0154320988 + 0.0007716049383$
$= 0.0162037037 \approx 0.016$

Example 4

A manufacturer of lightbulbs knows that the probability that a lightbulb is defective is 0.01. Out of 30 lightbulbs sold, find the probability (to the nearest ten-thousandth) that no more than 3 are defective.

Solution: No more than 3 defective bulbs means either 3, 2, 1, or 0 bulbs are defective. $P = {}_{30}C_0(0.01)^0(0.99)^{30} + {}_{30}C_1(0.01)^1(0.99)^{29} + {}_{30}C_2(0.01)^2(0.99)^{28}$ $+ {}_{30}C_3(0.01)^3(0.99)^{27} \approx 0.7397003734 + 0.2241516283 + 0.032830289$ $+ 0.0030951114 = 0.9997774021 \approx 0.9998$

Check Your Understanding of Section 12.4

1. In a binomial experiment, the probability of success is 0.35 and the probability of failure is 0.65. Which of the following expressions represents the probability of 6 successes out of 9 trials?
   (1) ${}_9C_6(0.65)^6(0.35)^3$
   (2) ${}_9C_6(0.65)^5(0.35)^4$
   (3) ${}_9C_6(0.35)^6(0.65)^3$
   (4) ${}_9C_6(0.65)^4(0.35)^5$

2. The probability that Tom will win a game anytime he plays chess with Joe is 0.7. If they play 5 games, to the nearest hundredth, what is the probability that Joe wins 1 game?
   (1) 0.36 (2) 0.03 (3) 0.15 (4) 0.24

3. An urn contains 5 white balls and 3 red balls. One ball is selected from the urn, and its color is recorded before returning it back to the urn. This is repeated 7 times. Which expression represents the probability that a red ball is recorded no more than twice?

   (1) ${}_7C_2\left(\frac{3}{8}\right)^2\left(\frac{5}{8}\right)^5$

   (2) ${}_7C_0\left(\frac{3}{8}\right)^0\left(\frac{5}{8}\right)^7 + {}_7C_1\left(\frac{3}{8}\right)^1\left(\frac{5}{8}\right)^6$

   (3) ${}_7C_0\left(\frac{5}{8}\right)^7 + {}_7C_1\left(\frac{3}{8}\right)^1\left(\frac{5}{8}\right)^6 + {}_7C_2\left(\frac{3}{8}\right)^2\left(\frac{5}{8}\right)^5$

   (4) ${}_7C_0\left(\frac{1}{2}\right)^7 + {}_7C_1\left(\frac{1}{2}\right)^1\left(\frac{1}{2}\right)^6 + {}_7C_2\left(\frac{1}{2}\right)^2\left(\frac{1}{2}\right)^5$

4. The probability that it rains any day in October in New York City is 0.3. A traveler plans to visit New York City for 10 days. What is the probability (to the nearest hundredth) that it rains less than 2 days?
   (1) 0.03 (2) 0.15 (3) 0.00 (4) 0.38

5. A fair coin is tossed 7 times. What is the probability that it lands heads up more than 4 times?

6. The adjacent spinner is spun 8 times. If it is equally likely to land on any one section, determine the probability (to the nearest thousandth) that:
   a. It lands on a 1 at least 6 times.
   b. It lands on a 1 no more than 3 times.
   c. It lands on a 3 more than 5 times.

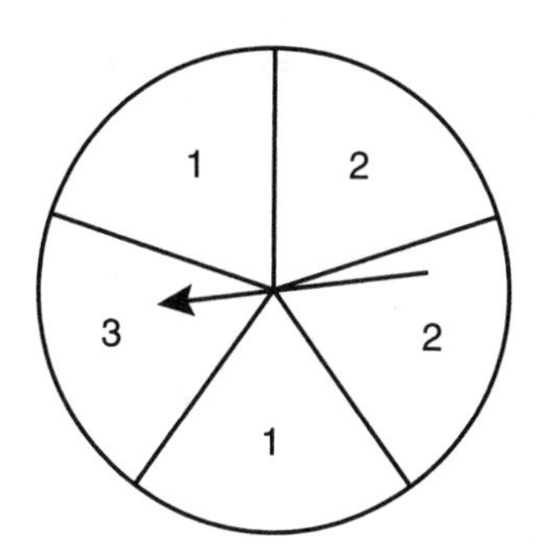

7. The adjacent spinner is spun 5 times. If it is equally likely to land on any one section, determine the exact probability that:
   a. It lands on a 7 more than twice.
   b. It lands on a number larger than 5 at least 4 times.
   c. It lands on a multiple of 3 exactly twice.

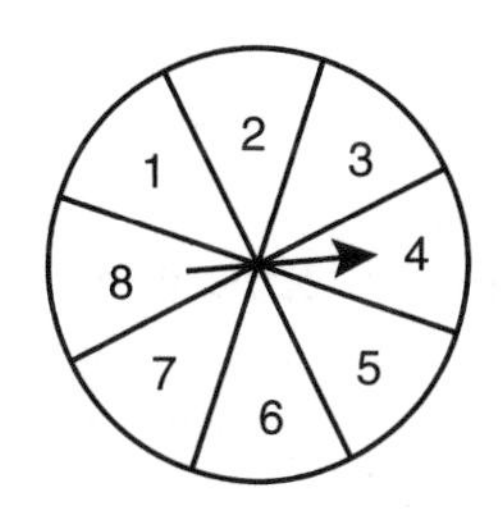

8. A woman is equally likely to give birth to a boy or a girl. If she has 3 children, what is the probability that the children consist of:
   a. All boys
   b. All girls
   c. At least 2 boys
   d. At most 2 girls

9. Whenever Gene goes to the local ice cream shop, the probability that he orders a vanilla ice cream cone is 0.7. If Gene goes to the ice cream shop 15 times over summer vacation, what is the probability (to the nearest ten-thousandth) that:
   a. He orders a vanilla ice cream cone at least 10 times?
   b. He orders a vanilla ice cream cone a majority of the time?

12.5 BINOMIAL EXPANSION

When a binomial is raised to an exponent, the result is a polynomial whose degree depends on the exponent the binomial is raised to as well as the degree of the binomial itself.

The Binomial Theorem

One way to expand $(x + y)^n$ would be to multiply out n factors of $(x + y)$. Here are some examples of this when $n = 1$, 2, and 3.

n	n **Factors Multiplied**	$(x + y)^n$
1	$(x + y)^1 =$	$x + y$
2	$(x + y)^2 = (x + y)(x + y) = x^2 + xy + xy + y^2 =$	$x^2 + 2xy + y^2$
3	$(x + y)^3 = (x + y)(x + y)(x + y) =$ $(x + y)^2(x + y) =$ $(x^2 + 2xy + y^2)(x + y) =$ $x^3 + 2x^2y + xy^2 + x^2y + 2xy^2 + y^3 =$	$x^3 + 3x^2y +$ $3xy^2 + y^3$

Notice that when $x + y$ is raised to the third power, the coefficient of x^2y is 3. That means that there were really three separate sets of the term x^2y in this product. If one were to consider taking either a factor of x or a factor of y from each parentheses of $(x + y)(x + y)(x + y)$ in such a way that an x is chosen from two of the parentheses and a y is chosen from one of the parentheses, then it becomes easy to see why the number of ways this can occur is a combination of successfully selecting a y one out of three times or ${}_3C_1$. In fact, the coefficients of each term of $(x + y)^3 = x^3 + 3x^2y + 3xy^2 + y^3$ can be expressed in terms of combinations as ${}_3C_0x^3 + {}_3C_1x^2\,y + {}_3C_2x\,y^2 + {}_3C_3y^3$.

MATH FACTS

The Binomial Theorem

For any positive integer n:

$$(x+y)^n = {}_nC_0x^n + {}_nC_1x^{n-1}y + {}_nC_2x^{n-2}y^2 + {}_nC_3x^{n-3}y^3 + \ldots + {}_nC_rx^{n-r}y^r + \ldots + {}_nC_{n-2}x^2y^{n-2} + {}_nC_{n-1}xy^{n-1} + {}_nC_ny^n$$

Example 1

Expand: $(a + b)^6$

Solution: $(a+b)^6 = {}_6C_0a^6 + {}_6C_1a^5b + {}_6C_2a^4b^2 + {}_6C_3a^3b^3 + {}_6C_4a^2b^4 + {}_6C_5ab^5 + {}_6C_6b^6 = a^6 + 6a^5b + 15a^4b^2 + 20a^3b^3 + 15a^2b^4 + 6ab^5 + b^6$

There is another way to expand a binomial. In 1655, the French mathematician Blaise Pascal assembled previous research into the relationship between the coefficients of the terms of a binomial after it has been expanded and the number of combinations of n things selected r at a time. Pascal's triangle is started by writing three 1's in a triangle. The numbers in each subsequent row are the sum of the two numbers diagonally above it on the left and the right. The first seven rows of Pascal's Triangle are shown below:

1

1 1

1 2 1

1 3 3 1

1 4 6 4 1

1 5 10 10 5 1

1 6 15 20 15 6 1

The second number in each row is the degree that the binomial is raised to, and each number in the row represents the coefficients of that binomial. In Example 1, the expansion of $(a + b)^6$ yielded $a^6 + 6a^5b + 15a^4b^2 + 20a^3b^3 + 15a^2b^4 + 6ab^5 + b^6$. Notice that these coefficients are the numbers in the seventh row in Pascal's triangle as illustrated above.

Example 2

Use Pascal's triangle to expand $(r + s)^4$.

Solution: The row associated with $(r + s)^4$is 1 4 6 4 1.

$$(r + s)^4 = r^4 + 4r^3s + 6r^2s^2 + 4rs^3 + s^4$$

Example 3

Expand: $(x + 2)^5$

Solution: In this problem, the binomial is $x + 2$, so y is the constant 2.

$$(x + 2)^5 = {}_5C_0x^5 + {}_5C_1x^4(2) + {}_5C_2x^3(2)^2 + {}_5C_3x^2(2)^3 + {}_5C_4x(2)^4 + {}_5C_5(2)^5$$
$$= 1x^5 + 5x^4(2) + 10x^3(4) + 10x^2(8) + 5x\,(16) + 1(32)$$
$$= x^5 + 10x^4 + 40x^3 + 80x^2 + 80x + 32$$

The next step in this process is to expand a binomial that has two variables with coefficients other than 1.

Example 4

Expand: $(3c + 4d)^5$

Solution: $(3c + 4d)^5 = {}_5C_0(3c)^5 + {}_5C_1(3c)^4(4d) + {}_5C_2(3c)^3(4d)^2 + {}_5C_3(3c)^2(4d)^3 + {}_5C_4(3c)(4d)^4 + {}_5C_5\,(4d)^5$

$$= 1 \cdot 243c^5 + 5 \cdot 81c^4 \cdot 4d + 10 \cdot 27\,c^3 16d^2 + 10 \cdot 9c^2 64d^3 + 5 \cdot 3c256d^4 + 1024d^5$$
$$= 243c^5 + 1620c^4d + 4320c^3d^2 + 5760c^2d^3 + 3840cd^4 + 1024d^5$$

All of the previous examples involved binomials that are the sum of two terms. If a binomial difference is expanded, then the coefficient of the second term can be treated as a negative number.

Example 5

Expand: $(2x - 3y)^4$

Solution: $(2x - 3y)^4$

$$= {}_4C_0(2x)^4 + {}_4C_1(2x)^3(-3y) + {}_4C_2(2x)^2(-3y)^2 + {}_4C_3(2x)^1(-3y)^3 + {}_4C_4(-3y)^4$$
$$= 1 \cdot 16x^4 + 4 \cdot 8x^3(-3y) + 6 \cdot 4x^2(9y^2) + 4 \cdot 2x(-27\,y^3) + 1 \cdot 81y^4$$
$$= 16x^4 - 96x^3y + 216x^2y^2 - 216xy^3 + 81y^4$$

r^{th} Term in a Binomial Expansion

In Example 1, it was determined that

$$(a + b)^6 = {}_6C_0a^6 + {}_6C_1a^5b + {}_6C_2a^4b^2 + {}_6C_3a^3b^3 + {}_6C_4a^2b^4 + {}_6C_5ab^5 + {}_6C_6b^6.$$

Notice that the third term of this expansion is ${}_6C_2a^4b^2$. This suggests that the exponent of the second term of the binomial is 1 less than the number of the term. Of course, that exponent is the r in the number of combinations of n items selected r at a time.

MATH FACTS

The rth Term in a Binomial Expansion

In the binomial expansion of $(x + y)^n$, the rth term in the expansion is equal to:

$${}_nC_{r-1}x^{n-(r-1)}y^{r-1}$$

Example 6

What is the fifth term in the expansion of $(x + y)^{11}$?

Solution: The fifth term is ${}_{11}C_4x^7y^4 = 330x^7y^4$.

Example 7

What is the third term in the expansion of $(x - 3)^8$?

Solution: The third term is ${}_8C_2x^6(-3)^2 = 28x^6(9) = 252x^6$.

Example 8

What is the sixth term in the expansion of $(2a - 5b)^7$?

Solution: The sixth term is

$${}_7C_5(2a)^2(-5b)^5 = 21 \cdot 4a^2 \cdot (-3125)b^5 = -262500a^2b^5.$$

Check Your Understanding of Section 12.5

1. What is the seventh term in the binomial expansion of $(r + s)^{12}$?
(1) $924r^5s^7$ (2) $792\ r^5s^7$ (3) $924r^6s^6$ (4) $495r^4s^8$

2. The fourth term in the binomial expansion of $(a - b)^9$ is:
(1) $-84a^6b^3$ (2) $84a^5b^4$ (3) $-126a^6b^3$ (4) $126a^5b^4$

3. What is the fifth term in the binomial expansion of $(3x - 2y)^7$?
(1) $9072x^3y^4$ (2) $15120\ x^3y^4$ (3) $-15120\ x^3y^4$ (4) $-9072x^2y^5$

4. The second number in a row of Pascal's triangle is 6. What is the fifth number in that same row?
(1) 10 (2) 15 (3) 20 (4) 6

5. The coefficient of x^5 in the expansion of $(x - 3)^7$?
(1) 9 (2) 21 (3) -189 (4) 189

6. Use the Binomial Theorem to expand $(x - y)^7$.

7. Use the Binomial Theorem to expand $(3a - 4)^4$.

8. Expand: $\left(x^2 + \frac{1}{x}\right)^5$

9. What is the fourth term of $\left(\sqrt{x} - \frac{2}{\sqrt{x}}\right)^8$?

10. Since 1.001 can be thought of as the binomial 1 + 0.001, use the Binomial Theorem to determine 1.001^4 to the nearest thousandth.

12.6 STATISTICAL STUDIES

Statistics are used to analyze and draw conclusions from data. In order to do a statistical study, understanding how to collect, organize, summarize, and analyze the data is important.

Descriptive Statistics

Descriptive statistics are concerned with the collection, organization, presentation, and summarization of data. The way data is collected may affect the resulting analysis.

When conducting a statistical study, it is necessary to understand the difference between the population and a sample.

MATH FACTS

- The population consists of all subjects that are being studied.
- A sample is a subset of the population that is being studied.

Collecting Data

How is data collected for a study? Researchers use many different techniques, including but not limited to those listed below.

When it is possible to study the entire population, a census is taken. For example, the U.S. census is conducted every 10 years to determine information about the population of our country.

One of the most popular forms of data collection is a survey. A survey can be in the form of a written questionnaire, a telephone survey or questions over the telephone, and personal interviews. For example, a researcher mails a questionnaire asking for your views on the need for wind-generated energy. A company may hire a telemarketer to call all the homes in a certain region to see if the residents would shop at a new store if it opened in the region.

Controlled experiments are structured studies that consist of two or more groups chosen from the population. One group, the experimental group, utilizes the subject of the study. At least one group, called a control group, does not utilize the subject of the study. For example, the principal of your school may provide an interactive whiteboard for one class of students and not provide another similar class with this technology. Test performance by both classes is compared.

Sometimes it is not possible or ethical to use a control group. In that case, an observational study is used. A researcher observes the consequences for portions of a population who have already been exposed to the treatment being studied. For example, researchers may observe the long-term effects of a medical procedure that was performed on people 10 years ago.

Factors Affecting the Outcome of a Survey

Many things can affect the outcome of a survey. For instance, the question asked may be too vague. An example of a vague question may be, "Did you perform well on the test?" This is vague because a "B" level of performance may be considered performing well by one person but another person may consider it very bad.

Compare the questions, "Should we build a new bridge from Long Island to Connecticut?" as opposed to "Should taxes be raised in order to build a new bridge from Long Island to Connecticut?" These two questions may get very different responses from the same people.

Examine the question, "Do you think that New York City should raise the subway fare?" Most will probably answer, "No." However, if this were rephrased to ask, "Is the subway fare in New York City too high, too low, or just right?" the response might be less biased.

It is also very important to make sure that if a sample is selected to conduct a study, the sample must be representative of the population. There are many different ways that samples can be selected. Random samples are selected by methods such as numbering each sample and picking numbers arbitrarily out of a hat or generating selections randomly with a computer. Samples can be selected systematically by numbering each sample and selecting perhaps every fourth subject from the population. Samples can be selected by picking subjects that have a specific characteristic or by geographic clusters. Sometimes samples are selected because of convenience. For instance, a researcher may go to a shopping mall to interview a sample of shoppers about the merits of a new product.

Example

An auto manufacturer wants to market a new luxury car. A market researcher decides to go to the parking lot of a discount store in a town that has an average family income of $40,000 a year. The researcher plans to conduct a survey for three hours during a weekday to ask people entering the discount store to look at this new car to get their feedback about potential pricing. Is this a good sample to use for this study? Why or why not?

Solution: It is not a good sample to use for this study. The sample does not represent the population of people likely to purchase a luxury vehicle.

Check Your Understanding of Section 12.6

1. For each of the following proposed studies, suggest a method of selecting a sample:
   a. What is the average weight of a hamburger at the Hamburger Hut?
   b. How many hours of homework are assigned to students who attend Binghamton High School?
   c. Who performed better on last year's English Regents Examination—boys or girls?

2. Which method of collecting data will most likely result in an unbiased random sample?
   (1) Selecting every fourth senior citizen leaving a bank to complete a survey about savings plans for college tuition
   (2) Surveying teachers as they enter their school about the merits of the testing program in the No Child Left Behind Act
   (3) Conducting a poll on the Internet about a tax rebate program in the state of Ohio
   (4) Selecting customers by the last two digits of their telephone number to determine if they have telephoned France

12.7 MEASURES OF CENTRAL TENDENCY

What constitutes the average value from data that has been collected? The collected data is called a distribution. An average data value represents the center of this distribution. There are four basic measures to describe the center of a distribution: mean, median, mode, and midrange.

The Mean

The mean (often called the arithmetic mean) is the measure of central tendency most frequently associated with the word *average*. It is calculated by adding all the data values and dividing by the number of pieces of data in the distribution.

MATH FACTS

The symbol $\overline{x}$ represents the mean of a sample distribution.

$$\overline{x} = \frac{x_1 + x_2 + x_3 + \ldots + x_n}{n} \text{ or } \frac{\sum x}{n},$$

where n is the number of pieces of data collected for the sample and the Greek letter Σ stands for sum.

The symbol μ represents the mean of the population.

$$\mu = \frac{x_1 + x_2 + x_3 + \ldots + x_n}{N} \text{ or } \frac{\sum x}{N},$$

where N is the size of the entire population and the Greek letter Σ stands for sum.

Example 1

Eleven people who work at Jim's Hardware were surveyed to find out how far they commute to work. Their responses in miles were:

11, 32, 14, 26, 14, 11, 22, 27, 21, 32, 14

Calculate the mean of this distribution.

Solution: $\overline{x} = \dfrac{11+32+14+26+14+11+22+27+21+32+14}{11}$

$= 20.\overline{36}36$

The Median

The median is middle number when the data in a distribution is arranged in numeric order.

Example 2

Eleven people who work at Jim's Hardware were surveyed to find out how far they commute to work. Their responses in miles were:

11, 32, 14, 26, 14, 11, 22, 27, 21, 32, 14

Calculate the median for this distribution.

Solution: First arrange these 11 numbers in order from smallest to largest: 11, 11, 14, 14, 14, 21, 22, 26, 27, 32, 32. Since there are 11 pieces of data, the middle number will be the sixth number in this arrangement of the data, which is 21. There are five pieces of data listed below the 21 and five pieces of data listed above it.

Example 3

Twelve people who work at Jim's Hardware were surveyed to find out how far they commute to work. Their responses in miles were:

11, 32, 14, 26, 14, 11, 22, 27, 21, 32, 14, 17

Calculate the median for this distribution.

Solution: The same process is used to start this problem. The data is arranged as 11, 11, 14, 14, 14, 17, 21, 22, 26, 27, 32, 32. Now there is an even number of pieces of data. To calculate the median, simply average the two middle pieces of data—17 and 21. They average to $\frac{17+21}{2}=19$. Even though 19 is not a number in the data set or distribution, it is the median of the data.

The Mode

The mode is the data value that occurs most often in the distribution. To determine the mode, once again, arranging the data in numerical order is helpful (but not necessary). The number of times a piece of data occurs in a distribution is called the frequency of that piece of data.

Example 4

Eleven people who work at Jim's Hardware were surveyed to find out how far they commute to work. Their responses in miles were:

11, 32, 14, 26, 14, 11, 22, 27, 21, 32, 14

Determine the mode for this distribution.

Solution: The table below shows the frequency of each data value for this example:

x	11	14	21	22	26	27	32
f	2	3	1	1	1	1	2

The mode is therefore 14.

Sometimes more than one data value occurs with greatest frequency, and sometimes every piece of data occurs the same number of times. When two data values occur the most frequently, the data distribution is called bimodal.

If there are more than two data values that have the highest frequency, the distribution is referred to as multimodal. If no one data value occurs with greater frequency than any other, the data set has no mode.

The Midrange

The midrange is calculated by averaging the highest data value and the lowest data value.

$$MR = \frac{\text{Highest data value} + \text{Lowest data value}}{2}$$

Example 5

Eleven people who work at Jim's Hardware were surveyed to find out how far they commute to work. Their responses in miles were:

11, 32, 14, 26, 14, 11, 22, 27, 21, 32, 14

Calculate the midrange for this distribution.

Solution: The lowest data value is 11 and the highest is 32.

$$MR = \frac{11+32}{2} = 21.5$$

Example 6

Twenty-two students at SUNY Albany were asked how many times they withdrew money from the ATM in the Campus Center over the last month. Their responses are shown below.

3 9 2 2 6 4 1 0 7
10 2 5 8 11 4 5 12
2 4 10 16 9

Determine the mean, median, mode, and midrange of this data.

Solution: Mean $= \frac{\sum x}{n}$

$$= \frac{3+9+2+2+6+4+1+0+7+10+2+5+8+11+4+5+12+2+4+10+16+9}{22}$$

$$= \frac{132}{22}$$

$$= 6$$

Median: Arrange the numbers in numerical order:

$$\underbrace{0\ 1\ 2\ 2\ 2\ 2\ 3\ 4\ 4\ 4}\ 5\ 5\ \underbrace{6\ 7\ 8\ 9\ 9\ 10\ 10\ 11\ 12\ 16}$$

There are two numbers in the middle. Average them to get the median, 5.

Mode: Once the numbers are arranged numerically, it is easy to see that the frequency of the number 2 is four and that it is higher than the frequency of any other number. Therefore, the mode is 2.

$$\text{Midrange} = \frac{16+0}{2} = 8$$

Grouped Frequency Data

Many times data is collected and reported in a table that groups the data. Sometimes it is simply because data is repeated, so the table records the frequency for each data value.

To find the mean in this case, use the formula $\bar{x} = \frac{\sum f \cdot x}{\sum f}$ or $\mu = \frac{\sum f \cdot x}{\sum f}$.

For the median, since $n = \Sigma f$, find the data value that contains the middle or the data value in the $\frac{n+1}{2}$,s place.

The mode is the data value with the highest frequency.

Example 7

Twenty-five people were asked how many first cousins they have. Their responses are reported in the following table.

Number of Cousins	0	1	2	3	4	5	6	7	8	9	10	11
Frequency	1	1	3	4	2	3	3	1	3	2	1	1

Determine the mean, median, and mode of this data.

Solution: Mean:

$$\bar{x} = \frac{\begin{array}{c}0\cdot1+1\cdot1+2\cdot3+3\cdot4+4\cdot2+5\cdot3+\\6\cdot3+7\cdot1+8\cdot3+9\cdot2+10\cdot1+11\cdot1\end{array}}{25} = \frac{130}{25} = 5.2$$

Median: $\frac{n+1}{2} = \frac{26}{2} = 13$. Now look for the thirteenth data point in the order from smallest to largest. Since 1 + 1 + 3 + 4 + 2 = 11 and 1 + 1 + 3 + 4 + 2 + 3 = 14, the thirteenth piece of data is 5 and the median is 5.

Mode: The number 3 occurs more often than any other, so the mode is 3.

Sometimes the data is collected and reported in a table that groups the data into an interval of data values. Since the exact value of each and every piece of data is unknown, use the midpoint of the interval to compute the value of the mean. The formula is now μ or $\overline{x} = \frac{\sum f \cdot x_m}{\sum f}$. For the median, find the interval in which the median occurs. Then estimate the median based on the proportion of data in that interval as opposed to how many additional data values are needed to reach the location of the median. The best that can be done for the mode is to determine the interval and call it the modal class.

Example 8

Your Spanish teacher gave a test to all 58 of her students. The results on this test are summarized in the table.

Intervals of Test Scores	Frequency, f
50–59	3
60–69	9
70–79	21
80–89	10
90–99	15

Calculate the mean, median, and mode of this data.

Solution: Add a column to this table for the midpoint of each interval, another column to multiply the midpoint by the frequency, and a box to find the sum of this product.

Intervals of Test Scores	Midpoint, x_m	Frequency, f	$f \cdot x_m$
50–59	54.5	3	163.5
60–69	64.5	9	580.5
70–79	74.5	21	1564.5
80–89	84.5	10	845
90–99	94.5	15	1417.5
Sum		58	4571

Mean: $\frac{4571}{58} \approx 78.8$

Median: Look for the 29th and 30th terms; the median is in the interval from 70–79. There are 12 data values in intervals less than 70–79 and another 21 data values in the interval from 70–79. To estimate the 29th data value, recognize that 29 – 12 = 17 and there are 10 integers between 70 and 79. Therefore, use $\frac{17}{21} \cdot 10 \approx 8.1$. So the 29th data value $\approx 70 + 8.1 = 78.1$. Similarly, the 30th data value is $70 + \frac{18}{21} \cdot 10 \approx 78.6$. The average of these two data values is 78.35, and that is the best estimate for the median.

Mode: The mode is in the interval between 90 and 99. The modal class is 90–99.

Check Your Understanding of Section 12.7

1. Response times for 9 emergency police calls in Rochester were reported to the nearest minute as 7, 10, 8, 5, 8, 6, 8, 9, and 6 Which of the following statements is true?
(1) mean = mode
(2) mode = median
(3) mean = median
(4) mean > mode

2. A data set of five numbers contains the numbers 4, 8, 8, 11, and 12. What number can be added to this data set so that the mean, median, and mode are equal?
(1) 4 (2) 5 (3) 8 (4) 11

3. The number of homeruns hit by Derek Jeter each year from 1995 to 2008 is 0, 10, 10, 19, 24, 15, 21, 18, 10, 23, 19, 14, 12, and 11. What was the mean number of homeruns (to the nearest tenth) hit by Derek Jeter from 1995 to 2008?
(1) 14.7 (2) 14.5 (3) 14 (4) 10

4. The excise tax on gasoline is assessed by individual states. The excise tax collected in the following states were reported as of January 2008.

State	Alabama	Alaska	Iowa	Kansas	Maine	Ohio	Utah
Gas excise tax	16	8	20.7	24	27.6	28	24.5

What is the mean tax rate (to the nearest tenth) for these seven states?
(1) 21.3 (2) 22 (3) 24 (4) 24.5

5. The data below represents the number of cars passing through an intersection in Watertown between 4 P.M. and 8 P.M. on 10 consecutive nights:

28 19 31 24 27 35 26 28 30 21

Which measure of central tendency is the greatest?
(1) mean (2) median (3) mode (4) midrange

6. Sixteen people were asked how many times they have exercised their right to vote in a presidential election. The results are displayed below.

7 5 3 1 3 0 4 6 2 1 6 4 2 5 2 0

Determine the mean, median, and mode for this data.

7. The weights (in pounds) of 11 newborn babies at University Hopsital were recorded:

6.8 7.5 8.2 7.1 9.4 8.6 6.7 7.3 8.1 7.8 6.9

a. Determine the mean, median, and mode of this data.
b. If the 9.4-pound baby was taken out of these recorded weights, which of these measures of central tendency is most affected? Support your answer with appropriate work.

8. Janice was asked to calculate the median of the data values 11, 15, 22, 16, and 19. She answered that the median is 22 because the median is the third piece of data out of the five. Is she correct? Why or why not?

9. On the last math test in your class, three students scored a 61, one scored a 73, two scored a 77, four scored an 82, two scored an 88, and one scored a 97. What is the median score for this test?

10. A student researcher polled 24 recent college graduates to find out how many weddings they attended in the first year after graduation. The data is displayed in the table below. Determine the mean, median, and mode of this data.

Number of Weddings	Frequency
1	3
2	5
3	6
4	9
5	1

11. A bank decided to record the number of transactions made at one of its branches over the past 32 days. The data are summarized in the table below. Calculate the mean, median, and mode of the number of daily transactions made during these past 32 days.

Number of Transactions	Frequency, f
45–59	5
60–74	12
75–89	13
90–104	2

12.8 MEASURES OF DISPERSION

Now that an approach to finding a center for a distribution is known, the next step is to explore how the data is spread out or dispersed from the center.

The Range

The range of a data set is the difference between the highest piece of data and the lowest piece of data.

Example 1

A student in Niskayuna High School asked all nine of her teachers how many years they have taught at Niskayuna. Her data is recorded below:

12 8 21 14 17 25 2 13 24

What is the range of the number of years of teaching experience at Niskayuna High School for her teachers?

Solution: Highest number of years = 25. Lowest number of years = 2. Range = 25 − 2 = 23 years.

Quartiles

Just as the median is the data value at the middle of all data values or the 50th percentile, the quartiles are at the 25th percentiles. To calculate these, first calculate the median and then take all the data values below the median (which is the second quartile) and find the median of those. This value is the first or lower quartile. Then take the median of all the data values above the median. This value is the third or upper quartile.

Example 2

There are eleven sections of Algebra and Trigonometry at Ossining High School. The class size for each section is listed below. Determine the three quartiles for these class sizes.

21 27 24 28 30 23 25 19 29 22 17

Solution: First arrange the data in numerical order.

17 19 21 22 23 24 25 27 28 29 30

The 24 has five data values below it and five data values above it. Therefore, the median or second quartile is 24.

Examine the five data values below 24: 17 19 21 22 23
Their median is 21, so 21 is the first or lower quartile.
Examine the five data values above 24: 25 27 28 29 30
Their median is 28, so 28 is the third or upper quartile.
This can be written or summarized as $Q_1 = 21$, $Q_2 = 24$, and $Q_3 = 28$.

The results of a problem such as this can be summarized in a Box-and-Whisker Plot. A Box-and-Whisker Plot uses what is called the five-number summary. It includes the values of the lowest data value or Q_1, the median or Q_2, Q_3, and the highest data value. In this case, these values are 17, 21, 24, 28, and 30. These values are then graphed as below:

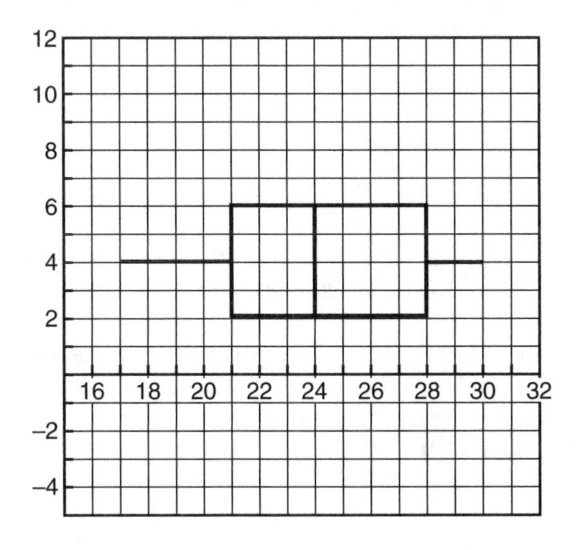

Example 3

Test results for the 24 students in your class are shown below. Determine the three quartiles for this data.

58	63	68	70	71	72	74	77	78	78	80	81
83	84	86	87	89	90	90	92	93	95	98	100

Solution: Since there are 24 data values, the median is the average of the 12th and 13th data value, $\frac{81+83}{2}=82$. The median divides the data values into two sets of data consisting of 12 data values. The median of the lower set of data values is the average of the 6th and 7th data values or $\frac{72+74}{2}=73$. The lower quartile is 73. The median of the upper set of data values is the average of the 18th and 19th data values or $\frac{90+90}{2}=90$. The upper quartile is 90. So $Q_1 = 73$, $Q_2 = 82$, and $Q_3 = 90$.

Range

The range of a set of data is the difference between the highest and lowest data values.

Interquartile Range

The interquartile range is the difference between the upper quartile and the lower quartile or $Q_3 - Q_1$.

Example 4

Test results for the 24 students in your class are shown below.

58	63	68	70	71	72	74	77	78	78	80	81
83	84	86	87	89	90	90	92	93	95	98	100

Determine the range and the interquartile range of scores for this test.

Solution: Range = 100 − 58 = 42
Interquartile range = 90 − 73 = 17

Variance

Variance is a measure of the dispersion of the data in a distribution. The variance is the average of the squared difference of each data value from the mean of the distribution.

MATH FACTS

$$\text{Variance for a population} = \frac{\sum_{i=1}^{n}(x_i - \overline{x})^2}{n}$$

$$\text{Variance for a sample} = \frac{\sum_{i=1}^{n}(x_i - \overline{x})^2}{n-1}$$

Example 5

The manager of the Jill's Boutique records the number of sales over a 5-day period. These sales records were 36, 58, 27, 41 and 33. Determine the variance for these sale records.

Solution: $\overline{x} = \dfrac{36+58+27+41+33}{5} = 39$

It is easier to put this information into a chart in order to calculate the variance. Notice that the headings of the chart provide a step-by-step approach to this calculation.

x_i	$x_i - \overline{x}$	$(x_i - \overline{x})^2$
27	$27 - 39 = -12$	$(-12)^2 = 144$
33	$33 - 39 = -6$	$(-6)^2 = 36$
36	$36 - 39 = -3$	$(-3)^2 = 9$
41	$41 - 39 = 2$	$2^2 = 4$
58	$58 - 39 = 19$	$19^2 = 361$
$\overline{x} = 39$		$\sum_{i=1}^{n}(x_i - \overline{x})^2 = 554$

$$\text{Variance} = \frac{\sum_{i=1}^{n}(x_i - \overline{x})^2}{n} = \frac{554}{5} = 110.8$$

Example 6

The manager of Smith's Nursery records the number of sales over a 5-day period of time. These sales records were 46, 58, 52, 49 and 53. Determine the variance for these sale records.

Solution: $\overline{x} = \dfrac{46 + 58 + 52 + 49 + 53}{5} = 51.6$

x_i	$x_i - \overline{x}$	$(x_i - \overline{x})^2$
46	$46 - 51.6 = -5.6$	$(-5.6)^2 = 31.36$
49	$49 - 51.6 = -2.6$	$(-2.6)^2 = 6.76$
52	$52 - 51.6 = 0.4$	$(0.4)^2 = 0.16$
53	$53 - 51.6 = 1.4$	$(1.4)^2 = 1.96$
58	$58 - 51.6 = 6.4$	$(6.4)^2 = 40.96$
$\overline{x} = 51.6$		$\sum_{i=1}^{n}(x_i - \overline{x})^2 = 81.2$

$$\text{Variance} = \frac{\sum_{i=1}^{n}(x_i - \overline{x})^2}{n} = \frac{81.2}{5} = 16.24$$

In a grouped frequency distribution, the mean of each interval is used in place of individual data values. In that case, the mean becomes $\dfrac{\sum f \cdot x_m}{\sum f}$ and the variance becomes $\dfrac{\sum (x_m - \overline{x})^2}{\sum f}$. Example 9 will explore this process.

Standard Deviation

Standard deviation is essentially found by taking the square root of the variance. It is the most commonly used statistical measure for the dispersion or spread of data.

MATH FACTS

The formula for the standard deviation of the population is

$$\sigma = \sqrt{\frac{\sum_{i=1}^{n}(x_i - \mu)^2}{n}}.$$

The formula for the standard deviation of a sample is

$$s = \sqrt{\frac{\sum_{i=1}^{n}(x_i - \overline{x})^2}{n-1}}.$$

Example 7

The weekly salaries of all 8 workers in a small office are listed below. Determine the standard deviation for the weekly salaries of these 8 workers.

$255, $265, $270, $275, $280, $285, $290, $300

Solution: It is best to organize the data in a table to calculate the standard deviation.

x_i	$x_i - \overline{x}$	$(x_i - \overline{x})^2$
255	–22.5	506.25
265	–12.5	156.25
270	–7.5	56.25
275	–2.5	6.25
280	2.5	6.25
285	7.5	56.25
290	12.5	156.25
300	22.5	506.25
$\mu = \frac{\sum x_i}{n} = \frac{2220}{8} = 277.5$		$\sigma = \sqrt{\frac{1450}{8}} \approx 13.46$

This work can also be done on a calculator when available. Press **STAT** **1** and enter the 8 pieces of data into L1. If L1 already has data, clear it out by pressing **DEL** on each entry in L1 or by going up to L1, pressing **CLEAR** and then pressing **ENTER**. The window should look like this after entering all 8 data values:

L1	L2	L3 1
270		
275		
280		
285		
290		
300		

L1(9)=

Now press **STAT** **CALC** **1** **L1** **ENTER**. A list of values will appear as shown below:

```
1-Var Stats
 x̄=277.5
 Σx=2220
 Σx²=617500
 Sx=14.39245834
 σx=13.46291202
↓n=8
```

The list includes a few lines that are not necessary to view, such as $\sum x^2$, which can also be used in an alternative approach to calculating the standard deviation. The symbol S_x is for the sample standard deviation and the symbol σ_x is for the population standard deviation, which is what was calculated in Example 7.

Example 8

Six students from the tenth grade at your school were asked how many minutes they spend doing their geometry homework each night. The number of minutes was reported to be 40, 45, 80, 20, 10 and 60. Determine the standard deviation for the number of hours spent on geometry homework by tenth graders in your school.

Solution: The table below is used to calculate the standard deviation.

x_i	$x_i - \overline{x}$	$(x_i - \overline{x})^2$
10	−32.5	1056.25
20	−22.5	506.25
40	−2.5	6.25
45	2.5	6.25
60	17.5	306.25
80	37.5	1406.25
$\mu = \frac{\sum x_i}{n} = \frac{255}{6} = 42.5$		$s = \sqrt{\frac{3287.5}{6-1}} \approx 25.64$

In a grouped frequency distribution, the mean of each interval is used in place of individual data values. In that case, the mean becomes $\frac{\sum f \cdot x_m}{\sum f}$ and the standard deviation becomes $\sqrt{\frac{\sum f \cdot (x_m - \overline{x})^2}{\sum f}}$.

Example 9

Thirty people were surveyed to find out how far they live from where they work. The results were recorded in the chart below.

# of Miles	0–7	8–15	16–23	24–31
Frequency	12	9	6	3

Determine the variance and standard deviation for the number of commuting miles.

Solution: The chart below is used to calculate both of these measures.

Miles Traveled	x_m	f	$f \cdot x_m$	$x_m - \overline{x}$	$(x_m - \overline{x})^2$
0–7	3.5	12	42	−8	64
8–15	11.5	9	103.5	0	0
16–23	19.5	6	117	8	64
24–31	27.5	3	82.5	16	256
			$\overline{x} = \frac{\sum f \cdot x_m}{\sum f} = \frac{345}{30} = 11.5$		$\text{Variance} = \frac{\sum f \cdot (x_m - \overline{x})^2}{\sum f} = \frac{1920}{30} = 64$

$$= \sqrt{\frac{\sum f \cdot (x_m - \overline{x})^2}{(\sum f)}} = \sqrt{\frac{1920}{30}} = 8$$

Check Your Understanding of Section 12.8

1. Seven data values have a range of 13. If six of these data values are 8, 12, 19, 15, 11, and 17, which of the following must be the seventh data value?
(1) 7 (2) 18 (3) 20 (4) 21

2. Nine teenagers went on a fishing trip. The number of fish each teenager caught were 3, 8, 0, 6, 2, 3, 5, 1, and 4. Identify the lower quartile associated with the number of fish caught by the teenagers on this trip.
(1) 6 (2) 2 (3) 1.5 (4) 5.5

3. Your school collected bids in order to get the best price for graphing calculators. The prices for a single graphing calculator from the bids were \$126, \$114, \$130, \$108, \$116, \$122, \$114, \$109, \$128, and \$110. Identify the interquartile range for the calculator prices from the bids submitted by vendors.
(1) 22 (2) 16 (3) 5 (4) 10

4. Your local video store recorded the number of videos rented each night for a week. The number of videos rented were 141, 173, 126, 98, 161, 174, and 165. The 5-number summary for the number of rentals per night from this data is:
(1) 98, 133.5, 161, 169, 174
(2) 98, 126, 161, 173, 174
(3) 98, 141, 161, 165, 174
(4) 98, 134, 161, 169, 174

5. Five senior citizens were asked how many grandchildren they have. They reported 6, 2, 9, 3, and 1 grandchildren. Calculate the variance for the number of grandchildren they have.
(1) 3.27 (2) 8.56 (3) 10.7 (4) 42.8

6. Dan wants to buy a car, and he decided to look at 10 different cars. He examined specifications on each of the 10 cars that he was interested in purchasing. The specifications claimed that the cars average 26, 21, 23, 24, 24, 26, 22, 24, 26, and 22 miles per gallon. Calculate the mean, range, variance, and standard deviations of the average miles per gallon for the cars Dan is considering purchasing.

7. A college admissions officer selects 30 of the high school transcripts at random and records the number of AP credits each student has as shown below.

4	7	3	2	5	2	1	5	6	2
4	3	5	2	0	4	1	3	3	2
5	1	0	3	6	1	4	5	2	1

Calculate the mean and standard deviation of the number of AP credits for high school students applying to this college.

8. An Internet site records the number of hits made to its site each day. One week after changing the looks of the site, the site's owners examine the number of hits made over the next 10 days. The number of hits recorded were 42, 75, 80, 99, 112, 114, 121, 122, 123, and 125. Calculate the quartiles and the interquartile range of this data.

9. The owner of the Internet site described in Exercise 8 decides that the first day after the changes were made to the site, it did not make enough of a difference to the number of hits and was not indicative of the changes made. He decided to repeat the study and examine the data values 75, 80, 99, 112, 114, 121, 122, 123, and 125. Calculate the quartiles and the interquartile range of this data.

10. The owner of the Internet site described in the last two examples wants to compare the data values 42, 75, 80, 99, 112, 114, 121, 122, 123, and 125 and the same data set without the 42. This time he wants to compare the standard deviation of the original set of data and the data set without the 42. Compute each.

12.9 NORMAL DISTRIBUTIONS

Some data sets are distributed normally. The data will fit a bell-shaped curve where the data is symmetric to the mean. For instance, the mean human temperature is 98.6° F. Temperatures of all humans are not exactly 98.6° F. If the temperatures of all the human beings on Earth were measured, the data gathered would be evenly distributed from the mean temperature. Abnormal temperatures could be identified by how far away from the mean they are.

The Normal Distribution and the Bell-Shaped Curve

Look at a histogram created for the data below:

1, 2, 2, 3, 3, 3, 4, 4, 4, 4, 5, 5, 5, 5, 5, 6, 6, 6, 6, 6, 6, 7, 7, 7, 7, 7, 8, 8, 8, 8, 9, 9, 9, 10, 10, 11

Notice that this data is symmetric to the mean = the median = 6.

The display of data in a frequency histogram clearly shows how the data is spread out evenly from the mean. Notice also that the data graphed in the histogram has symmetry to a the mean. The data seems to conform to a bell-shaped curve. The graph that follows shows a bell-shaped curve that appears to fit this data closely. The bell-shaped curve is shown alone in the second diagram.

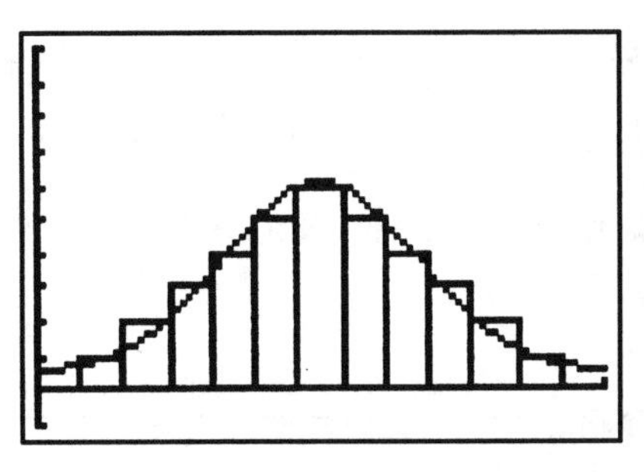

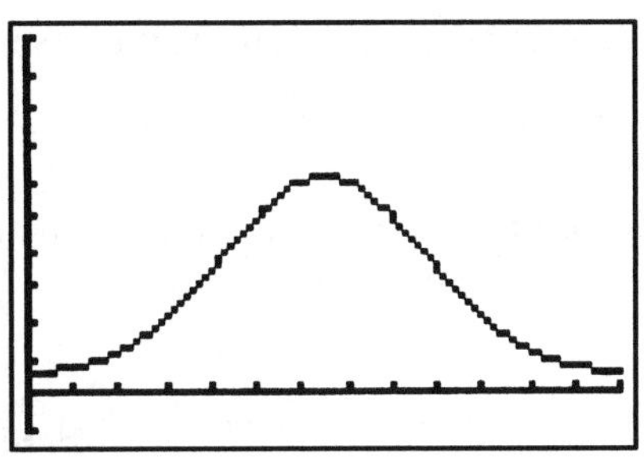

The frequency histogram represents the actual data values that are all integers. This type of data is called discrete data. Discrete variables are variables whose values are counting numbers. The bell-shaped curve assumes the data is continuous data. Continuous variables are variables that can have an infinite number of values between any two counting numbers.

The Empirical Rule

In a normal distribution, the amount of data one standard deviation on either side of the mean consists of approximately 68.2% of the data values. Two standard deviations consist of approximately 95.4% of the data values, and three standard deviations consist of approximately 99.8% of the data values.

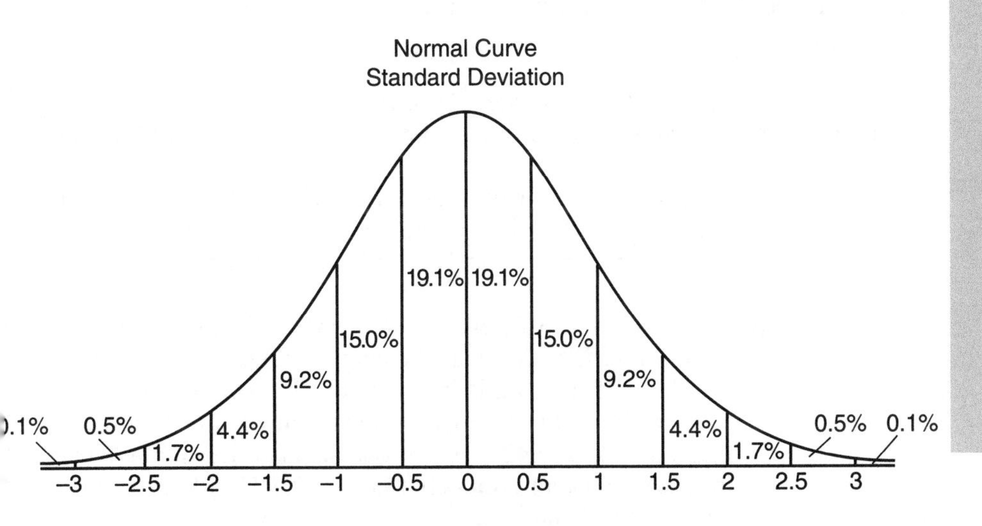

A z-score is a characteristic measured from a data point that is calculated by the formula $z = \frac{x - \overline{x}}{s}$ or $\frac{x - \mu}{\sigma}$. A z-score measures the number of standard deviations that a data value is from the mean. Notice that if $x = \overline{x}$, then $z = 0$ and if $x = s + \overline{x}$, then $z = 1$.

MATH FACTS

The Empirical Rule

- Approximately 68.2% of data in a normal distribution fall within one standard deviation from the mean. Approximately 34.1% of the data in a normal distribution lie between $z = 0$ to $z = 1$, or $z = -1$ to $z = 0$.
- Approximately 95.4% of data in a normal distribution fall within two standard deviations from the mean. Approximately 13.6% of the data in a normal distribution lie between $z = 1$ to $z = 2$, or $z = -2$ to $z = 1$.
- Approximately 99.8% of data in a normal distribution fall within three standard deviations from the mean. Approximately 2.2% of the data in a normal distribution lie between $z = 2$ to $z = 3$, or $z = -3$ to $z = 2$.

Example 1

In a distribution whose mean = 42 and whose standard deviation = 3.2, determine the approximate percentage of data values that lie between:

a. 38.2 and 45.2 b. 42 and 48.4 c. 35.6 and 45.2

Solution: In each part of this example, it first is necessary to determine the number of standard deviations away from the mean (or z-score) for each value.

a. 38.2 = 42 – 3.2 or one standard deviation below the mean and 45.2 = 42 + 3.2 or one standard deviation above the mean. So approximately 68.2% of the data is in this interval.

b. 48.4 = 42 + 2(3.2) or two standard deviations above the mean while 42 is the mean. So approximately (34.1 + 13.6)% or 47.7% of the data is in this interval.

c. 35.6 = 42 – 2(3.2) or two standard deviations below the mean and 45.2 = 42 + 3.2 or one standard deviation above the mean. So approximately (34.1 + 47.7)% or 81.8% of the data is in this interval.

Normal Approximation for a Binomial Distribution

In Section 12.4, the concept of binomial probability was developed where p represented the probability of success and q represented the probability of failure.

In a binomial distribution of n trials, the mean = np and the standard deviation = $\sqrt{np(1-p)}$ or $\sqrt{npq}$. Let r represent the number of successes in n trials. Since the data in a binomial distribution is discrete rather than continuous, to estimate the probability of at least r successes in n trials, it is necessary to subtract 0.5 from r.

MATH FACTS

Normal Approximation for a Binomial Probability

- $P(x = r) = P(r - 0.5 < x < r + 05)$
- $P(x > r) = P(x > r + 0.5)$
- $P(x \geq r) = P(x > r - 0.5)$
- $P(x < r) = P(x < r - 0.5)$
- $P(x \leq r) = P(x < r + 0.5)$

Example 2

A manufacturer of lightbulbs knows that the probability that a lightbulb is defective is 0.01. Out of 30 lightbulbs sold, use a normal distribution to approximate the probability that no more than 3 are defective.

Solution: In this binomial distribution, the mean = 30(0.01) = 0.3 and the standard deviation = $\sqrt{npq} = \sqrt{30(0.01)(0.99)} \approx 0.545$. $P(r \leq 3) = P(r < 3 + 0.5) = P(r < 3.5)$. The graphing calculator can be used to determine this result using a low value of 0.5 lower than the least-possible number 0, or –0.5, and 3.5 as the highest number, mean = 0.3, and standard deviation of 0.545. Press **2ND** **VARS** **2** **–0.5** **,** **3.5** **,** **0.3** **,** **0.545** **)** **ENTER**.

The result is displayed below.

```
normalcdf(-0.5,3
.5,0.3,0.545)
      .9289328967
```

This same example was calculated in Example 4 on page 327 as follows:

$$P = {}_{30}C_0(0.01)^0(0.99)^{30} + {}_{30}C_1(0.01)^1(0.99)^{29} + {}_{30}C_2(0.01)^2(0.99)^{28} +$$

$${}_{30}C_3(0.01)^3(0.99)^{27} \approx 0.7397 + 0.2242 + 0.0328 + 0.0031 = 0.9998$$

Notice that the actual answer is greater than the one calculated in the approximation, but it is only a little more than 0.07 too small. So it is a reasonable approximation.

Example 3

Any time Gary plays James in a game of chess, he has a 70% probability of winning the game. If they play 10 chess games, use a normal distribution to approximate the probability that Gary wins at least 8 games.

Solution: In this binomial distribution, the mean = 10(0.7) = 7 and the standard deviation = $\sqrt{npq} = \sqrt{10(0.7)(0.3)} \approx 1.45$. $P(r \geq 8) = P(r > 8 - 0.5)$ = $P(r > 7.5)$. Using the calculator, the answer is revealed as:

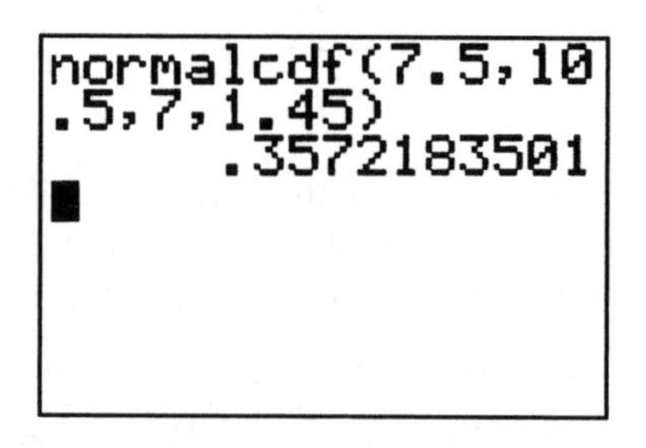

Actual calculations would result in an answer of 0.382, approximately 0.025 larger than the approximation.

Check Your Understanding of Section 12.9

1. In a distribution whose mean = 15 and whose standard deviation = 2.7, determine the approximate percentage of data values that lie between 12.3 and 17.7.
(1) 25% (2) 34% (3) 68% (4) 32%

2. The mean score on the last test given in your math class was 83 with a standard deviation of 9. What proportion of the class scored less than 65 on this test?
(1) 2.3% (2) 13.6% (3) 47.7% (4) 97.8%

3. In a normal distribution, the mean is 322 with a standard deviation of 11.6. Determine the approximate percentage of data values that lie between:
a. 322 and 345.2 b. 310.4 and 333.6 c. 333.6 and 345.2
d. 298.8 and 310.4 e. 287.2 and 298.8

4. Every time a single die is tossed, the probability that the die lands on an even number is 0.5. Use a normal distribution to approximate the probability that if a die is tossed 12 times, it lands on an even number more than 8 times. Include your calculation for the mean and standard deviation of this distribution in your work.

5. Sara takes a multiple-choice test (where each question has four choices to choose from) on a topic that she did not study. She randomly guesses on all 35 questions on the test. Use a normal distribution to approximate the probability that she gets between 10 and 15 questions correct inclusively by pure guessing. Be sure to include your answer for the mean and standard deviation for this problem.

Chapter Thirteen

REGRESSION

13.1 TYPES OF REGRESSION

One way to examine statistical data is to make a scatter plot. Sometimes the data values appear to conform closely to a line or a curve. Regression analysis attempts to fit the data to a function.

Scatter Plots and Regression

A scatter plot for the values in the chart below can be created on a graphing calculator by inputting the values of r and s in the data lists L1 and L2.

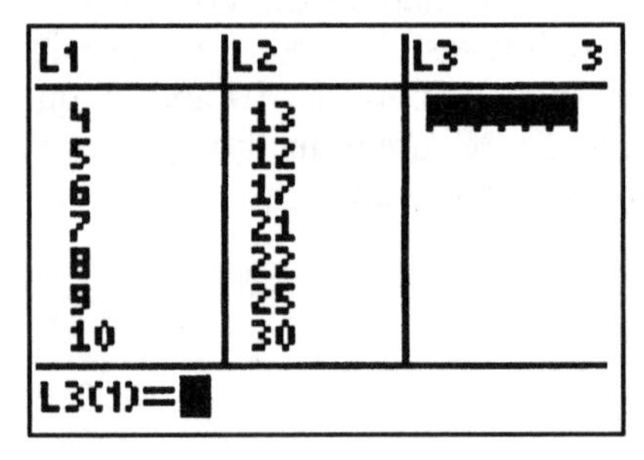

L1	L2	L3 3
4	13	
5	12	
6	17	
7	21	
8	22	
9	25	
10	30	

L3(1)=

The scatter plot is displayed below.

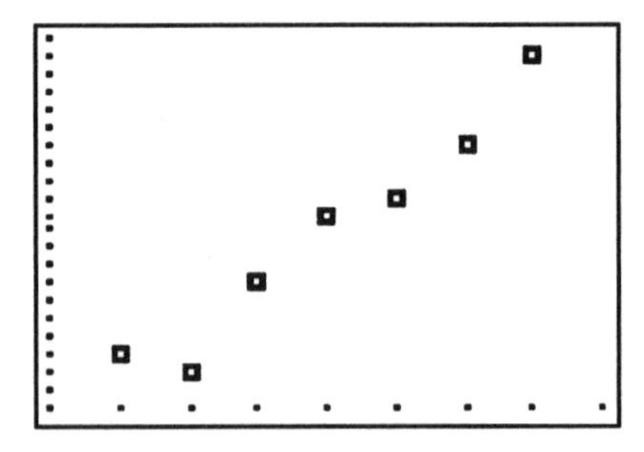

Notice that the data lies in a path that appears to be close to a line. In fact, the next display shows a line that may model these ordered pairs.

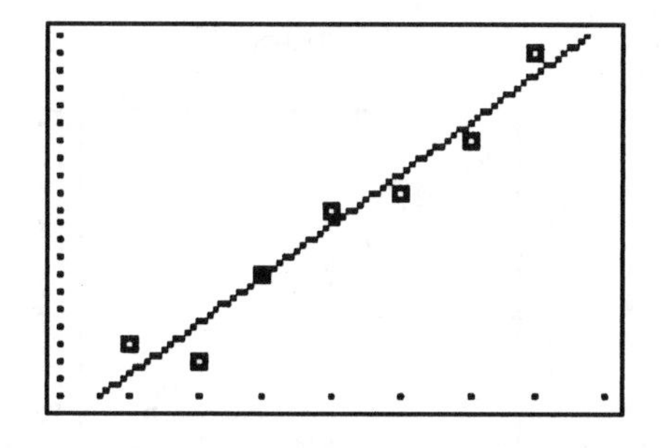

Linear Regression on a Graphing Calculator

To determine the equation of the regression line for the data, press

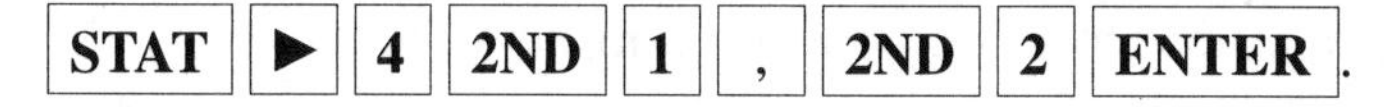

.

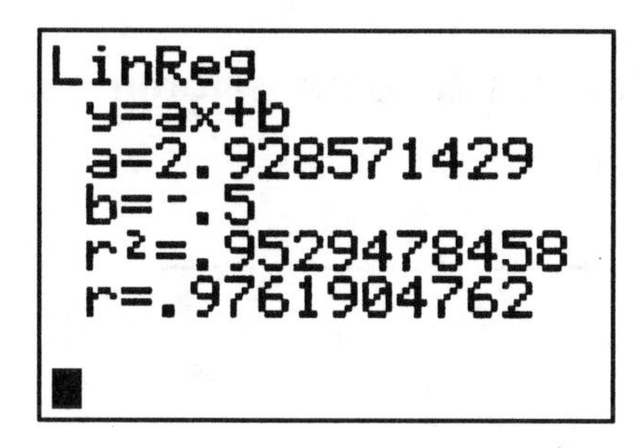

The equation of this line of best fit is $y = 2.928571429x - 0.5$. The calculator also displays the value of r, which is called the correlation coefficient of the regression line.

If $|r|$ is close to 1, the line fits the data well.

If your calculator does not display r^2 and r, then press 2ND 0 X^{-1} and scroll down to DiagnosticOn and press ENTER. This enables the displaying of the correlation coefficient.

The diagrams that follow show what the correlation coefficient indicates about a data set.

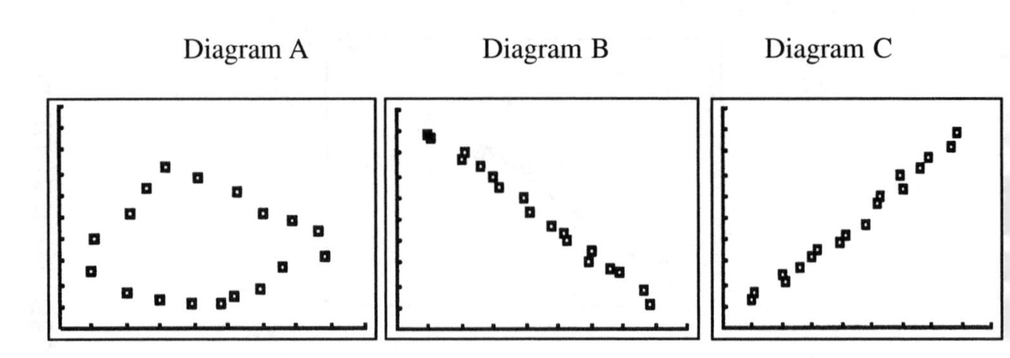

Diagram A shows a data set with no correlation or $r = 0$.
Diagram B shows a data set with a negative correlation or $-1 < r < 0$.
Diagram C shows a data set with a positive correlation or $0 < r < 1$.
The closer $|r|$ is to 1, the closer the data conforms to a line.

Fitting Other Curves to Data—Nonlinear Regression

Observe the scatter plot below.

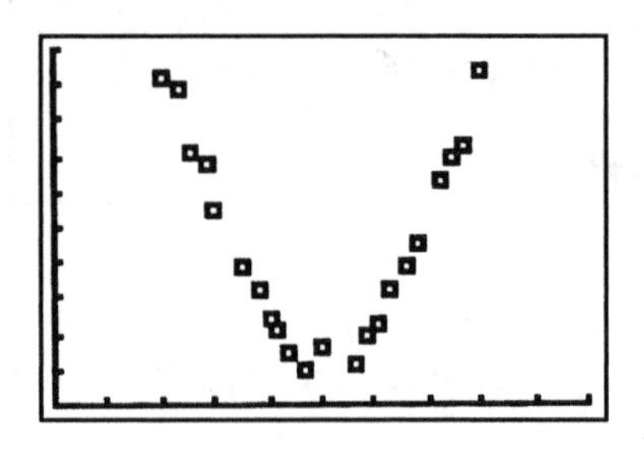

This data would obviously not fit a linear model. In fact it appears very close to a quadratic function. The actual data is {(0, 27.1), (0.2, 25.1), (0.5, 22.2), (0.9, 18.2), (1.1, 16.1), (1.4, 14.8), (1.7, 13.8), (2, 9.2), (2.3, 8.9), (2.6, 7.1), (2.9, 6.8), (3, 5.5), (3.5, 3.9), (3.8, 3.3), (4, 2.4), (4.2, 2.1), (4.4, 1.4), (4.7, 1), (5, 1.6), (5.6, 1.2), (5.8, 2), (6.1, 2.3), (6.3, 3.2), (6.6, 3.8), (6.8, 4.5), (7.2, 6.3), (7.4, 6.9), (7.7, 7.2), (8, 9.3), (8.3, 11.7), (8.6, 13.6), (8.8, 15.2), (9, 16.2), (9.2, 18.3), (9.6, 22), (10, 26.8), (10.4, 29.1), (11.1, 37.8), (11.5, 42.7)}. To perform a quadratic regression, press **STAT** **▶** **5** **2ND** **1** **,** **2ND** **2** **ENTER**.

The resulting quadratic equation is $y \approx 0.9994679445x^2 - 10.12266249x + 26.89140207$. To show this quadratic function graphed on top of the scatter plot, press STAT ▶ 5 VARS ▶ 1 1 ENTER. Then press GRAPH. This display is shown below.

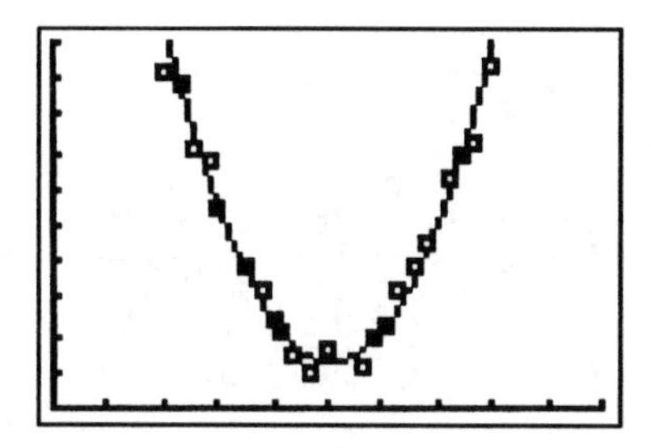

Some data best fits a logarithmic curve.

Example 1

For her science project, Anna measures the height of a plant each day for 10 days after she feeds it a chemical she hypothesizes will increase its growth. The data are summarized in the chart below.

Day	1	2	3	4	5	6	7	8	9	10
Height	17.3	23.4	26.1	26.9	27.3	27.8	28.2	28.9	28.7	29.1

Create a scatter plot for this data. What type of function does this data appear to fit best? Determine the regression equation for this.

Solution:

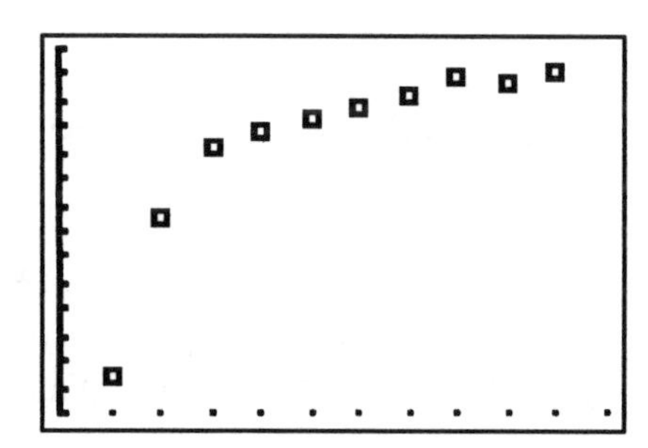

The data seems close to a logarithmic function. To perform a logarithmic regression, press **STAT** **►** **9** **2ND** **1** **,** **2ND** **2** **ENTER**. This yields

```
LnReg
 y=a+blnx
 a=19.29052674
 b=4.687023229
 r²=.9081500704
 r=.9529690815
```

or approximately $y = 19.3 + 4.7 \ln x$.

To show this logarithmic function graphed on top of the scatter plot, press **STAT** **►** **9** **VARS** **►** **1** **1** **ENTER**. Then press **GRAPH**. This display is shown below.

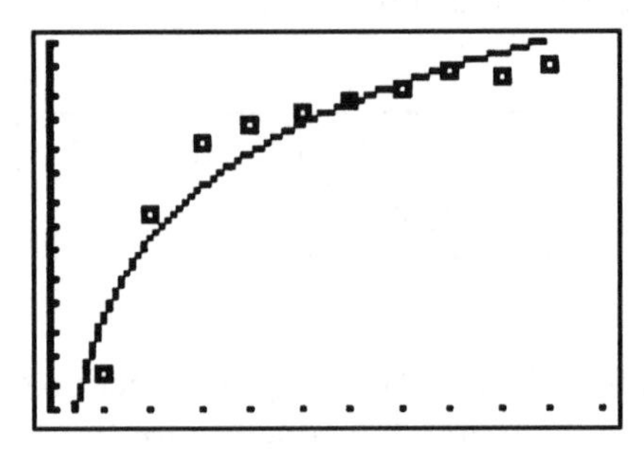

Example 2

Gino measures the height of the liquid (in centimeters) in a beaker each second after he adds a new chemical to the beaker. The data are displayed in the table below.

Seconds	1	2	3	4	5	6	7	8	9	10
Height	0.3	0.6	0.7	0.6	0.8	1.1	2.4	3.4	4.8	5.9

Create a scatter plot for this data. What type of function does this data appear to fit best? Determine the regression equation for this.

Solution:

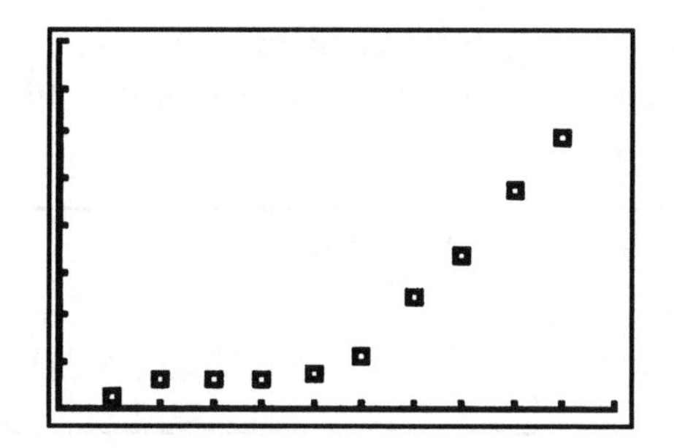

The data seems close to a exponential function. To perform an exponential regression, press **STAT** **▶** **0** **2ND** **1** **,** **2ND** **2** **ENTER**.

This yields

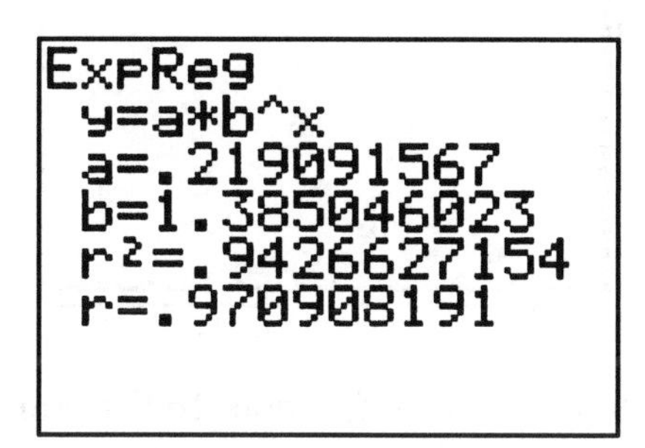

or approximately $y = 0.22(1.4)^x$.

To show this exponential function graphed on top of the scatter plot, press **STAT** **▶** **0** **VARS** **▶** **1** **1** **ENTER**. Then press **GRAPH**. This display is shown below.

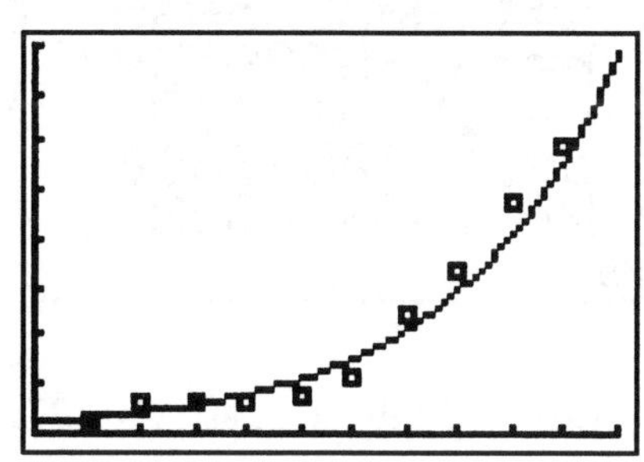

Example 3

A report that a company is going to split created a surge in stock prices. The chart below shows the value of this stock a specific number of seconds after the report was released.

# of Seconds	0.2	0.5	1.2	1.9	2.6	3.1	3.4	4.3
Value of Stock	1.015	2.145	2.985	3.285	4.925	8.041	13.545	29.865

Create a scatter plot for this data. What type of function does this data appear to fit best? Determine the regression equation for this.

Solution:

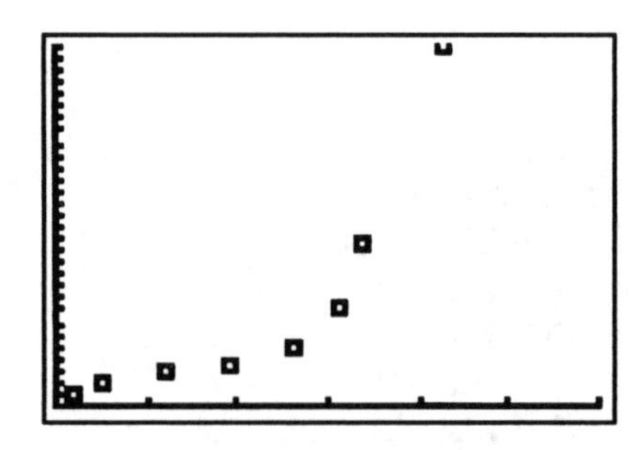

The data seems close to a cubic function. To perform a cubic regression, press **STAT** **►** **6** **2ND** **1** **,** **2ND** **2** **ENTER**.

This yields

```
CubicReg
 y=ax³+bx²+cx+d
 a=1.003535635
 b=-3.986293304
 c=5.53510139
 d=.1770527616
 R²=.9963032895
```

or approximately $y = 1.0x^3 - 3.99x^2 + 5.54x + 0.177$. To show this cubic function graphed on top of the scatter plot, press **STAT** **►** **6** **VARS**

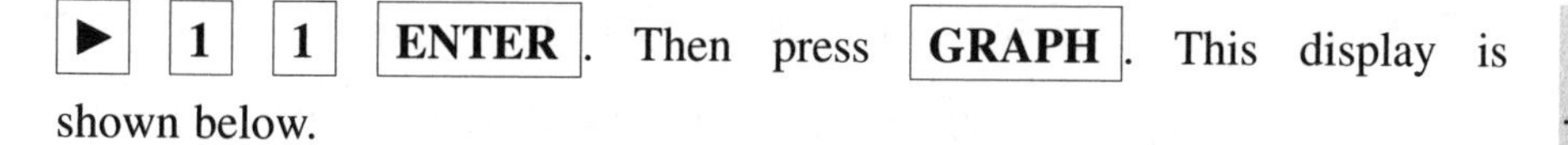
▶ 1 1 ENTER. Then press GRAPH. This display is shown below.

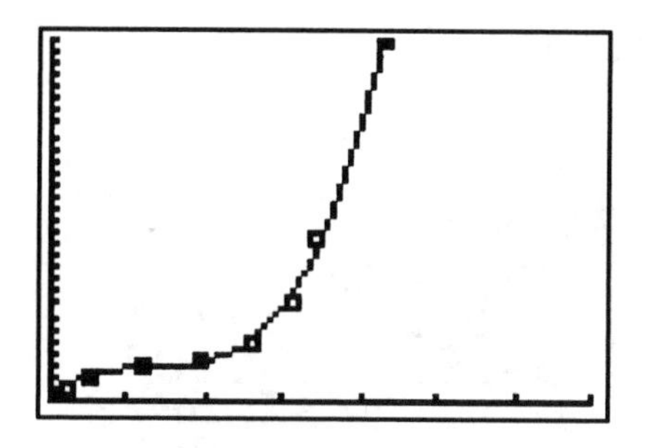

Example 4

The number of sales of CDs for a rock group dramatically increased over a period of 20 years. The chart below shows the number of sales in the thousands for the group's CDs. The table below indicates the number of years, x, after 1985.

Year	**1986**	**1988**	**1990**	**1994**	**1998**	**2000**	**2003**	**2005**
x	1	3	5	9	13	15	18	20
Number of sales	1.28	2.45	3.82	5.57	7.35	7.63	9.22	9.84

Create a scatter plot for this data. What type of function does this data appear to fit best? Determine the regression equation for this.

Solution:

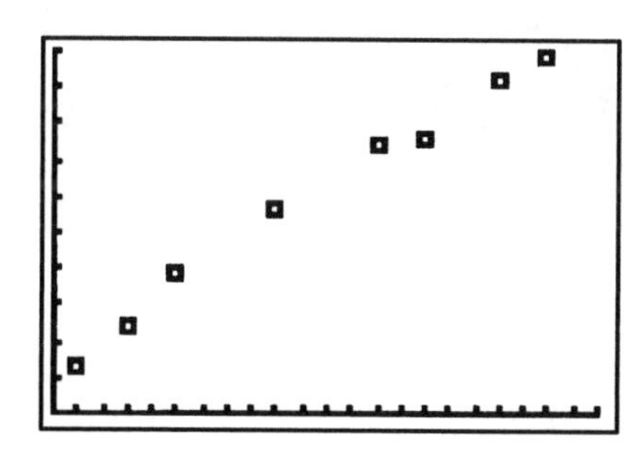

The data seems close to that of a power function, $y = a \cdot x^b$. To perform a power regression, press **STAT** **▶** **ALPHA** **MATH** **2ND** **1** **,** **2ND** **2** **ENTER**.

This yields

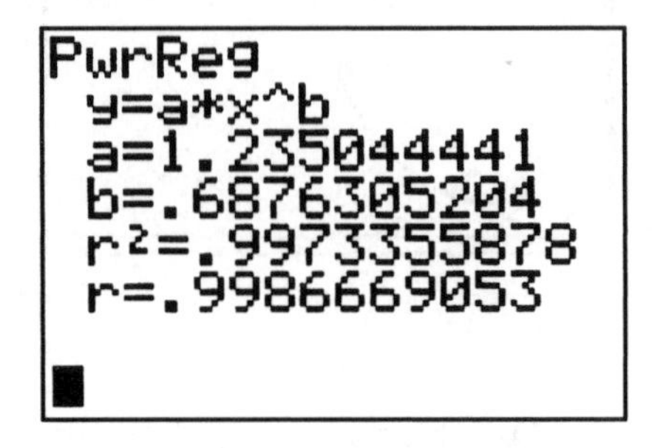

or approximately $y = 1.24x^{0.688}$. To show this power function graphed on top of the scatter plot, press **STAT** **▶** **ALPHA** **MATH** **VARS** **▶** **1** **1** **ENTER**.

Then press **GRAPH**. This display is shown below.

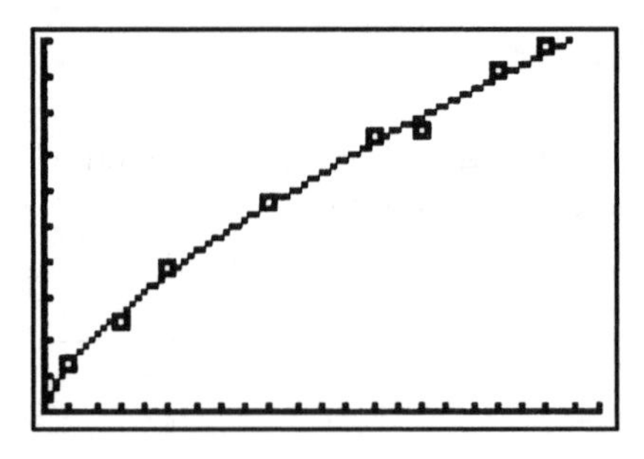

Check Your Understanding of Section 13.1

1. Which of the following scatter plots represents data used in a linear regression that has a correlation coefficient closer to −1?

(1)

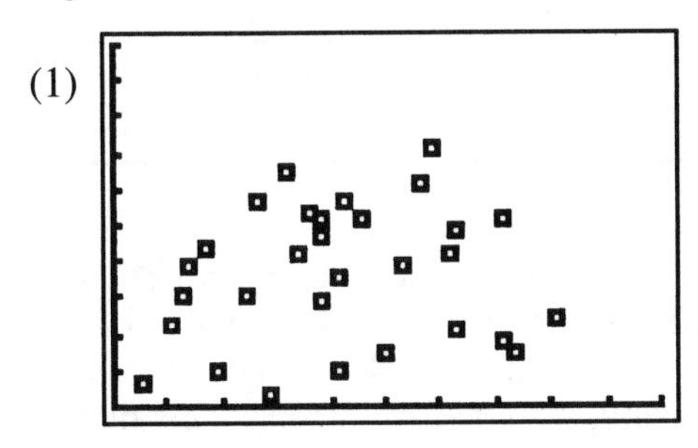

(2)

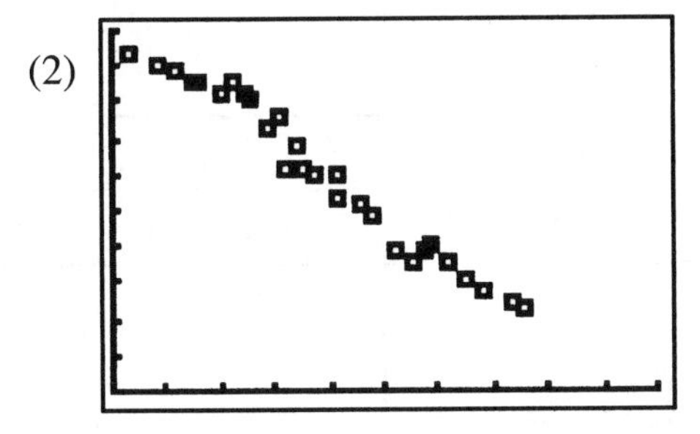

(3)

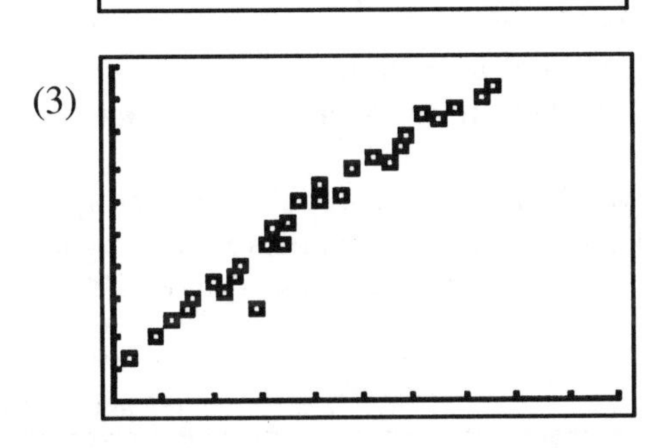

(4)

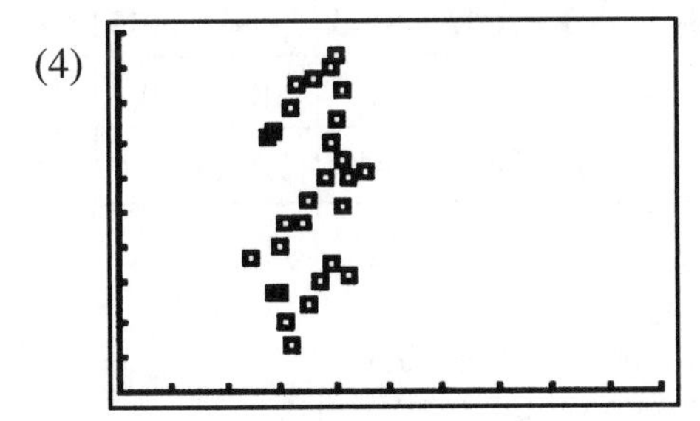

2. Seven babies were born at University Hospital on the same day. The chart below shows their length and weight at birth. What is the linear correlation for this relationship rounded to the nearest thousandth?

Length in Inches	20	18.5	19	21.5	20.5	21	19.5
Weight in Ounces	115	110	111	118	116	117	112

3. Tara has nine pieces of candy. She weighs the first one, adds the second to the scale, weighs it, adds the third, and so on. The weights of one to all nine together are summarized in the table below, where x represents the number of pieces of candy being weighed and y represents their total weight.

x	1	2	3	4	5	6	7	8	9
y	2.8	6.4	10	11.7	15.2	17.1	21.4	23.8	28.6

a. Create a scatter plot for the data.
b. Perform a linear regression on this data.
c. Graph the linear regression on the scatter plot.
d. Approximate the correlation coefficient to the nearest thousandth.

4. In 1997, a new organization was formed with initial membership of 5 people. The chart below shows the number of members in this organization by year. The table indicates the number of years, x, after 1997 and the number of members, y.

Year	**1997**	**1998**	**1999**	**2000**	**2001**	**2002**	**2003**	**2004**	**2005**	**2006**
x	1	2	3	4	5	6	7	8	9	10
y	5	7	11	18	28	43	86	105	164	256

a. Create a scatter plot for the data.
b. Perform an exponential regression on this data.
c. Graph the exponential regression on the scatter plot.
d. Approximate the correlation coefficient to the nearest thousandth.

5. The number of hits per day on a new web site is recorded in the table below.

Day	1	2	3	4	5	6	7	8	9	10
Hits	5	12	35	51	82	92	131	168	204	234

a. Perform a quadratic regression on this data.
b. Perform an exponential regression on this data.
c. Perform a cubic regression on this data.
d. Which type of regression fits the data best?

13.2 REGRESSION AND PREDICTION

When the correlation is strong, the regression model can be used to predict values of y for new values of x.

Interpolation and Extrapolation

Examine the data in the data set used to perform the regression. Predicting a value when the input or x-value is within the range of the x-values in the data set is called an *interpolation*. Predicting an x-value that is not within the range of the x-values in the data set is called an *extrapolation*.

TIP

Be careful when performing an extrapolation! Since the prediction is for values outside the range of the original data set, the regression model may not fit these numbers.

Example 1

Integrated Algebra	Algebra 2/Trigonometry
70	65
87	83
96	92
75	68
81	72
77	70
84	80

Your guidance counselor decided to compare student scores on the Integrated Algebra Regents Examination with the same students' scores on the Algebra 2/Trigonometry Regents Examination. She arbitrarily looked up scores for seven students. Their results are recorded in the table above. Perform a linear regression.

a. Use the linear regression to predict the Algebra 2/Trigonometry Regents score for a student in her school who scored a 90 on the Integrated Algebra Regents.
b. Use the linear regression to predict the Algebra 2/Trigonometry Regents score for a student in her school who scored a 50 on the Integrated Algebra Regents.

Solution:

```
LinReg
 y=ax+b
 a=1.104463131
 b=-14.22056921
 r²=.9666175272
 r=.9831670902
```

```
Plot1 Plot2 Plot3
\Y1=1.1044631306
597X+-14.2205692
10861
\Y2=
\Y3=
\Y4=
\Y5=
```

X	Y1	
90	85.181	
50	41.003	

X=

a. Approximately 85
b. Approximately 41

Example 2

For her science project, Anna measures the height of a plant each day for 10 days after she feeds it a chemical she hypothesizes will increase its growth. The data are summarized in the chart below.

Day	1	2	3	4	5	6	7	8	9	10
Height	17.3	23.4	26.1	26.9	27.3	27.8	28.2	28.9	28.7	29.1

A logarithmic regression was performed for this data in Example 1 on pages 363–364. Use this to estimate the height of the plant on day 15.

Solution:

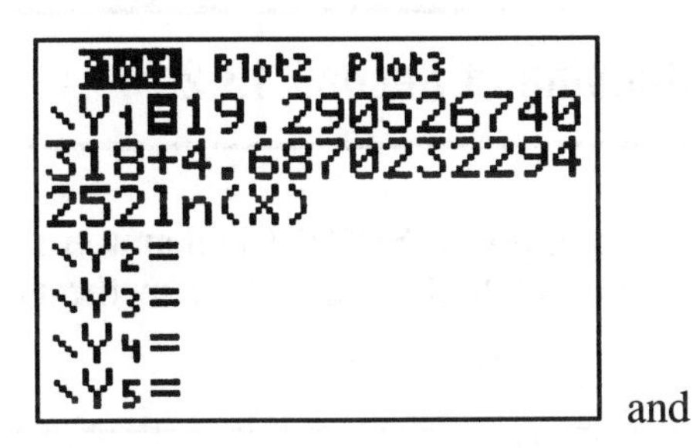

and

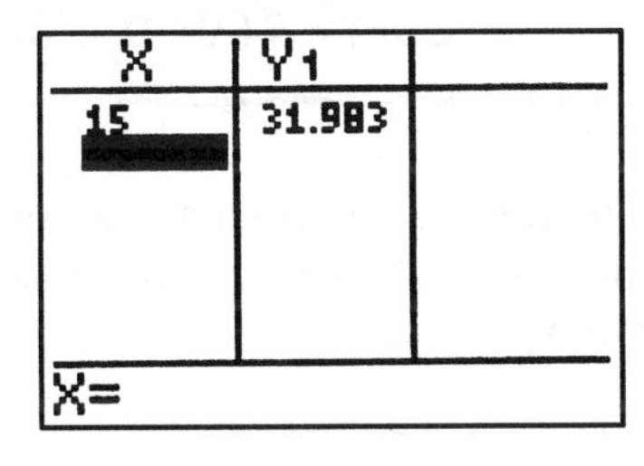

The estimated height of the plant on day 15 is approximately 32.0.

Example 3

Gino measures the height of the liquid (in centimeters) in a beaker each second after he adds a new chemical to the beaker. The data are displayed in the table that follows.

Seconds	1	2	3	4	5	6	7	8	9	10
Height	0.3	0.6	0.7	0.6	0.8	1.1	2.4	3.4	4.8	5.9

In Example 2 on pages 364–365 an exponential regression was performed. Use this to estimate the height of the liquid:

a. After 4.5 seconds
b. After 18 seconds

Solution:

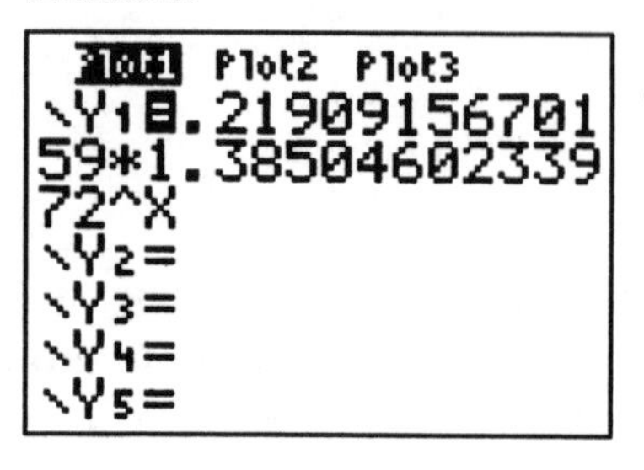

and

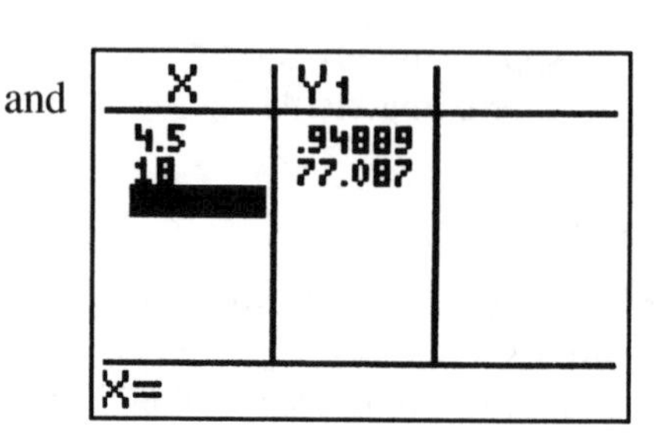

After 4.5 seconds, the height in the beaker is approximately 0.9 centimeters. After 18 seconds, the height in the beaker is approximately 77 centimeters. Note that 18 seconds is out of the range of number of seconds for which data has been observed, and 77 centimeters may not be an accurate estimate.

Check Your Understanding of Section 13.2

1. The number of students choosing to apply to SUNY Binghamton from Homer High School each year from 2004–2008 are listed in the table below.

Year	2004	2005	2006	2007	2008
#Applicants	45	42	37	32	25

Write a linear equation to model the data in the table. Estimate approximately how many students from Homer High School will apply to SUNY Binghamton in 2010.

2. The population of a town has been growing steadily since 1995. If t represents the number of years since 1995, the population, p, is displayed in the table below.

t	0	1	2	4	6	8	10
p	14,000	16,500	17,900	22,000	24,700	29,100	31,500

Write a linear equation to model the data in the table. Estimate the population of the town in 2025.

3. The amount of a radioactive material decays exponentially. The table below shows how much 1000 units of this material remains after t years.

t	0	1	2	3	4	6	10
Amount	1000	730	533	389	284	151	43

Write an exponential equation to model this data. Use that to approximate the number of units present (to the nearest whole number) after
a. 5 years b. 8 years c. 15 years

4. A company earns a profit, p from the sales of x items according to the table below.

x	1	3	6	10	15
p	$24	$26	$89	$437	$734

Perform a cubic regression on this data to estimate the profit for selling
a. 4 items b. 8 items c. 22 items

5. A company estimates that after a new product is released, the number of inquiries they will receive by phone will increase dramatically. The table below shows the number of calls, c, they receive each hour, h hours after the product is released.

h	1	2	3	6	9	12	15
c	6	13	21	49	81	114	151

The company decides to use a power regression to estimate the number of phone calls. How many phone calls (to the nearest whole number of phone calls) do they expect after
a. 4 hours b. 11 hours c. 20 hours

6. Todd invested $850 in stocks. The value of his stocks over the first 6 years of his investment is shown in the table below.

Years Since Investment	Value in $
0	$850
1	$910
2	$945
3	$987
4	$1032
5	$1094
6	$1152

Write the exponential regression equation for this set of data, rounding all values to the nearest hundredth. Using this equation, estimate the value of his stock after 12 years.

7. The chart below shows the average height (in inches) for children according to their age in years.

Age	1	2	3	4	5	6	7
Height	28	31	33	37	40	42	45

Using a linear regression, estimate the average height of a 12 year old to the nearest tenth of an inch.

Chapter Fourteen

SEQUENCES AND SERIES

14.1 SEQUENCES

KEY IDEAS

Think of a sequence as a function whose domain is limited to the positive integers. Instead of using functional notation, such as $f(1)$, $f(2)$, $f(3)$, . . . $f(n)$, symbols such as a_1, a_2, a_3, . . ., a_n are used.

Ordered Lists, Patterns

Patterns of numbers are certainly not a new concept. Look at the pattern of numbers 1, 4, 9, 16, 25, 36, 49, . . . This is probably recognizable as the perfect squares of the counting numbers (or positive integers). This pattern is a one-to-one function mapping the counting numbers to their perfect squares, as illustrated in the chart.

x	1	2	3	4	5	6	7
x^2	1	4	9	16	25	36	49

The function rule is, of course, $f(x) = x^2$. Each of the values just indicated can also be written in terms of this function rule as $f(1) = 1$, $f(2) = 4$, $f(3) = 9$, $f(4) = 16$, $f(5) = 25$, $f(6) = 36$, and $f(7) = 49$.

A sequence can simply be defined as a set of ordered numbers or a number pattern. A more rigorous definition of a sequence would be that a sequence is a one-to-one function whose domain is the set of counting numbers (or a subset of the counting numbers). The terms of a sequence are named a_1, a_2, a_3, a_4, and so on. The subscripts in this notation represent the counting number in the domain corresponding to the position or place that term has in the sequence. These correspond to the function values $f(1)$, $f(2)$, $f(3)$, $f(4)$, and so on. In the previous example, it would be appropriate to say that $a_n = n^2$ defines the nth term of this sequence. It is best to use the function rule in the form a_n to find each term of a sequence. The symbol a_n can also be referred to as the sequence generator.

Example 1

Find the first three terms of the sequence defined by $a_n = \frac{2n}{n^3}$.

$$Solution: \; a_1 = \frac{2\cdot 1}{1^3} = \frac{2}{1} = 2$$

$$a_2 = \frac{2\cdot 2}{2^3} = \frac{4}{8} = \frac{1}{2}$$

$$a_3 = \frac{2\cdot 3}{3^3} = \frac{6}{27} = \frac{2}{9}$$

Notice that even though the domain of a sequence is the counting numbers, the range is not limited to the counting numbers.

Sequences of numbers occur in the real world. One famous sequence that has a real-world application is named the Fibonacci Sequence. Its terms can be identified with such examples as measuring the fraction of a turn between successive leaves on the stalk of a plant.

Example 2

Identify the *n*th term of the sequence: 1, 3, 5, 7, 9, . . .

Solution: These numbers represent the odd numbers. Every even number is a multiple of 2. To get an odd number it is necessary to subtract 1 from the even numbers. Therefore, $a_n = 2n - 1$.

Example 3

Identify the *n*th term of the sequence: 14, 15, 16, 17, 18, . . .

Solution: These numbers are consecutive and start at 14. So $a_n = n + 13$.

Finite and Infinite Sequences

What is the difference between the following two sequences?

3, 7, 11, 15, 19, 23, 27 and 3, 7, 11, 15, . . .

The first sequence has exactly 7 terms. The second sequence has an infinite number of terms, but the first 4 of them are the ones listed in the first sequence. The first sequence is called a *finite sequence*, and the second one is called an *infinite sequence*. A finite sequence has its domain limited to a subset of the counting numbers, still starting with 1. An infinite sequence is defined on the entire set of counting numbers or on the positive integers.

Recursive Definitions and Terms of a Sequence

Sometimes a sequence can be defined recursively. That means that the first term or terms are assigned specific values and the rest of the terms are defined as a function of previous terms in the sequence.

Example 4

If $a_1 = 4$ and $a_n = 3a_{n-1} + 2$, find the first 5 terms of the sequence.

Solution: $a_1 = 4$

$a_2 = 3(4) + 2 = 14$

$a_3 = 3(14) + 2 = 44$

$a_4 = 3(44) + 2 = 134$

$a_5 = 3(134) + 2 = 404$

The sequence is 4, 14, 44, 134, 404, . . .

The Fibonacci Sequence is defined recursively as $a_1 = 1$, $a_2 = 1$, and $a_n = a_{n-2} + a_{n-1}$. Examine how this generates the terms of the Fibonacci Sequence:

$a_1 = 1$

$a_2 = 1$

$a_3 = a_1 + a_2 = 1 + 1 = 2$

$a_4 = a_2 + a_3 = 1 + 2 = 3$

$a_5 = a_3 + a_4 = 2 + 3 = 5$

$a_6 = a_4 + a_3 = 3 + 5 = 8$

This is the sequence: 1, 1, 2, 3, 5, 8, . . .

Example 5

Write the first 5 terms of the recursively defined sequence $a_1 = -2$ and $a_n = -2a_{n-1} + 6$.

Solution: $a_1 = -2$

$a_2 = -2(-2) + 6 = 10$

$a_3 = -2(10) + 6 = -14$

$a_4 = -2(-14) + 6 = 34$

$a_5 = -2(34) + 6 = -62$

Example 6

In Example 3, the task was to identify the *n*th term of the sequence: 14, 15, 16, 17, 18, . . . Now determine how to define this sequence recursively.

Solution: Since these numbers are consecutive and start at 14, then let $a_1 = 14$ and define each subsequent term as 1 more than the last one by assigning $a_n = a_{n-1} + 1$.

Check Your Understanding of Section 14.1

1. What are the first 4 terms of the sequence: $a_n = 5n - 1$?

2. What are the first 4 terms of the sequence: $a_n = n^3 + n$?

3. What are the first 4 terms of the sequence: $a_n = -2n + 4$?

4. What are the first 4 terms of the sequence: $a_n = 2^n$?

5. What are the first 4 terms of the sequence: $a_n = n!$?

6. What are the first 4 terms of the sequence: $a_n = \frac{n}{n+1}$?

7. If $a_n = n(n + 2)$, what is a_9?

8. If $a_n = \frac{5n-3}{n^2}$, what is a_7?

(1) 15 (2) $\frac{20}{49}$ (3) $\frac{32}{49}$ (4) $\frac{35}{49}$

9. What is the rule for the *n*th term, a_n, when the first 4 terms of the sequence are −8, −6, −4, −2?

(1) $a_n = 10 - 2n$
(2) $a_n = -4n - 4$
(3) $a_n = 2n - 10$
(4) $a_n = 4 - 4n$

10. What is the rule for the *n*th term, a_n, when the first 4 terms of the sequence are 3, 9, 27, 81?

(1) $a_n = (-3)^{n-1}$
(2) $a_n = (-3)^n$ 4
(3) $a_n = 3n$
(4) $a_n = 3^n$

11. What is the rule for the nth term, a_n, when the first 4 terms of the sequence are $\frac{1}{2}, \frac{1}{4}, \frac{1}{6}, \frac{1}{8}$?

(1) $a_n = \left(\frac{1}{n}\right)^2$ (2) $a_n = \frac{1}{2n}$ (3) $a_n = \left(\frac{1}{2n}\right)^2$ (4) $a_n = 2n^{-1}$

12. What is the rule for the nth term, a_n, when the first 4 terms of the sequence are 6, 9, 12, 15?

(1) $a_n = 3(n + 1)$ (3) $a_n = 3n^2$
(2) $a_n = 3n - 3$ (4) $a_n = 2(n + 2)$

13. Write the next 3 terms of the sequence: 2, 6, 10, 14, . . .

14. Write the next 3 terms of the sequence: 3, –6, 12, –24, . . .

15. Write the next 3 terms of the sequence: $\frac{3}{2}$, 1, $\frac{5}{6}$, $\frac{3}{4}$, . . . (Hint: $1 = \frac{4}{4}$ and $\frac{3}{4} = \frac{6}{8}$).

16. Write the first 5 terms of the recursively defined sequence $a_1 = 3$ and $a_n = 4a_{n-1} - 7$.

17. For the sequence whose first 5 terms are 0.1, 0.01, 0.001, 0.0001, 0.00001, identify the sequence generator, a_n as well as define the sequence recursively.

14.2 ARITHMETIC SEQUENCES

Some sequences have a pattern that results from adding the same amount between terms. This type of sequence is called an Arithmetic Sequence.

Common Difference and *n*th Term

One example referred to in the last section was the infinite sequence 3, 7, 11, 15, . . . Notice that by starting with the 3 and adding 4, the result is the next term of the sequence, 7. Similarly, by adding 4 to the 7, the result is the third term of this sequence, 11. Adding 4 to the 11 yields the 15. The amount added each time, 4, is the common difference, *d*. In the following chart, the second row shows the difference between consecutive terms:

n	1		2		3		4
a_n	3		7		11		15
d		+4		+4		+4	

Let's examine how to generate, in general form, the first 5 terms of an arithmetic sequence:

$$a_1$$

$$a_2 = a_1 + d$$

$$a_3 = a_2 + d = (a_1 + d) + d = a_1 + 2d$$

$$a_4 = a_3 + d = (a_1 + 2d) + d = a_1 + 3d$$

$$a_5 = a_4 + d = (a_1 + 3d) + d = a_1 + 4d$$

Notice that this pattern suggests a formula for the *n*th term of an arithmetic sequence. Examine the simplified form in each of the lines above. From this, observe that the subscript for the term in the sequence is always 1 more than the coefficient of *d*. Therefore, it can be concluded that the formula for the *n*th term of an arithmetic sequence is $a_n = a_1 + (n - 1)d$.

MATH FACTS

The *n*th term in an arithmetic sequence is $a_n = a_1 + (n - 1)d$.

Example 1

Identify the common difference in the arithmetic sequence, 5, 11, 17, 23, . . . and determine the 17th term in the sequence.

Solution: Since $11 - 5 = 6$, $17 - 11 = 6$, and $23 - 17 = 6$, $d = 6$. Using the formula $a_n = a_1 + (n - 1)d$ with $a_1 = 5$ and $d = 6$ yields $a_n = 5 + (n - 1)6 = 5 + 6n - 6 = 6n - 1$. Substituting $n = 17$, results in $a_n = 6(17) - 1 = 101$.

Once the values of a_1 and d have been determined, it is possible to use the graphing calculator to finish this exercise. To do this, press **MODE** and on the FUNC line scroll over and change the option to SEQ. This is what it will look like on the calculator screen:

Now press the **Y =** key and see that the screen looks different than it usually does. Below is the type of screen that comes up now. Notice that this screen allows definition of 3 sequences, *u*, *v*, and *w*.

```
 Plot1 Plot2 Plot3
 nMin=1
·.u(n)=
 u(nMin)=
·.v(n)=
 v(nMin)=
·.w(n)=
 w(nMin)=
```

To enter the arithmetic sequence where $a_1 = 5$ and $d = 6$, begin by setting *n*Min to 1 because the subscript of our first term is 1 as in a_1. For the purposes of using the calculator for this work, the notation $u(n)$ will substitute for the notation a_n. In the $u(n)$ row on this screen type in $u(n - 1) + 6$ and then set $u(n\text{Min})$ to 5 as shown in the window below.

```
 Plot1 Plot2 Plot3
 nMin=1
·.u(n)=u(n-1)+6
 u(nMin)={5}
·.v(n)=
 v(nMin)=
·.w(n)=
 w(nMin)=
```

Notice that this is using a recursive form of the definition of the nth term of the sequence. Of course, $a_n = a_1 + (n - 1)d$. It is possible to see that recursively, the first term is defined to be 5 and then each subsequent term is defined to be 6 more than the previous term. Examine the recursively defined sequence that was just entered into the calculator and see that both $6n - 1$ and $u(n - 1) + 6$ will generate the same sequence:

n	$6n - 1$	$u(n - 1) + 6$
2	$6(2) - 1 = 12 - 1 = 11$	$5 + 6 = 11$
3	$6(3) - 1 = 18 - 1 = 17$	$11 + 6 = 17$
4	$6(4) - 1 = 24 - 1 = 23$	$17 + 6 = 23$
5	$6(5) - 1 = 30 - 1 = 29$	$23 + 6 = 29$

Now press **2ND** **GRAPH** to generate a table containing the terms of this sequence. To determine the 17th term, scroll down under the n column until $n = 17$ to see that this yields the 17th term, 101, as was just calculated in this example.

n	u(n)
11	65
12	71
13	77
14	83
15	89
16	95
17	101

n=17

Example 2

What is the 8th term in the arithmetic sequence whose first four terms are 19, 15, 11, 7?

Solution: $15 - 19 = -4 = d$, $a_1 = 19$, $n = 8$. Substituting these values into $a_n = a_1 + (n - 1)d$ yields $a_8 = 19 + (8 - 1)(-4) = 19 + 7(-4) = 19 - 28 = -9$.

Example 3

In an arithmetic sequence, if $a_4 = 13$ and $a_9 = 28$, determine a_1, d, and a_n.

Solution:

Method I	Method II
Since the common difference, d, is constant,	$a_4 = a_1 + (n - 1)d = 13$
$d = \frac{28-13}{9-4} = \frac{15}{3} = 3$	$a_1 + (4 - 1)d = 13 \rightarrow a_1 + 3d = 13$
$a_4 = a_1 + (n - 1)d = 13$	$a_1 + (9 - 1)d = 28 \rightarrow a_1 + 8d = 28$
$13 = a_1 + 3(3)$	subtracting
$13 = a_1 + 9$	$a_1 + 8d = 28$
$4 = a_1$	$\underline{a_1 + 3d = 13}$
$a_n = 4 + (n - 1)d$	$5d = 15 \rightarrow d = 3$
$= 4 + (n - 1)3$	$a_1 + 3d = 13$
$= 4 + 3n - 3$	$a_1 + 3(3) = 13$
$a_n = 3n + 1$	$a_1 + 9 = 13$
	$a_1 = 4$
	$a_n = 4 + (n - 1)d$
	$= 4 + (n - 1)3$
	$= 4 + 3n - 3$
	$a_n = 3n + 1$

Sum of the First *n* Terms

Now that the method to determine the common difference d and the nth term a_n is known, it is easy to determine how to find the sum of the first n terms of an arithmetic sequence. In fact, it is reported that when the famous mathematician Carl Gauss (1777–1855) was in elementary school, his teacher gave his class a busywork assignment asking the students to add all the counting numbers from 1 to 100. Carl answered the question almost immediately because he paired numbers to see that $1 + 100 = 101$, and $2 + 99 = 101$, and so on. He easily saw that there were 50 such pairs of numbers adding to 101. To his teacher's dismay, he quickly answered 5050.

In general, adding the first n terms of an arithmetic sequence yields the sum

$$S_n = a_1 + a_2 + a_3 + \ldots + a_{n-2} + a_{n-1} + a_n.$$

Therefore,

$$S_n = a_1 + (a_1 + d) + (a_1 + 2d) + \ldots + (a_n - 2d) + (a_n - d) + a_n.$$

Reversing the terms, results in

$S_n = a_n + (a_n - d) + (a_n - 2d) + \ldots + (a_1 + 2d) + (a_1 + d) + a_1.$

Pairing the terms in order from both lines of work, yields

$2S_n = (a_1 + a_n) + (a_1 + a_n) + (a_1 + a_n) + \ldots + (a_1 + a_n) + (a_1 + a_n) + (a_1 + a_n).$

This can be simplified as $2S_n = n(a_1 + a_n)$ or $S_n = \frac{n}{2}(a_1 + a_n)$ or $S_n = \frac{n}{2}(a_1 + a_1 + (n-1)d)$, yielding $S_n = \frac{n}{2}(2a_1 + (n-1)d)$. This is the formula for the sum of the first n terms of an arithmetic sequence.

MATH FACTS

The sum of *n* terms in an arithmetic sequence is either

$$S_n = \frac{n}{2}(a_1 + a_n) \quad \text{or} \quad S_n = \frac{n}{2}(2a_1 + (n-1)d)$$

Example 4

Determine the sum of the first 20 terms of the sequence 2, 5, 8, 11, . . .

Solution:

$S_n = \frac{n}{2}(2a_1 + (n-1)d)$ where $n = 20$, $d = 3$, and $a_1 = 2$.

$$S_{20} = \frac{20}{2}[2(2) + (20-1)3] = 10[4 + 19(3)] = 10(4 + 57) = 10(61) = 610$$

Example 5

Determine the sum of the first 15 terms of the sequence 13, 9, 5, 1, . . .

Solution: $S_n = \frac{n}{2}(2a_1 + (n-1)d)$ and $n = 15$, $d = -4$, and $a_1 = 13$.

$$S_{15} = \frac{15}{2}[2(13) + (15-1)(-4)] = 7.5[26 + 14(-4)]$$
$$= 7.5(26 - 56) = 7.5(-30)$$
$$= -225$$

Once again, a graphing calculator can be used to determine the sum of any number of terms in an arithmetic sequence. To do this, press **MODE** and highlight SEQ as was done in Example 1. Press **Y =** and enter the recursive form for the nth term, just as it was done in Example 1. Then set $u(n\text{Min})$ to the first term of the sequence, 13. To define $u(n)$ recursively so that the calculator will add –4 to the previous term in the sequence, enter it as $u(n) = u(n-1) - 4$. This alone will generate the sequence. However, to generate the sum of the first 15 terms, it is also necessary to use $v(n)$. Here simply enter a formula recursively that creates a sequence where each term is the sum from the previous term plus the new term. In $v(n)$ just enter $v(n-1) + u(n-1) - 4$. Now enter 13 as $v(n\text{Min})$ to provide the first term. This is what the calculator screen should look like:

```
Plot1 Plot2 Plot3
nMin=1
·.u(n)=u(n-1)-4
 u(nMin)={13}
·.v(n)=v(n-1)+u(n
-1)-4
 v(nMin)={13}
·.w(n)=
```

Now press **2ND** **GRAPH** to generate a table containing the terms of the sequence. The display now has three columns. To find the sum of the first 15 terms, scroll down the n column to 15 and find the answer –225 under the $v(n)$ column in that row as shown below:

n	u(n)	v(n)
9	-19	-27
10	-23	-50
11	-27	-77
12	-31	-108
13	-35	-143
14	-39	-182
15	-43	[illegible]

v(n)=-225

Now see that the fifteenth term of the sequence is –43 and the sum of all fifteen terms is –225.

Example 6

Determine the sum of $4 + 7 + 10 + \ldots + 37$.

Solution: $37 - 4 = 33$ and $d = 3$, so $n-1=\dfrac{33}{3}=11$, $n = 12$.

Note that $a_n - a_1$ tells how many times d has been added to the first term, a_1. Therefore, the number of the last term (the n in a_n) is 1 more than this difference.

$$S_n = \frac{n}{2}(a_1 + a_n) = \frac{12}{2}(4+37) = 6(41) = 246$$

Check Your Understanding of Section 14.2

1. What is the common difference in each of the following arithmetic sequences?
a. −8, −1, 6, 13, . . .
b. 120, 108, 96, 84, . . .
c. 5, 9, 13, 17, . . .
d. 3, 31, 59, 87, . . .

2. What is the 12th term in the arithmetic sequence whose first 4 terms are 14, 19, 24, 29?
(1) 5 (2) 64 (3) 69 (4) 74

3. What is the 19th term in the arithmetic sequence whose first 4 terms are 3, 9, 15, 21?
(1) 111 (2) 105 (3) 117 (4) 115

4. What is the 53rd term in the arithmetic sequence whose first 4 terms are −8, −4, 0, 4?
(1) −216 (2) −212 (3) 204 (4) 200

5. What is the 30th term in the arithmetic sequence whose first 4 terms are 34, 27, 20, 13?
(1) −50 (2) −169 (3) 237 (4) −176

6. What is the 27th term in the arithmetic sequence whose first 4 terms are 2, 11, 20, 29?
(1) 288 (2) 254 (3) 245 (4) 236

7. Write a recursive formula that generates the sequence 21, 25, 29, 33, . . .

8. Determine the sum of the first 8 terms of the arithmetic sequence whose first 4 terms are 8, 11, 14, 17.

9. Determine the sum of the first 17 terms of the arithmetic sequence whose first 4 terms are −15, −9, −3, 3.

10. Determine the sum of the first 40 terms of the arithmetic sequence whose first 4 terms are 6, 17, 28, 39.
(1) 435 (2) 446 (3) 8820 (4) 8385

11. Determine the sum of $22 + 16 + 10 + \ldots + -80$.
(1) −261 (2) −493 (3) −522 (4) −551

12. Determine the sum of all the even integers from 2 to 50.

13. Determine the sum of all the integers that are multiples of 3 from 12 to 81.

14. Determine a_1 in an arithmetic sequence in which $a_8 = 42$ and $d = 8$.

15. Ted takes a job that pays $3000 the first month and increases $100 each month for the first 2 years. What will his salary be at the end of the 2 years? How much did he earn over these 2 years?

14.3 GEOMETRIC SEQUENCES

Some sequences have a pattern that results from multiplying by the same amount between terms. This type of sequence is called a Geometric Sequence.

Common Ratio and *n*th Term

Look at the sequence 2, 6, 18, 54, . . . Notice by multiplying each term by 3, the result is the next term. This type of sequence is called a Geometric Sequence. The ratio between one term of a Geometric Sequence and the previous term is called the common ratio, r. This can also be written as $r = \frac{a_{n+1}}{a_n}$.

n	1		2		3		4
a_n	2		6		18		54
r		× 3		× 3		× 3	

Let's examine how to generate, in general form, the first 5 terms of a geometric sequence:

$$a_1$$

$$a_2 = a_1 r$$

$$a_3 = a_2\, r = (a_1 r)r = a_1 r^2$$

$$a_4 = a_3 r = (a_1 r^2)r = a_1 r^3$$

$$a_5 = a_4 r = (a_1 r^3)r = a_1 r^4$$

Notice that this pattern suggests a formula for the nth term of a geometric sequence. Examine the simplified form in each of the lines above and notice that the subscript for the term in the sequence is always 1 more than the superscript (exponent) of r. Therefore, it can be deduced that the formula for the nth term of a geometric sequence is $a_n = a_1 r^{n-1}$.

MATH FACTS

The nth term in a geometric sequence is $a_n = a_1 r^{n-1}$.

Example 1

Identify the common ratio in the geometric sequence, 3, 9, 27, 81, . . . and determine the 12th term in the sequence.

Solution: $r = \frac{9}{3} = 3$

Using the formula $a_n = a_1 r^{n-1}$, the result is $a_{12} = 3(3)^{11} = 3(177{,}147) = 531{,}441$.

Check that the graphing calculator is in the correct mode by pressing **MODE** and then highlighting SEQ. Now press **Y =**. In the $u(n)$ row, type $u(n-1) * 3$. Set the $u(n\text{Min})$ row equal to 3. Then press **2ND** **GRAPH**.

```
Plot1 Plot2 Plot3
nMin=1
\u(n)=u(n-1)*3
 u(nMin)={3}
\v(n)=
 v(nMin)=
\w(n)=
 w(nMin)=
```

and

n	u(n)
6	729
7	2187
8	6561
9	19683
10	59049
11	177147
12	[illegible]

u(n)=531441

The same answer for a_{12} is determined, 531,441.

Example 2

What is the 10th term in the geometric sequence whose first four terms are 2, 3, $\frac{9}{2}$, $\frac{27}{4}$? Write a recursive formula that will generate the same sequence.

Solution: $r = \frac{3}{2}$, $a_1 = 2$, $n = 10$. Substituting these values into

$$a_n = a_1 r^{n-1} \text{ yields } a_{10} = 2\left(\frac{3}{2}\right)^{(10-1)} = 2\left(\frac{3}{2}\right)^9 = 2\left(\frac{19683}{512}\right) = \frac{19683}{256}.$$

Recursively defined, $a_1 = 2$ and $a_n = (a_n - 1)\left(\frac{3}{2}\right)$.

Example 3

In a geometric sequence, if $a_3 = 9$ and $a_6 = \frac{9}{8}$, determine a_1, r, and a_n.

Solution: $a_3 = 9 = a_1 r^2$

$$a_6 = \frac{9}{8} = a_1 r^5 = (a_1 r^2) r^3 = 9r^3$$

$$9r^3 = \frac{9}{8} \rightarrow r^3 = \frac{1}{8} \rightarrow r = \frac{1}{2}$$

$$a_3 = 9 = a_1 r^2$$

$$9 = a_1\left(\frac{1}{2}\right)^2$$

$$\frac{1}{4} a_1 = 9 \rightarrow a_1 = 36$$

By substituting the solved values of r and a_1 into the formula $a_n = a_1r^{n-1}$, the result is $a_n = 36\left(\frac{1}{2}\right)^{n-1}$.

Sum of the First *n* Terms

Now let's find the common ratio r and the nth term, a_n. It is easy to determine how to find the sum of the first n terms of an arithmetic sequence.

In general, if the first n terms of a geometric sequence are added, then the sum is $S_n = a_1 + a_2 + a_3 + \ldots + a_{n-2} + a_{n-1} + a_n$.

Therefore, $S_n = a_1 + a_1r + a_1r^2 + \ldots + a_1r^{n-3} + a_1r^{n-2} + a_1r^{n-1}$.

If both sides of this equation are multiplied by r, the result is,

$$rS_n = a_1r + a_1r^2 + a_1r^3 + \ldots + a_1r^{n-2} + a_1r^{n-1} + a_1r^n.$$

Now subtract $S_n - rS_n$.

$$S_n - rS_n = (a_1 + a_1r + a_1r^2 + \ldots + a_1r^{n-3} + a_1r^{n-2} + a_1r^{n-1})$$
$$- (a_1r + a_1r^2 + a_1r^3 + \ldots + a_1r^{n-2} + a_1r^{n-1} + a_1r^n)$$

Notice that all the terms in the first parentheses, except a_1, are also in the second parentheses. In the second parentheses, there is an additional term, a_1r^n. Therefore, by subtracting, all these terms cancel out except for the a_1 and the a_1r^n. The result is $S_n - rS_n = (a_1 - a_1r^n)$. Factoring the left side of this equation results in $S_n(1 - r)$. Dividing by $1 - r$ on both sides of the equation yields the formula for the sum of the first n terms of a geometric sequence.

MATH FACTS

The sum of *n* terms in a geometric sequence:

$$S_n = \frac{a_1 - a_1r^n}{1-r} \quad or \quad \frac{a_1(1-r^n)}{1-r} \text{ where } r \neq 1$$

Example 4

Determine the sum of the first 15 terms of the geometric sequence 1, 2, 4, 8, . . .

Solution: $S_n = \dfrac{a_1(1-r^n)}{1-r}$ where $n = 15$, $r = 2$, and $a_1 = 1$

$$S_n = \frac{1(1-2^{15})}{1-2} = \frac{1-32768}{-1} = 32{,}767$$

Once again, this problem can be solved using a graphing calculator. Just follow that same process for setting the mode. Then press **Y =** and insert $u(n-1) * 2$ in the $u(n)$ row, 1 in the $u(n\text{Min})$ row, $v(n-1) + u(n-1) * 2$ in the $v(n)$ row, and 1 in the $v(n\text{Min})$ row. Then press **2ND** **GRAPH**.

```
Plot1 Plot2 Plot3
nMin=1
\u(n)=u(n-1)*2
 u(nMin)={1}
\v(n)=v(n-1)+u(n
-1)*2
 v(nMin)={1}
\w(n)=
```

and

n	u(n)	v(n)
9	256	511
10	512	1023
11	1024	2047
12	2048	4095
13	4096	8191
14	8192	16383
15	16384	32767

n=15

Once again, this results in the same answer for S_{15}, 32767.

Example 5

Determine the sum of the first 11 terms of the geometric sequence 2, –6, 18, –54, . . .

Solution: $S_n = \dfrac{a_1(1-r^n)}{1-r}$ where $n = 11$, $r = -3$, and $a_1 = 2$.

$$S_n = \frac{2\left(1-(-3)^{11}\right)}{1-(-3)} = \frac{2(1-(-177147))}{4} = \frac{(177148)}{2} = 88{,}574$$

Example 6

Determine the sum of the first 6 terms of the geometric sequence 1000, 200, 40, 8, . . .

Solution: $S_n = \dfrac{a_1(1-r^n)}{1-r}$ where $n = 6$, $r = 0.2$, and $a_1 = 1000$

$$S_7 = \frac{1000\left(1-(0.2)^6\right)}{1-0.2} = \frac{1000(1-(0.000064))}{0.8}$$
$$= \frac{1000(0.999936)}{0.8} = \frac{999.936}{0.8} = 1{,}249.92$$

Example 7

Determine the sum of $36 + -18 + 9 + \ldots + \dfrac{-9}{8}$.

Solution: Since $a_n = a_1 * r^{n-1}$ and $a_1 = 36$, $r = \dfrac{36}{-18} = \dfrac{-1}{2}$, $\dfrac{-9}{8} = 36 * \left(\dfrac{-1}{2}\right)^{n-1}$.

That implies that $\dfrac{-1}{32} = \left(\dfrac{-1}{2}\right)^{n-1}$. This can be solved to determine that $n - 1$

$= 5$ and $n = 6$. Since $S_n = \dfrac{a_1(1-r^n)}{1-r}$, substitute to get

$$S_n = \frac{36\left(1-\left(\frac{-1}{2}\right)^6\right)}{1-\frac{-1}{2}} = \frac{36\left(1-\frac{1}{64}\right)}{\frac{3}{2}} = \frac{36\left(\frac{63}{64}\right)}{\frac{3}{2}} = 36\left(\frac{63}{64}\right)\frac{2}{3} = \frac{189}{8} = 23.625$$

or $23\frac{5}{8}$.

Check Your Understanding of Section 14.3

1. In each of the following geometric sequences, identify the common ratio, write the sequence generator a_n, and define the sequence with a recursive formula:
   a. −8, 4, −2, 1, . . .
   b. 2, 8, 32, 128, . . .
   c. −100, −10, −1, −0.1, . . .
   d. 1, 6, 36, 216, . . .

2. Determine the 9th term in the geometric sequence whose first 4 terms are 20, 40, 80, 160.
   (1) 10,240 (2) 5,120 (3) 2560 (4) 180

3. Determine the 20th term in the geometric sequence whose first 4 terms are –6400, 3200, –1600, 800.

(1) $\frac{-25}{4096}$ (2) $\frac{25}{4096}$ (3) $\frac{25}{2048}$ (4) $\frac{25}{8192}$

4. Determine the 10th term in the geometric sequence whose first 4 terms are 3, 15, 75, 375.

(1) 98,415 (2) 5,859,375 (3) 29,296,875 (4) 146,484,375

5. What is the 12th term in the geometric sequence whose first 4 terms are 40, 4, 0.4, 0.04?

6. Determine the sum of the first 12 terms of the geometric sequence 2, 6, 18, 54, . . .

(1) 531,440 (2) 177,146 (3) 1,594,322 (4) 354,294

7. Determine the sum of the first 9 terms of the geometric sequence 6250, 25,000, 100,000, 400,000, . . .

(1) 136,531,250 (3) 546,131,250
(2) 409,600,000 (4) 2,184,531,250

8. Determine the sum of the first 11 terms of the geometric sequence –64, 32, –16, 8, . . .

(1) $\frac{-1}{16}$ (2) $\frac{-341}{8}$ (3) $\frac{-1365}{32}$ (4) $\frac{-2047}{48}$

9. Determine the sum of $\frac{1}{100}+\frac{1}{5}+4+...+640{,}000$.

10. In a geometric sequence, if $a_1 = 5000$ and $r = 0.01$, determine S_5.

11. In a geometric sequence, if $a_4 = 64$ and $a_7 = 4096$, determine S_{10}.

12. In a geometric sequence, if $a_4 = 12$ and $a_7 = -96$, determine a_1, r, and a_n.

13. A certain bacteria triples in a culture every hour. If the original culture had 120 bacteria present, how many will be present after 7 hours?

14. Tina sent out a chain e-mail to 5 people that asked each of them to send the e-mail to 5 other people. Assume that no one breaks the chain and that no one person receives the letter more than once. If the process is repeated 6 times, how many people will receive Tina's chain e-mail?

14.4 SERIES

A series is the sum of the terms of a sequence. Without realizing it, this concept was as already been introduced when the sum of the first n terms of a sequence was found.

Sum of the Terms

In Section 14.2, the sum of n terms of an Arithmetic Sequence was determined using the formula $S_n = \frac{n}{2}(2a_1 + (n-1)d)$. For instance, in Example 6 on pages 387–388, the sum of 4 + 7 + 10 + . . . + 37 was determined to be 246. Notice that in this example, rather than asking to find the sum of the first 12 terms of the sequence 4, 7, 10, . . . the question put each term of the sequence into an equation. Similarly, in Section 14.3 on page 394, the same type of situation was encountered in Example 7 in which the question was to calculate the sum of $36 + (-18) + 9 + ... + \frac{-9}{8}$. The only difference now is that the sequence whose terms were being added was a geometric sequence and the formula used was $S_n = \frac{a_1(1-r^n)}{1-r}$.

Finite or Infinite

All of the series examined in Sections 14.2 and 14.3 were sums of finite sequences. A finite sequence is a sequence that has a finite number of terms and is represented by a function whose domain consists of the first n positive integers.

An infinite series represents the sum of the terms of an infinite sequence. In Example 4 on page 386, the task was to determine the sum of the first 20 terms of the sequence 2, 5, 8, 11, . . . Although the question was adding only the first 20 terms, the three dots after the term 11 indicates that on page 386, the sequence itself has an infinite number of terms. The domain of an infinite sequence is the entire set of positive integers or counting numbers.

Can the sum of a series associated with an arithmetic sequence be calculated? For each of the examples already examined, relatively small values of d, the common difference, were used. For example, if $d = 2$, think about how large the terms will eventually get by adding 2 to a_1 many times. If a_1 were equal to 5, the 100th term of the sequence would be 205. By adding 2 a total of 10,000,000 times, the result now starts to approach a pretty large number. By putting these terms in a series rather than in a sequence, the sum will be even larger. It doesn't take too long to realize that in an infinite arithmetic series, the sum of its terms will be infinite.

Notice that in all the arithmetic sequences that were previously studied, the sequence generator, a_n, was always in the form $a_n = dn + c$, where d is the common difference and $c = a_1 - d$. For instance, in Example 3 on pages 384–385, $a_n = 3n + 1$. In functional notation, that would compare with $f(x) = 3x + 1$. Notice that this is a linear function. In a linear function, as x increases without bound, y will either increase or decrease without bound. The sum of these terms will be infinite.

What about a geometric series? Think about the examples previously examined for geometric sequences. Some gave very large numbers, and others resulted in relatively small answers. Let's explore a few on the calculator.

Let's input values for a_1 and r for a geometric series and examine the sum of a few large values of n. First check the calculator's mode. Press **MODE**, scroll down to FUNC, and highlight FUNC by pressing **ENTER**. Then press **3** **STO>** **ALPHA** **A** **ENTER** and **2** **STO>** **ALPHA** **R** **ENTER**.

Now enter the function rule for Y= as indicated in the window below:

```
 Plot1  Plot2  Plot3
\Y1=A(1-R^X)/(1-
R)
\Y2=
\Y3=
\Y4=
\Y5=
\Y6=
```

Now press **2ND** **WINDOW** which goes to TABLE SETUP. Set TblStart to 1 and ΔTbl to 1. On the line for Indpnt, choose the Ask command as shown below:

Now press **2ND** **GRAPH**, which is TABLE. Type in 10 for the value of x and enter and then type in **5** **0** for x followed by **ENTER**, **1** **0** **0** for x followed by **ENTER**, and finally **1** **0** **0** **0** for x and then **ENTER**. Each time a value is input for x, the corresponding value for Y_1 (the sum of that many terms of the series) also appears. This is what should now be displayed on the calculator:

X	Y1	
10	3069	
50	3.4E15	
100	3.8E30	
200	4.8E60	
1000	ERROR	

X=

Even at $n = 10$, the y-value displayed is 3,069. At $n = 50$, the sum is 3.4E15. This means 3.4×10^{15} or 3,400,000,000,000,000. When $n = 1000$ is input, the calculator indicates an error because the number is extremely large.

Now let's repeat this process but change the values of a_1 and r. This time, lets change r to $\frac{1}{2}$. Go to the main screen and enter **1** **÷** **2** **STO>** **ALPHA** **R** **ENTER**. Keep the **Y =** entry as it is the formula for the sum of a series. Now press **2ND** **GRAPH**, which is TABLE. The calculator has adjusted the function Y_1 for the new value of r with the following result:

X	Y1	
10	5.9941	
50	6	
100	6	
200	6	
1000	6	

X=10

This display may be misleading because the calculator shows the Y_1 values for 50, 100, 200 and 1000 to be exactly 6 when, in fact, the values are actually just so close to 6 that the calculator was unable to display the differences between the actual answers and 6. Clearly, for this value of r, the series fairly quickly approaches a sum of 6 (or converges to 6) and does not get larger.

Change the values of a_1 and r again several times and examine what happens. Now it is possible to notice that the sum of a geometric series gets extremely large (or diverges) for values of r that are greater than 1. If r is less than −1, the sum becomes extremely small (or diverges). When r is a fraction between −1 and 1, the sum tapers off to some specific value. Of course, remember that when $r = 1$, the terms of the sequence remain constant and the series diverges. So the only arithmetic or geometric series whose sum you can evaluate is a geometric series with $|r| < 1$.

MATH FACTS

The sum of an infinite geometric series:

$$S_n = \frac{a_1}{1-r} \quad \textbf{where} \quad |r| < 1$$

Example 1

Determine the sum of the infinite geometric series 4 + 0.4 + 0.004 +. . .+ $4(0.1)^{n-1}$ + . . .

Solution: In this series, $a_1 = 4$ and $r = 0.1$. Since $|r| < 1$, use the formula, $S_n = \frac{a_1}{1-r}$ and substitute the values to get $S_n = \frac{4}{1-0.1} = \frac{4}{0.9} = 4.44\ \ldots$ or $4\frac{4}{9}$.

Sigma Notation

Sums of consecutive terms of a function can be written in a notation, called sigma notation. Sigma is a Greek letter, Σ, that is commonly used in mathematics to represent summation.

Sigma notation for the sum of n terms of a function can be written as $f(1) + f(2) + f(3) + \ldots + f(n) = \sum_{i=1}^{n} f(i)$. Notice that on the bottom of the sigma, the notation tells you that the first value of $i = 1$. The notation above the sigma tells us the last value of $i = n$.

Example 2

Write the following summation using sigma notation: 1! + 2! + 3! + . . . + 7!

Solution: $\sum_{i=1}^{7} i!$

Example 3

Write the following summation using sigma notation: $\frac{1}{3}+\frac{1}{4}+\frac{1}{5}+\ldots+\frac{1}{38}$

Solution: $\sum_{i=3}^{38} \frac{1}{i}$

Notice that this can also be written as $\sum_{i=1}^{35} \frac{1}{i+2}$.

Example 4

Rewrite $\sum_{i=5}^{8} (4-i)^2$ without the sigma notation. Then calculate the sum.

Solution: $(4-5)^2 + (4-6)^2 + (4-7)^2 + (4-8)^2$

$= (-1)^2 + (-2)^2 + (-3)^2 + (-4)^2$

$= 1 + 4 + 9 + 16 = 30$

There are some interesting properties for summations. For instance, think of a finite geometric series whose terms are from a geometric sequence whose common ratio is 1. How can such a series be represented in sigma notation? Let's make the first term of the series a. And with $r = 1$, the series would look like $\underbrace{a+a+a+\ldots+a}_{n\ terms}$. From this, it is possible to realize that the sum of the series is $n \cdot a$.

Can the answers to $\sum_{i=1}^{5} i^2$ and $\sum_{i=1}^{5} 3i$ be used to calculate $\sum_{i=1}^{5} (i^2 + 3i)$?

MATH FACTS

Properties of Summations in Sigma Notation

$$\sum_{i=1}^{n} c = n \times c \text{, where } c \text{ is a constant}$$

$$\sum_{i=1}^{n}(f(i)+g(i)) = \sum_{i=1}^{n} f(i) + \sum_{i=1}^{n} g(i)$$

$$\sum_{i=1}^{n}(f(i)-g(i)) = \sum_{i=1}^{n} f(i) - \sum_{i=1}^{n} g(i)$$

$$\sum_{i=1}^{n} f(i) = \sum_{i=1}^{j} f(i) + \sum_{i=j+1}^{n} f(i)$$

Let's just examine one of these formulas to see why these facts are true:

$$\sum_{i=1}^{n}(f(i)+g(i)) = \sum_{i=1}^{n} f(i) + \sum_{i=1}^{n} g(i)$$

$$\begin{aligned}
&\sum_{i=1}^{n}(f(i)+g(i)) \\
&= (f(1)+g(1)) + (f(2)+g(2)) + (f(3)+g(3)) + \ldots (f(n)+g(n)) \\
&= f(1)+g(1)+f(2)+g(2)+f(3)+g(3)+\ldots+f(n)+g(n) \\
&= f(1)+f(2)+f(3)+\ldots+f(n)+g(1)+g(2)+g(3)+\ldots+g(n) \\
&= (f(1)+f(2)+f(3)+\ldots+f(n)) + (g(1)+g(2)+g(3)+\ldots+g(n)) \\
&= \sum_{i=1}^{n}(f(i)) + \sum_{i=1}^{n}(g(i))
\end{aligned}$$

Example 5

$\sum_{n=1}^{6} a_n = 25$ and $\sum_{n=1}^{6} b_n = 10$, use the properties of sigma notation to evaluate each of the following:

a. $\sum_{n=1}^{6} 4 \cdot a_n$

b. $\sum_{n=1}^{6} (a_n + b_n)$

c. $\sum_{n=1}^{6} (2 \cdot a_n - 3 \cdot b_n)$

d. If $\sum_{n=1}^{3} b_n = 4$, determine $\sum_{n=4}^{6} b_n$

Solution: a. $\sum_{n=1}^{6} 4 \cdot a_n = 4(25) = 100$

b. $\sum_{n=1}^{6} (a_n + b_n) = 25 + 10 = 35$

c. $\sum_{n=1}^{6} (2 \cdot a_n - 3 \cdot b_n) = 2(25) - 3(10) = 20$

d. If $\sum_{n=1}^{3} b_n = 4$ and $\sum_{n=1}^{6} b_n = 10$, then $\sum_{n=4}^{6} b_n = 10 - 4 = 6$.

Representing the Sum of a Series Using Sigma Notation

Example 6

Write the summation 15 + 18 + 21 + . . . + 42 using sigma notation. Determine the indicated sum.

Solution: Notice that the terms of this series are part of an arithmetic sequence with $d = 3$ and $a_1 = 15$. Therefore, determine the sequence generator and rephrase the example in sigma notation. Since $a_n = a_1 + (n - 1)d = 15 + (n - 1)3 = 15 + 3n - 3$ or $3n + 12$. Additionally, $3n + 12 = 42$ when $n = 10$. In sigma notation, the sequence is $\sum_{i=1}^{10} (3i + 12)$. Applying the formula $S_n = \frac{n}{2}(a_1 + a_n)$ results in $S_n = \frac{10}{2}(15 + 42) = 5(57) = 285$.

Example 7

Write the summation 256 + 128 + 64 + . . . + 2 using sigma notation. Determine the indicated sum.

Solution: Notice that the terms of this series are part of a geometric sequence with $r = 0.5$ and $a_1 = 256$. Since the sequence is finite, $a_n = a_1 r^{n-1}$.

$$2 = 256(0.5)^{n-1}$$

$$\frac{2}{256} = \left(\frac{1}{2}\right)^{n-1}$$

$$\frac{1}{128} = \frac{1}{2^{n-1}}$$

$$128 = 2^{n-1}$$

$$2^7 = 2^{n-1}$$

$$n - 1 = 7$$

$$n = 8$$

Use the formula for the sum of a finite geometric series.

$$\begin{aligned} S_n &= \sum_{n=1}^{8} 256(0.5)^{n-1} \\ &= \frac{a_1(1-r^n)}{1-r} \\ &= \frac{256(1-(0.5)^8)}{1-0.5} \\ &= \frac{256(1-0.00390625)}{0.5} \\ &= \frac{265(0.99609375)}{0.5} \\ &= \frac{255}{0.5} \\ &= 510 \end{aligned}$$

Example 8

Write the summation $100 + 10 + 1 + 0.1 + \ldots + 100(0.1)^{n-1} + \ldots$ using sigma notation. Determine the indicated sum.

Solution: Notice that the terms of this series are part of a geometric sequence with $r = 0.1$ and $a_1 = 100$. Since the sequence is infinite and $|r| < 1$, $S_n = \sum_{n=1}^{\infty}\left(100(0.1)^{n-1}\right) = \frac{a_1}{1-r} = \frac{100}{1-0.1} = \frac{100}{0.9} = 111.11\ldots \text{ or } 111\frac{1}{9}$ or $111\frac{1}{9}$.

Example 9

Determine: $\sum_{n=1}^{\infty}\left(6(-0.3)^{n-1}\right)$

Solution: Notice that the series generator is that of an infinite geometric series with $a_1 = 6$ and $r = -0.3$. It is also important to note that $|r| < 1$. Using the formula, you get $S_n = \frac{a_1}{1-r} = \frac{6}{1-(-0.3)} = \frac{6}{1.3} \approx 4.615$.

Example 10

Determine: $\sum_{n=1}^{\infty}\left(24(2.5)^{n-1}\right)$.

Solution: Notice that the series generator is that of an infinite geometric series with $a_1 = 24$ and $r = 2.5$. Since $|r| > 1$, the series diverges.

Check Your Understanding of Section 14.4

1. Determine: $\sum_{i=1}^{6} i^2$

2. Determine: $\sum_{i=1}^{6} 5i$

3. Determine: $\sum_{i=1}^{6} 6$

4. Using your answers to questions 1–3 above, determine each of the following:

a. $\sum_{i=1}^{6}(i^2+5i)$ b. $\sum_{i=1}^{6} 3i^2$ c. $\sum_{i=1}^{6}(5i-6)$

d. $\sum_{i=1}^{6}(2i^2+10i)$ e. $\sum_{i=1}^{6}(5i+12)$ f. $\sum_{i=1}^{6}(4i^2+15i+24)$

5. Write the summation 800 + 160 + 32 + . . . + 1.28 using sigma notation. Determine the indicated sum.

6. Write the summation (–30) + (–35) + (–40) + . . . + (–80) using sigma notation. Determine the indicated sum.

7. Write the summation $12+4+\frac{4}{3}+\ldots+12\left(\frac{1}{3}\right)^{n-1}$ +. . . using sigma notation. Determine the indicated sum.

8. Write the summation 25 + 5 + 1 + . . . + $25(0.2)^{n-1}$ + . . . using sigma notation. Determine the indicated sum.

9. Determine the sum of an infinite geometric series –27 + 9 + –3 + 1 + . . .

(1) $\frac{81}{4}$ (2) $\frac{81}{2}$ (3) $\frac{-81}{4}$ (4) $\frac{-81}{2}$

10. Martha has a choice of receiving $100 each day for the next 25 days or a penny on day 1 and doubling the amount she receives each day for the each of the remaining 24 days. Which offer should she take? (Hint: Compare $\sum_{n=1}^{25} 100$ with $\sum_{n=1}^{25}\left(0.01(2)^{n-1}\right)$.)

Answers and Solution Hints to Practice Exercises

CHAPTER 1

Section 1.1

1. (3)
2. (3)
3. (2)
4. (2)
5. (1)
6. (4)
7. (1)
8. (3)
9. (4)
10. (4)
11. (2)
12. $4y^4 - \frac{7}{15}y^3 + \frac{34}{5}y^2 + \frac{7}{3}y - \frac{3}{8}$
13. $\frac{5}{6}s + \frac{1}{4}$ cm

Section 1.2

1. (4)
2. (3)
3. (4)
4. (1)
5. (4)
6. (3)
7. (2)
8. (1)
9. (4)
10. (2)
11. (2)
12. (4)
13. $(a - 4)(a - 7)$
14. $(x + 5y)(5x + y)$
15. $(r - 2s)(3r + 4s)$
16. $(p + 8)(p - 1)$
17. $(t^2 + 5)(t^2 - 5)$
18. $(y - 14)(y + 2)$
19. $(5a + 3b)(4a + 7b)$
20. $\left(\frac{2}{3}x + \frac{4}{5}y\right)\left(\frac{2}{3}x - \frac{4}{5}y\right)$
21. $(r^2x - 5s)(r^2x - 3s)$
22. $3p(p + 4)(p + 2)$
23. $(3j + 5k)(3j - 5k)(9j^2 + 25k^2)$
24. $5y(2x + 3)(x - 4)$
25. $3a^2b(5a + 2b)(5a - 2b)$
26. $6mn(m - 3)(m - 15)$
27. $3ax(x + 8)(x + 12)$

Section 1.3

1. (3)
2. (1)
3. (2)
4. (3)
5. (2)
6. (4)
7. (4)
8. (1)
9. (2)
10. $\frac{6-k}{5}$
11. $\frac{2x-30}{x^2+2x-15}$
12. $\frac{2x+5}{x-3}$
13. $\frac{-2x+y}{x-y}$
14. $\frac{-2x-5}{x+3}$
15. $\frac{1}{c^4}$
16. $\frac{7z-47}{-z^2+8z}$
17. $\frac{3a^2+29a+60}{8a^2+58a+90}$
18. $\frac{2r+4}{2r+1}$
19. $\frac{x^3-x^2-11x-33}{x^2-9}$
20. $\frac{1-20a^2b^3}{20a^2b^2}$

Section 1.4

1. (2) **5.** (1) **8.** (4) **11.** $\frac{x^4-2x^3+2x-1}{x^2}$

2. (4) **6.** (2) **9.** $\frac{346}{64}$ **12.** $\frac{1-y}{y^2}$

3. (1) **7.** (3) **10.** $\frac{3s^4}{5r^{11}}$ **13.** $\frac{48q^5}{p^7}$

4. (3)

Section 1.5

1. (3) **6.** (3) **11.** (2) **16.** $-35-6\sqrt{2}$

2. (1) **7.** (3) **12.** (2) **17.** $20\sqrt[6]{243}$

3. (4) **8.** (3) **13.** (3) **18.** $s^{\frac{13}{12}}$

4. (1) **9.** (4) **14.** $\frac{16+5\sqrt{6}}{53}$ **19.** $-\frac{6\sqrt{2}+18}{7}$

5. (2) **10.** (2) **15.** $\frac{5\sqrt{15}-23}{77}$

Section 1.6

1. (2) **8.** (4) **15.** $x=-3$ **21.** 36 and 100

2. (3) **9.** (1) **16.** $\{z: z<-2$ or $2<z\le 8\}$ **22.** $x=17$

3. (4) **10.** (4)

4. (1) **11.** (2) **17.** $x>18$ **23.** 12 feet

5. (2) **12.** (1) **18.** 8 mph **24.** 25 questions

6. (1) **13.** (3) **19.** 5 **25.** 20 batches of cookies

7. (4) **14.** (3) **20.** 60 mph **26.** No solution

CHAPTER 2

Section 2.1

1. (3) **2.** (1) **3.** Domain = {1, 2}, Range = {4, 5, 6}

Section 2.2

1. (3) **3.** (4) **5.** (1) **7.** (3) **9.** (4)

2. (2) **4.** (3) **6.** (1) **8.** (3) **10.** (2)

11. (4) **12.** a. $h(1)=0$ b. $h(-1)=-8$ c. $h(s)=\frac{(s-1)^3}{s^2}$

d. $h(3)=\frac{8}{9}$ e. $h\left(\frac{1}{3}\right)=\frac{-8}{3}$ f. $h(y+1)=\frac{y^3}{(y+1)^2}$

13.

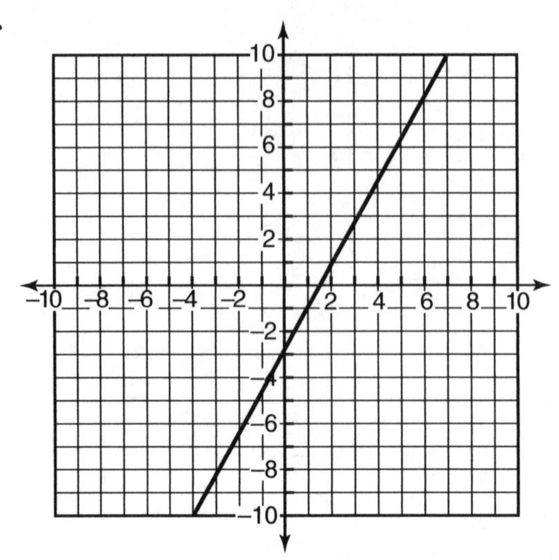

Explanation: The graph passes both the horizontal and vertical line tests.

14. a. $g(1) = 1$ b. $g(2) = 1$ c. $g(0) + 2 = 5$
d. Domain = all real numbers e. Range = $\{y: y \geq 0\}$
f. It is not one-to-one g. It is not onto.

15. a. It is a function because it passes the vertical line test
b. $\{x: -3 \leq x \leq 4\}$ or [– [−3, 4]
c. $\{y: 0 \leq y \leq 8\}$ or [0, 8]
d. It is not one-to-one because it does not pass the horizontal line test.

Section 2.3

1.

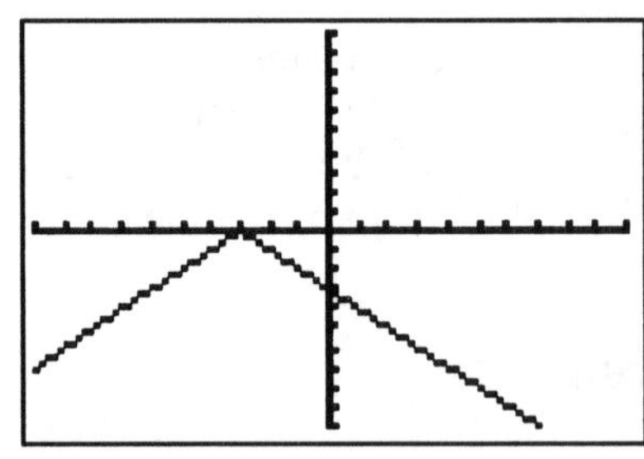

and a possible table

X	Y1
1	-4
-2	-1
-.3	-2.7
10000	-10003
-20000	-19997
-125	-122
480	-483

X=480

Based on these, Domain = all real numbers or $\mathscr{R}$, Range = $\{y: y \leq 0\}$.

2. Change your Y-Max to 15 and calculate the intersection as below:

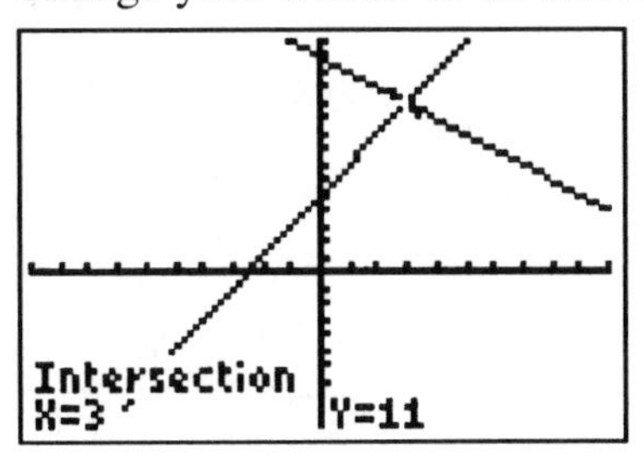

Solution $x = 3$

3.

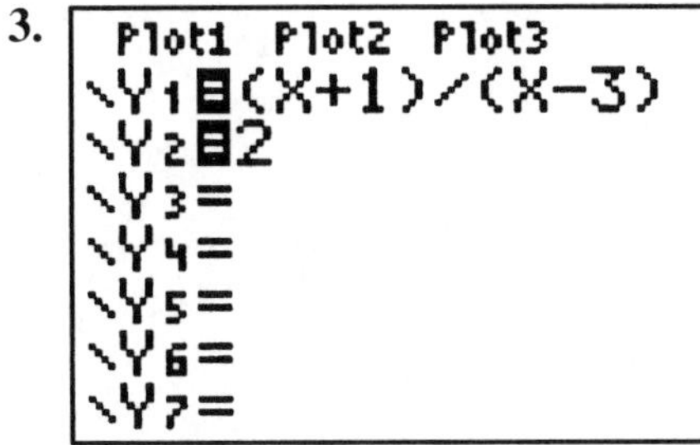

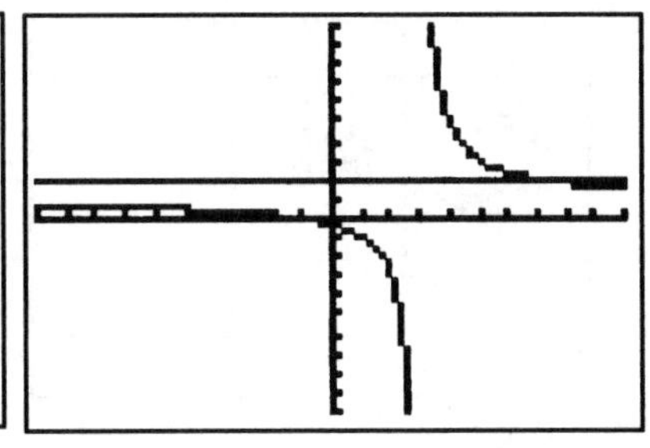

Use Zoom Box for a clearer picture to get

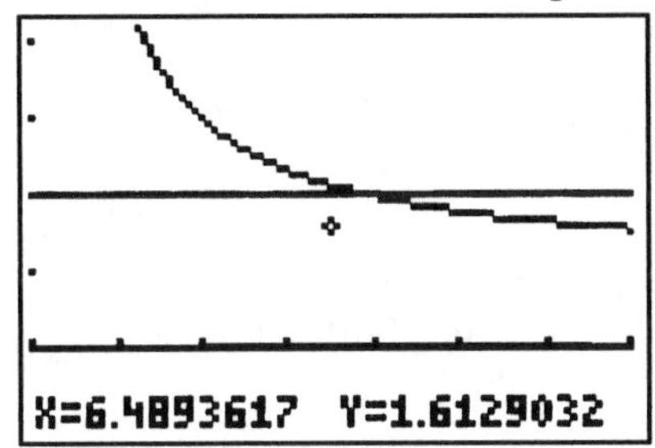

Find the intersection by **2nd** **TRACE** **5** **ENTER** **ENTER** **ENTER** to get:

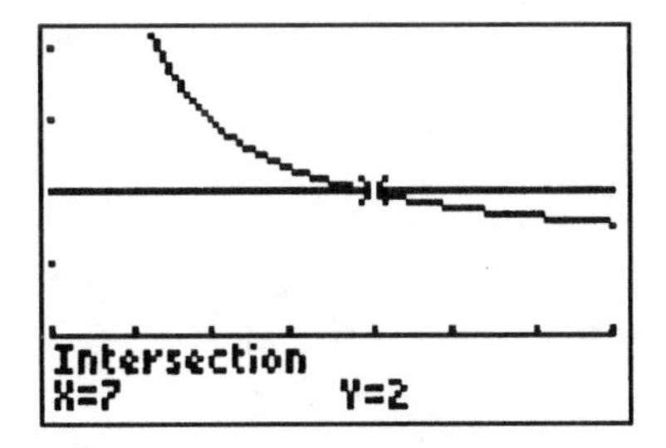

The screen shows that $x = 7$.

4. Set up the function in Y_1:

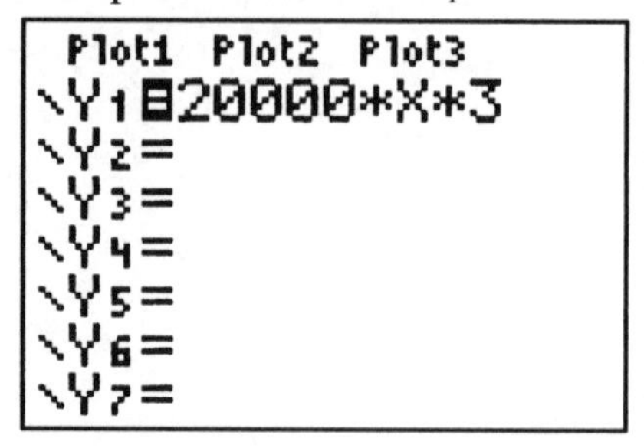

The answers are displayed in the table:

X	Y1	
.02	1200	
.024	1440	
.031	1860	

X=

At 2%, she earns \$1200 interest. At 2.4%, she earns \$1440 interest. At 3.1%, she earns \$1860 interest.

CHAPTER 3

Section 3.1

1. $f(x) = x - 12$

2. $f(x) = \frac{-3}{2}x + 6$

3. $f(x) = \frac{1}{4}x - 4$

4. a. $f(2) = -1$ b. $f(5) = 5$ c. $f(-3) = -11$ d. $f(10) = 15$

5. (4)

Section 3.2

1. (3)
2. (1)
3. (2)
4. (4)
5. (2)
6. (2)

7. $\{x: x = 2 \text{ or } x = 1\frac{1}{2}\}$

8. $\{y: y = 3 \text{ or } y = 9\}$

9. $\{x: x = \frac{1}{2} \text{ or } x = -3\}$

10. $\{z: 0 \le z \le 1\}$

0 1

11. $\{c: c \le -1 \text{ or } c \ge 8.5\}$

12. $|x - 9.25| \le 1.05$ The range of prices for a hamburger platter will be from \$8.25 to \$10.30.

13. $|10x - 28.90| \le 8$ The price for a single can of soup will range from \$2.09 to \$3.69.

Section 3.3

1. (2)
2. (4)
3. (1)
4. (4)
5. $\frac{7 \pm \sqrt{13}}{3}$
6. $x = \frac{-11}{9}$ or $x = -2$
7. $y = -6$
8. The ball will be 36 feet above the ground at 0.5 seconds and at 4.5 seconds.

Section 3.4

1. (3) 2. (1) 3. (2)

4\.

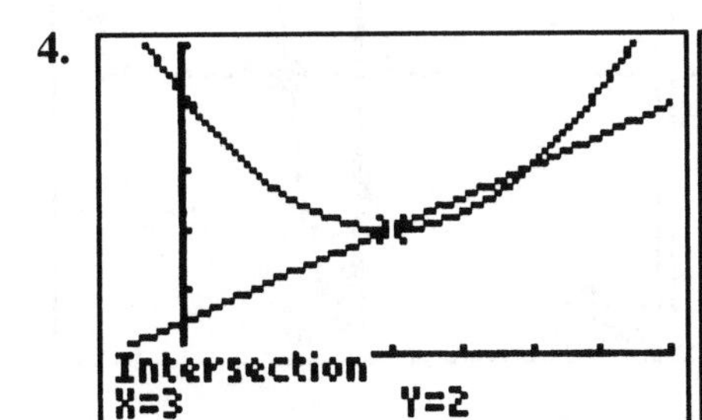

5. {(3, 12) and (2, −7)}
6. {(0, −14) and (2, 20)}
7. {(1, 4) and (4, −35)}
8. {(0.7, −15.84) and (0.2, −15.64)}
9. {(2, 21) and (3, 16)}
10. {(1, 6) and (3, 60)}
11. {(0, 2)}
12. {(0.5, 4)}
13. $\left\{\left(\frac{7+3\sqrt{5}}{2}, 6+3\sqrt{5}\right), \left(\frac{7-3\sqrt{5}}{2}, 6-3\sqrt{5}\right)\right\}$
14. $\left\{\left(\frac{2+\sqrt{6}}{2}, 2+2\sqrt{6}\right), \left(\frac{2-\sqrt{6}}{2}, 2-2\sqrt{6}\right)\right\}$
15. No. {(−1,16) (3, −112)} The first ordered pair represents 1 second before the bullet is shot, and the second ordered pair represents hitting the food relief package when it is 112 feet below ground.

Section 3.5

1. (1) 2. (2) 3. (4) 4. (2) 5. (3)

6.

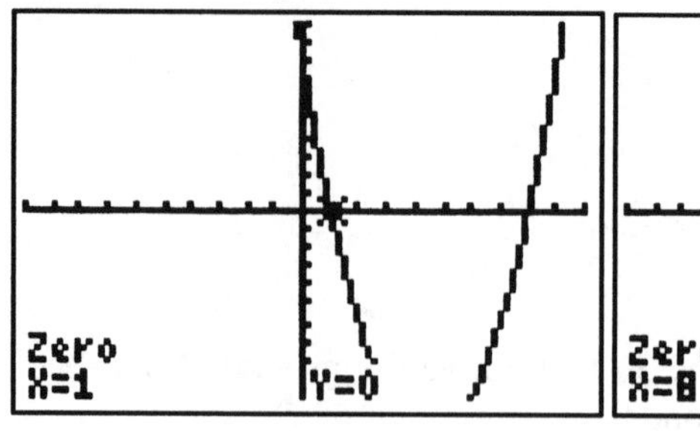

$\{x: 1 \le x \le 8\}$

7.

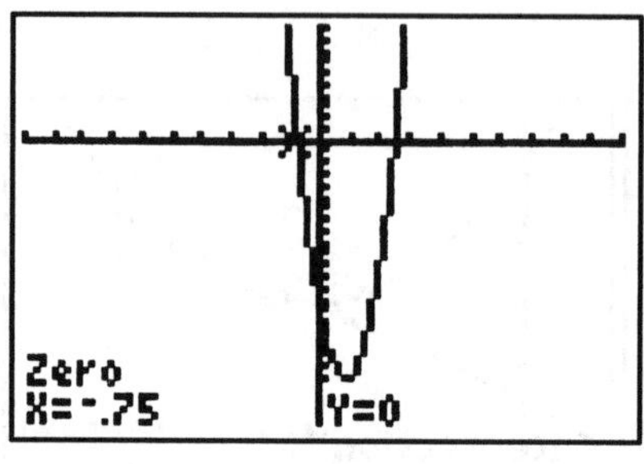

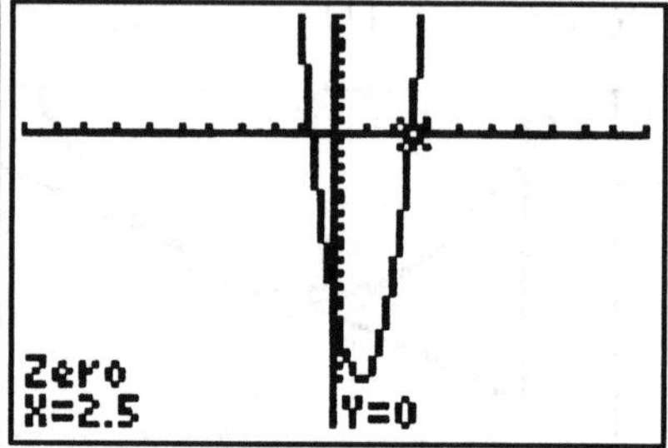

$\left\{x: x < \frac{-3}{4} \text{ or } x > \frac{5}{2}\right\}$

8.

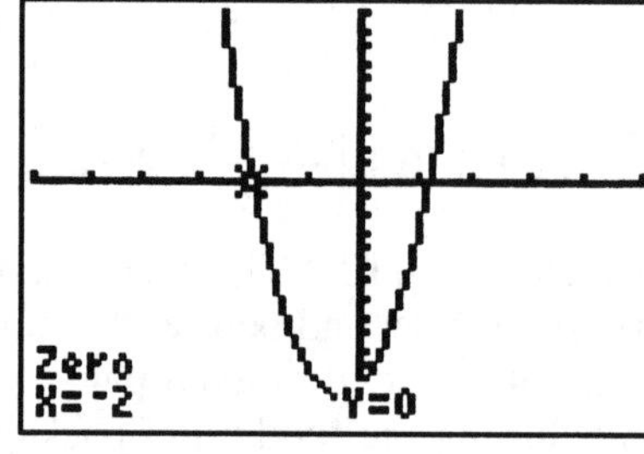

$\{x: -2 < x < 1.2\}$

9. $\left\{c: c \le \frac{3}{2} \text{ or } c \ge 5\right\}$

10. $\left\{t: t \le -\frac{3}{4} \text{ or } t \ge \frac{2}{5}\right\}$

11. $\left\{z: z \le -\frac{1}{4} \text{ or } z \ge \frac{3}{2}\right\}$

12. The ball is higher than 48 feet from the ground between 1 and 3 seconds, $1 < t < 3$.

13. The length of the rectangle must be at least 5 feet long and no larger than 45 feet long, $5 \le l \le 45$.

14. He must sell between 31 and 41 tickets to cover the cost. $9 < 22 - t < 19$ or if x = number of tickets in excess of 22, $9 < x < 19$.

Section 3.6

1. (4)
2. (2)
3. (1)
4. (3)
5. (3)
6. (2)
7. $a = \frac{-3}{5}$ and $b = 2$
8. $\left\{\frac{7}{3}, \sqrt{2}, -\sqrt{2}\right\}$
9. {3, –3, 7}
10. {3, –3, 6}
11. The edge was 15 cm long.

CHAPTER 4

Section 4.1

1. (3)
2. (4)
3. (1)
4. (2)
5. (3)
6. (4)
7. (2)
8. (1)
9. (4)
10. (3)

11. a. $h(f(x)) = 2 + x^2$
b. $(g \circ f)(x) = \frac{3}{x^4}$
c. $((h \circ g) \circ f)(x) = 2 + \frac{\sqrt{3}}{x^2}$
d. $(g \circ f)(1) = 3$
e. $(h(g(\frac{1}{3}))) = 5$
f. $(h \circ f)(2) = 6$
g. $(g \circ h)(25) = \frac{3}{7}$
h. $(f \circ g)(-6) = \frac{1}{16}$
i. $((g \circ f) \circ h)(1) = \frac{1}{27}$

12. $(f \circ g)(4) = 16$ and $(g \circ f)(4) = 16$. One example that verifies the commutative property is not enough is not enough to prove that composition of functions is commutative. In fact, composition is not commutative. This can be seen in a counterexample, such as when each composition is evaluated at –4.

13. $g(x) = \frac{x-1}{2}$ and $(g \circ f)(x) = x$

Section 4.2

1. (2)
2. (4)
3. (1)
4. (3)
5. (3)
6. (4)
7. (2)
8. (2)

9. $g^{-1}(x)=\frac{-1}{7}x+\frac{3}{7}$

$$(g\circ g^{-1})(x)=g(g^{-1}(x))=g\left(\frac{-1}{7}x+\frac{3}{7}\right)=-7\left(\frac{-1}{7}x+\frac{3}{7}\right)+3=x-3+3=x$$

$$(g^{-1}\circ g)(x)=g^{-1}(g(x))=g^{-1}(-7x+3)=\frac{-1}{7}(-7x+3)+\frac{3}{7}=x-\frac{3}{7}+\frac{3}{7}=x$$

10. $h^{-1}(x)=\frac{2x}{x-1}$

$$(h\circ h^{-1})(x)=h(h^{-1}(x))=g\left(\frac{2x}{x-1}\right)=\frac{\frac{2x}{x-1}}{\frac{2x}{x-1}-2}$$

$$=\frac{\frac{2x}{x-1}}{\frac{2x}{x-1}-\frac{2(x-1)}{x-1}}=\frac{\frac{2x}{x-1}}{\frac{2x}{x-1}-\frac{2x-2}{x-1}}=\frac{\frac{2x}{x-1}}{\frac{2x}{x-1}+\frac{-2x+2}{x-1}}$$

$$=\frac{\frac{2x}{x-1}}{\frac{2x-2x+2}{x-1}}=\frac{\frac{2x}{x-1}}{\frac{2}{x-1}}=\frac{\cancel{2}x}{\cancel{x-1}}\cdot\frac{\cancel{x-1}}{\cancel{2}}=x$$

$$(h^{-1}\circ h)(x)=h^{-1}(h(x))=h^{-1}\left(\frac{x}{x-2}\right)=\frac{2\left(\frac{x}{x-2}\right)}{\left(\frac{x}{x-2}\right)-1}$$

$$=\frac{2\left(\frac{x}{x-2}\right)}{\left(\frac{x}{x-2}\right)-\left(\frac{x-2}{x-2}\right)}=\frac{\frac{2x}{x-2}}{\frac{x-(x-2)}{x-2}}=\frac{\frac{2x}{x-2}}{\frac{2}{x-2}}=\frac{\cancel{2}x}{\cancel{x-2}}\cdot\frac{\cancel{x-2}}{\cancel{2}}=x$$

CHAPTER 5

Section 5.1

1. (4)
2. (4)
3. (1)
4. (2)
5. (3)
6. (3)
7. (1)

8.

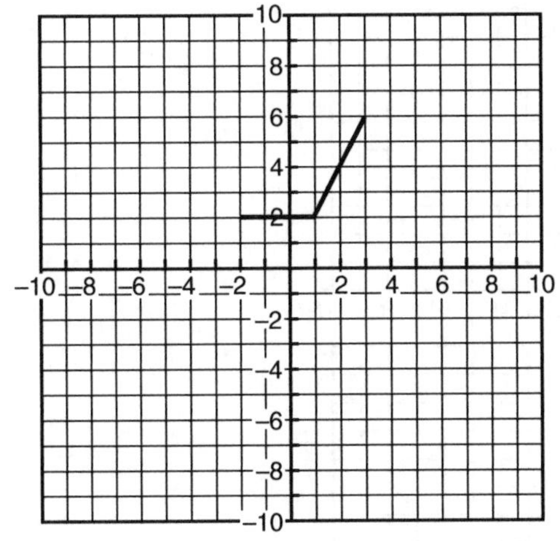
10
8
6
4
2
−10 −8 −6 −4 −2 2 4 6 8 10
−2
−4
−6
−8
−10

$h(x + 2)$

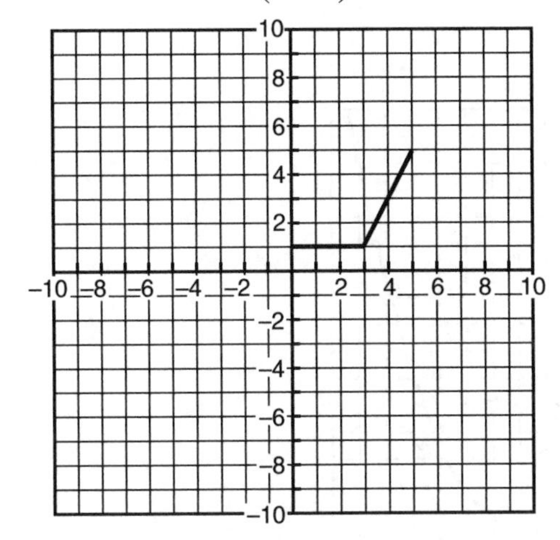
10
8
6
4
2
−10 −8 −6 −4 −2 2 4 6 8 10
−2
−4
−6
−8
−10

$h(x) - 1$

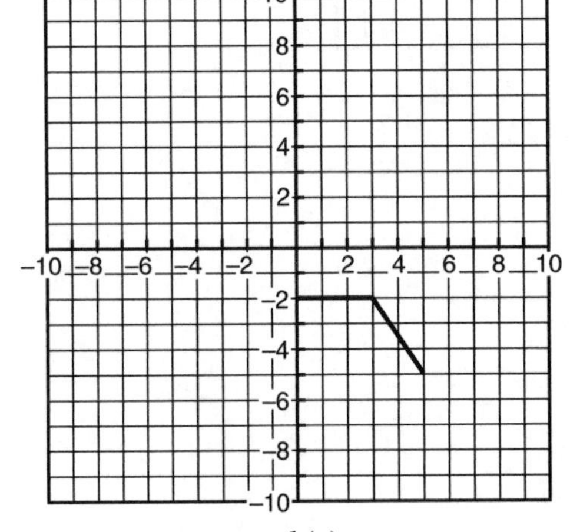
10
8
6
4
2
−10 −8 −6 −4 −2 2 4 6 8 10
−2
−4
−6
−8
−10

$-h(x)$

9.

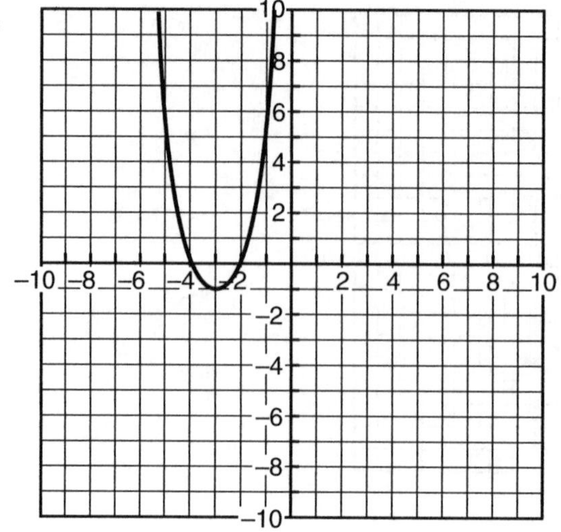

$f(x) = 2(x + 3)^2 - 1$

10.

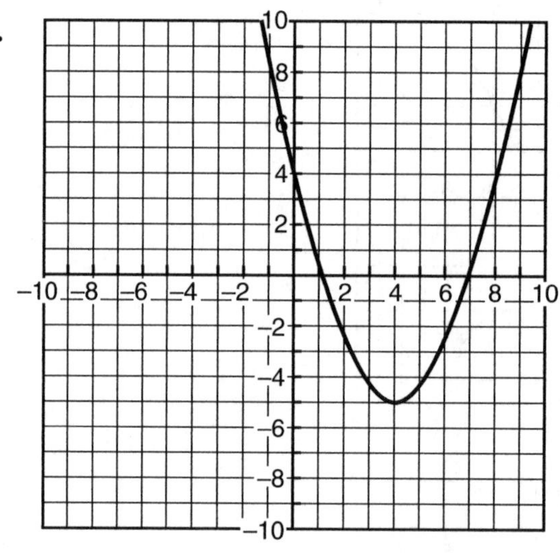

$f(x) = \frac{1}{2}(x - 4)^2 - 5$

11.

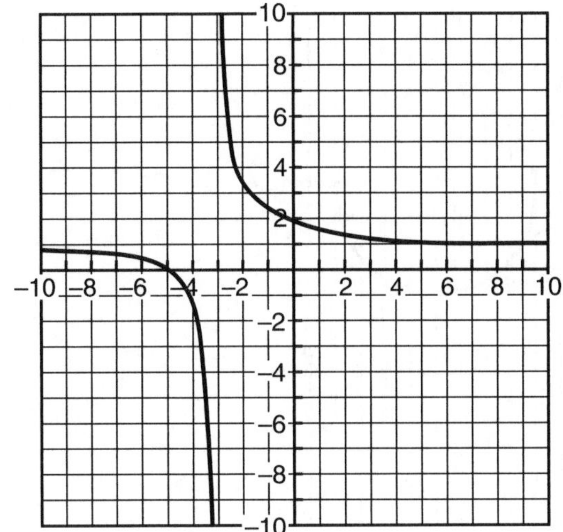

$$g(x)=\frac{2}{x+3}+1$$

12.

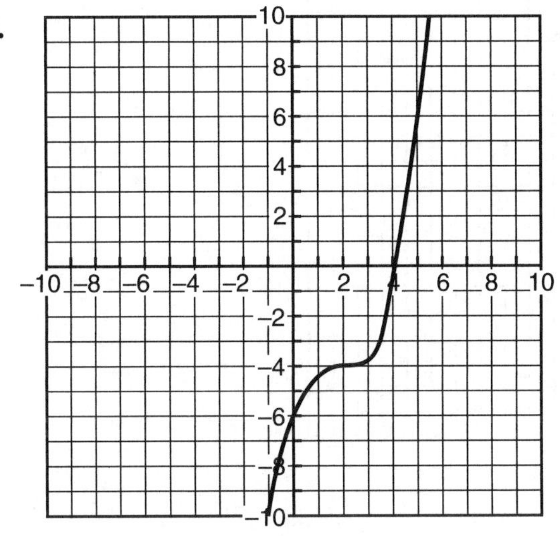

$$h(x)=\frac{1}{3}(x-2)^3-4$$

13. $f(x) = 4|x - 2| + 2$ **14.** $f(x) = 2(x - 2)^3 - 3$

15. $h(x)=\frac{1}{2}h(x+1)-3$ is a horizontal expansion with factor $\frac{1}{2}$, a horizontal shift of 1 unit to the left, and a vertical shift of 3 units down.

Section 5.2

1. (4) **3.** (4) **5.** (3) **7.** (1)
2. (1) **4.** (2) **6.** (1) **8.** center (–6, 2), radius 1

9.

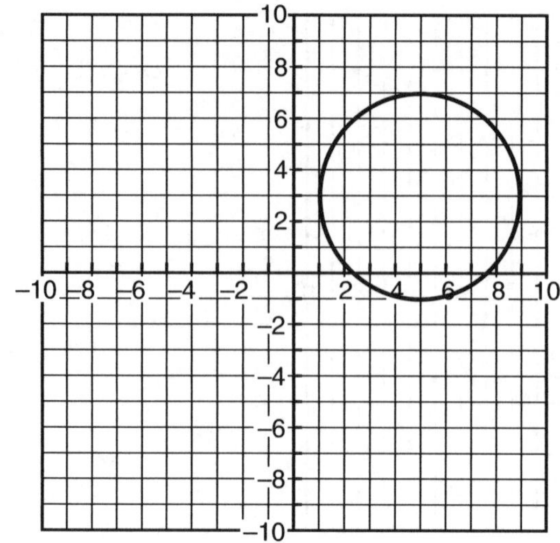

10.

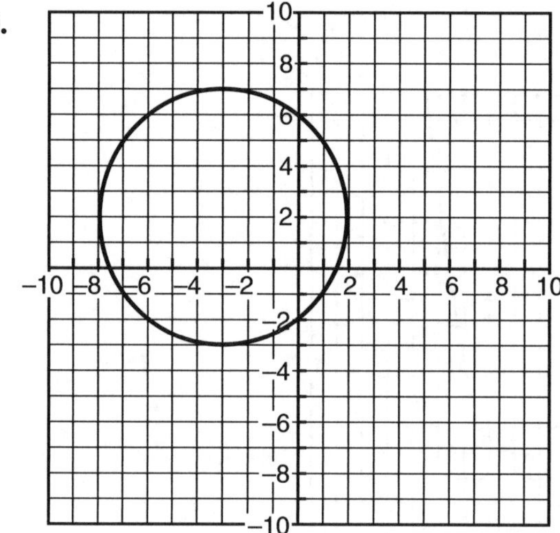

11. $(x-2)^2+(y-3)^2=25$

12. $(x-5)^2+(y+2)^2=130$

13. $(x-2)^2+(y-4)^2=25$

14. $(x-4)^2+(y+3)^2=9$

CHAPTER 6

Section 6.1

1. (2)
2. (1)
3. (4)
4. (4)
5. (3)
6. (2)
7. (3)
8. $14bi\sqrt{3b}$
9. $16i\sqrt{5}$
10. $5i\sqrt{6}$

Section 6.2

1. (2)
2. (4)
3. (1)
4. (2)
5. (4)
6. (4)
7. (1)
8. (3)
9. (3)

Section 6.3

1. (3)
2. (4)
3. (1)
4. (4)
5. (3)
6. (1)
7. (1)
8. (1)
9. (2)
10. (3)
11. $-41 + 13i$
12. $\frac{11}{13} - \frac{4}{13}i$
13. 41
14. $4 - 10i$
15. 3
16. $42 + 291i$

Section 6.4

1. (4)
2. (3)
3. (2)
4. (2)
5. (3)
6. (3)
7. (1)
8. (4)
9. (1)
10. (2)
11. (2)
12. (3)
13. $c = -\frac{3}{2} \pm \frac{\sqrt{11}}{2}i$
14. $t = \frac{1}{4} \pm \frac{\sqrt{11}}{4}i$
15. $x = 3$ or $x = 1$
16. $\frac{5}{4} \pm \frac{\sqrt{39}}{4}i$

Section 6.5

1. (2)
2. (2)
3. (3)
4. (1)
5. (4)
6. $f(x) = x^2 - 6x + 25$
7. $f(x) = 225x^2 - 300x + 73$
8. $f(x) = 25x^2 - 15x + 2,\ r_1 = \frac{1}{5}$ and $r_2 = \frac{2}{5}$
9. $r_1 = \frac{3}{2}$ and $r_2 = -4,\ 2x^2 + 7x - 49 = 0$

CHAPTER 7

Section 7.1

1. (3)
2. (4)
3. (4)
4. (4)
5. (2)
6. (2)
7. (1)
8. $y = \frac{4x - 15}{3}$
9. a. 28
b. 34.751
c. 82.451
d. 714.94
e. 1.19×10^7
10. $\left\{x: x = \frac{5}{2} \text{ or } x = -2\right\}$

11.

x	$f(x)$
–2	2.25
–1	1.5
0	1
1	.66667
2	.44444
3	.2963

12.

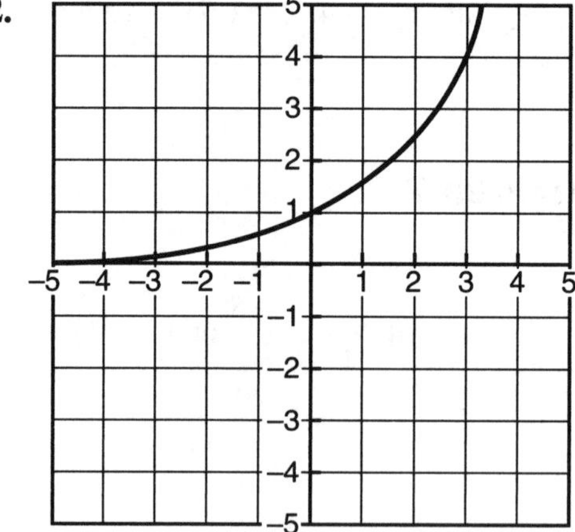

Section 7.2

1. (3)
2. (1)
3. (2)
4. (4)
5. (3)
6. (1)
7. (2)
8. 5
9. 8
10. 1.77
11. $\{x: x > 2 \text{ or } x < -2\}$
12. $2 - x$

13.

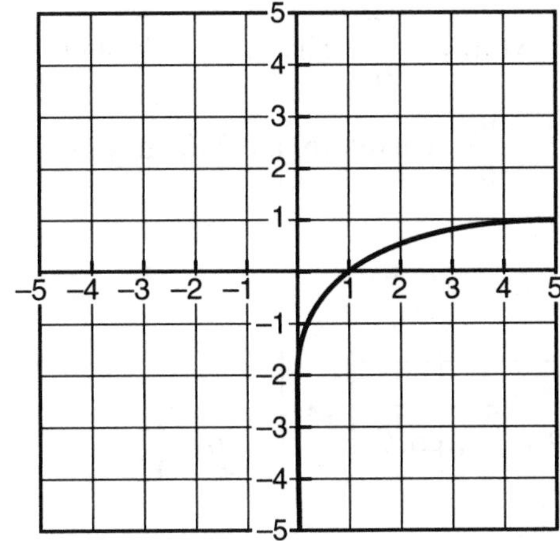

Section 7.3

1. (3) **3.** (1) **5.** (2)

2. (2) **4.** (3) **6.** $\dfrac{\sqrt[3]{18}}{864}$

7. $\log_b \dfrac{5x^3y}{\sqrt[4]{2z}} = \log_b 5 + 3\log_b x + \log_b y - \dfrac{1}{4}(\log_b 2 + \log_b z)$

8. $\log_b \sqrt{\dfrac{13(x+5)}{(2x-3)^3}} = \dfrac{1}{2}(\log_b 13 + \log_b (x+5) - 3\log_b (2x-3))$

Section 7.4

1. (2) **5.** (3) **9.** (1)

2. (2) **6.** (1) **10.** 478.63 parsecs and 1560.3 light-years

3. (3) **7.** (1) **11.** 3.98 decibels

4. (4) **8.** (3)

Section 7.5

1. (2) **3.** (4) **5.** (4) **7.** $x = 1.302$ **9.** $m = -1.59$

2. (3) **4.** (1) **6.** $x = 0.35$ **8.** $y = 1.13$ **10.** 1.1 hours

CHAPTER 8

Section 8.1

1. (2) **3.** (1) **5.** (3) **7.** (2)

2. (1) **4.** (4) **6.** (1) **8.** $\dfrac{50\pi}{3}$ feet

9. $240° = \frac{4\pi}{3}, \frac{3}{5}$ rotation $= \frac{6\pi}{5}$, $S(\text{arc}) = \frac{7\pi}{6}$. The LCD is 30 and $\frac{4\pi}{3} = \frac{40\pi}{30}$, $\frac{6\pi}{5} = \frac{36\pi}{30}$, $\frac{7\pi}{6} = \frac{35\pi}{30}$. Therefore, the 240° has the largest radian measure.

10. $\frac{120}{\pi}$

Section 8.2

1. $\sin\theta = \frac{5}{13}$, $\cos\theta = \frac{12}{13}$, $\tan\theta = \frac{5}{12}$, $\cot\theta = \frac{12}{5}$, $\sec\theta = \frac{13}{12}$, $\csc\theta = \frac{13}{5}$

2. $\sin\theta = \frac{2\sqrt{13}}{13}$, $\cos\theta = \frac{3\sqrt{13}}{13}$, $\tan\theta = \frac{2}{3}$, $\cot\theta = \frac{3}{2}$, $\sec\theta = \frac{\sqrt{13}}{3}$, $\csc\theta = \frac{\sqrt{13}}{2}$

3. a. 0.4245 b. –3.8637 c. –0.7071 d. 1.7321 e. 8.2055 f. 0.9962

4. a. $\frac{5}{3}$ b. 0.8 c. 0.75 d. 0.8 e. $\frac{4}{3}$ f. 1.25 g. $\frac{3}{5}$ h. $\frac{5}{3}$

5. 2.4571 **6.** $\frac{5\pi}{12}$ **7.** 12.8 feet

Section 8.3

1. (4)

2. (3)

3. (1)

4. (3)

5. (4)

6. (2)

7. a. $\frac{17}{15}$
b. $-\frac{15}{8}$
c. $-\frac{8}{17}$
d. 1
e. $-\frac{120}{289}$
f. 1

8. $\frac{2}{9}$

9. $-\frac{\sqrt{x^2-16}}{x}$

10. $\frac{7}{9}$

Section 8.4

1. a. $-\frac{\sqrt{3}}{3}$ b. $-\sqrt{2}$ c. $\frac{1}{2}$ d. $-\frac{1}{2}$ e. undefined f. 2

g. $\sqrt{2}$ h. $-\frac{2\sqrt{3}}{3}$

2. a. $-\sin 28°$ b. $-\tan\frac{\pi}{5}$ c. $\csc 41°$ d. $-\cos 41°$ e. $-\csc\frac{\pi}{14}$

f. $\cot 16°$ g. $-\cot 42°$ h. $-\cot 43°$

3. (3) **4.** (2)

CHAPTER 9

Section 9.1

1. (2) **2.** (4) **3.** (3) **4.** (4) **5.** (1) **6.** (3) **7.** (3) **8.** (2) **9.** (1) **10.** (3)

11.

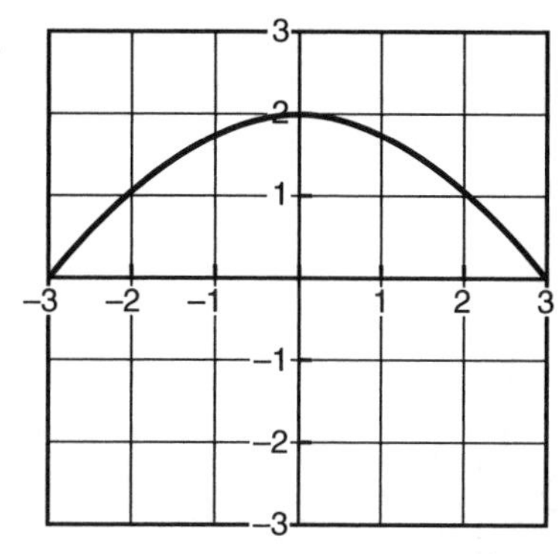

12.

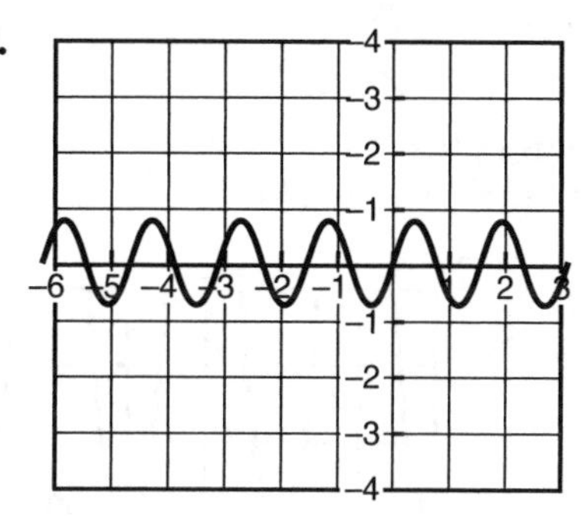

13.

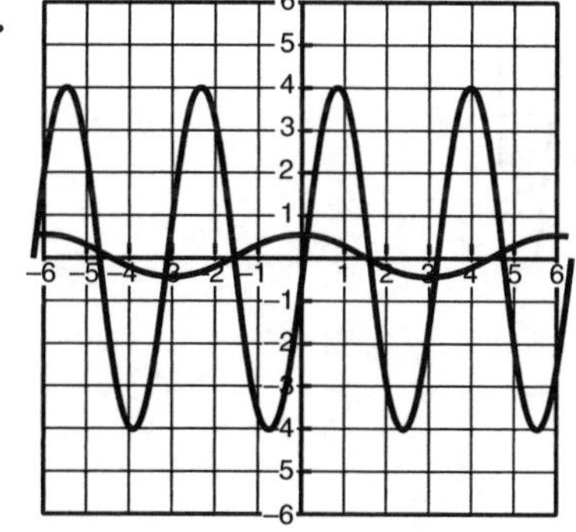

8 points of intersection

14.

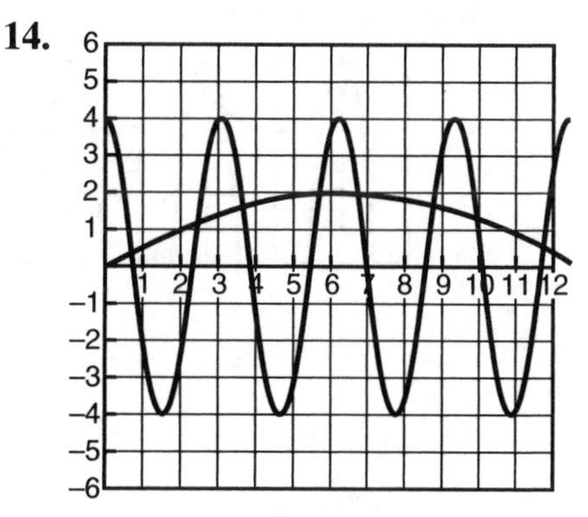

8 points of intersection

15.

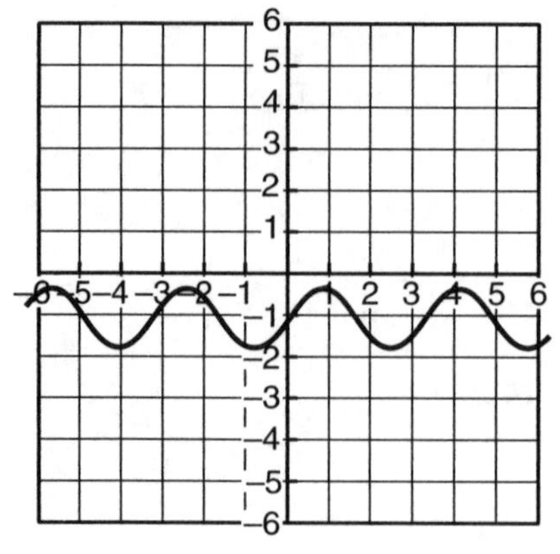

amplitude $= \frac{2}{3}$
frequency = 2
period $= \pi$
equation of midline: $y = -1$

16. $A = 2,\ B = \frac{1}{4},\ C = 3$

Section 9.2

1. (4) **2.** (3)

3.

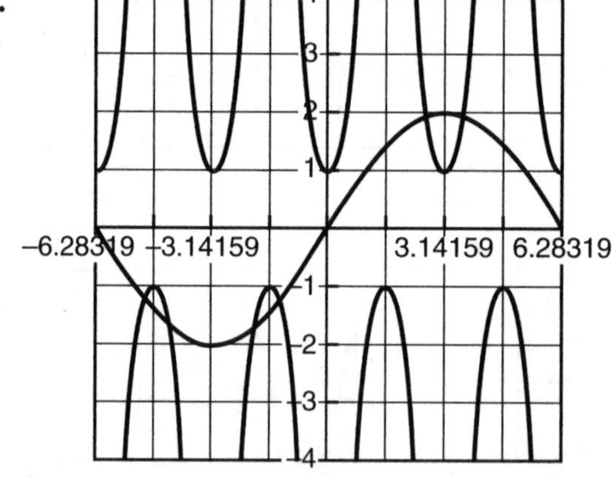

4.

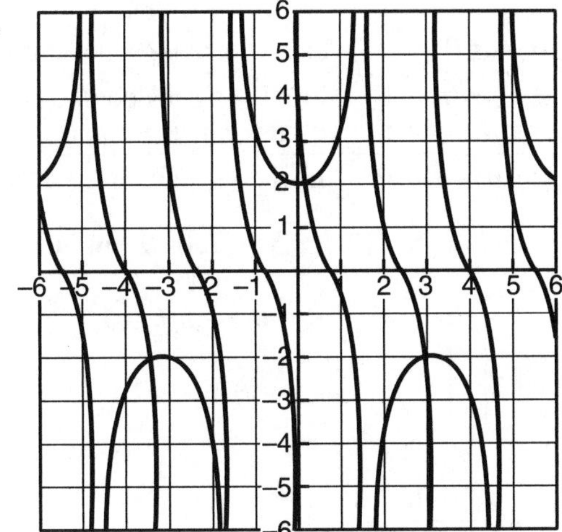

5.

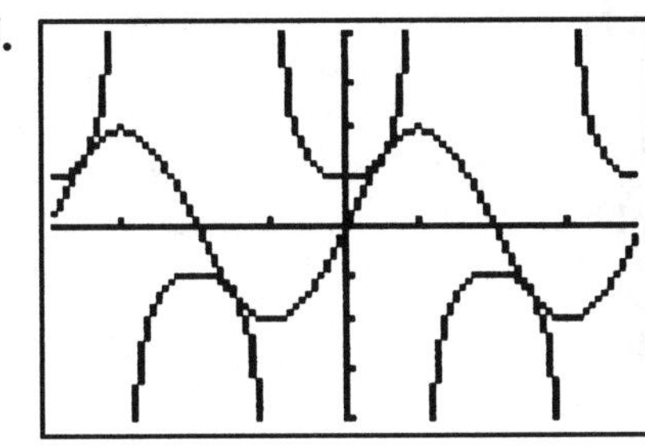

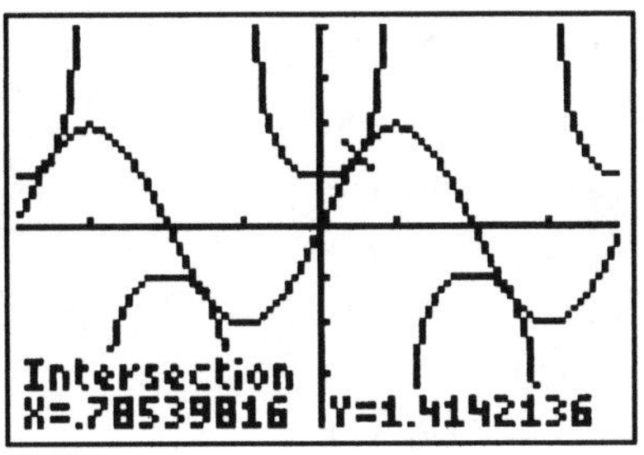

or $\left(\frac{\pi}{4}, \sqrt{2}\right)$

Section 9.3

1. (1)
2. (3)
3. (4)
4. (2)
5. (4)
6. (1)
7. (2)
8. (1)
9. (2)
10. $\frac{\sqrt{x^2+y^2}}{x}$
11. No, the domain of the cosecant function is $\{x: |x| \geq 1\}$.

Chapter 10

Section 10.1

1. (2)
2. (4)
3. (1)
4. (3)
5. (4)
6. (4)
7. (2)
8. $\frac{\sin A+1}{\sin A}$
9. $\frac{1-\sin^2 x}{\sin^2 x}$
10. $\frac{\sec A+2}{\sec^2 A-1}$

11.

$$\cot\theta(\tan\theta+\cos\theta\csc\theta)=\frac{\cos\theta}{\sin\theta}\left(\frac{\sin\theta}{\cos\theta}+\cos\theta\frac{1}{\sin\theta}\right)=\frac{\cos\theta}{\sin\theta}\left(\frac{\sin\theta}{\cos\theta}+\frac{\cos\theta}{\sin\theta}\right)$$
$$=1+\frac{\cos^2\theta}{\sin^2\theta}=\frac{\sin^2\theta}{\sin^2\theta}+\frac{\cos^2\theta}{\sin^2\theta}=\frac{\sin^2\theta+\cos^2\theta}{\sin^2\theta}=\frac{1}{\sin^2\theta}$$

12. $\sin^2\theta + \cos^2\theta = 1$

$$\frac{\sin^2\theta}{\sin^2\theta}+\frac{\cos^2\theta}{\sin^2\theta}=\frac{1}{\sin^2\theta}$$

$1 + \cot^2\theta = \csc^2\theta$

13. $(x, y) = (\cos\theta, \sin\theta)$
$x^2 + y^2 = 1$
$\cos^2\theta + \sin^2\theta = 1$

Section 10.2

1. (4)
2. (1)
3. (2)
4. (3)
5. (3)
6. (2)
7. (1)
8. (3)
9. (3)
10. (1)

11. a. $-\frac{140}{221}$

b. $-\frac{171}{221}$

c. $\frac{140}{171}$

d. $\frac{171}{140}$

12. $\cos(x + y)\cdot\cos(x - y) = (\cos x\cos y - \sin x\sin y)(\cos x\cos y + \sin x\sin y)$
$= \cos^2x\cos^2y - \sin^2x\sin^2y = (1 - \sin^2x)\cos^2y - \sin^2x\sin^2y$
$= \cos^2y - \sin^2x\cos^2y - \sin^2x\sin^2y = \cos^2y - \sin^2x(\cos^2y + \sin^2y)$
$= \cos^2y - \sin^2x\cdot 1 = \cos^2y - \sin^2x$

Section 10.3

1. (2)
2. (3)
3. (2)
4. (1)
5. (4)
6. (3)

7. $\sin 3A = \sin(A + 2A) = \sin A\cos 2A + \sin 2A\cos A$
$= \sin A(\cos^2A - \sin^2A) + 2\sin A\cos A\cos A = \sin A\cos^2A - \sin^3A +$
$2\sin A\cos^2A = 3\sin A\cos^2A - \sin^3A$

8. $\frac{\sqrt{117}}{13}$

9. $\cos 2\theta = 1 - 2\sin^2\theta \rightarrow 2\sin^2\theta = 1 - \cos 2\theta \rightarrow \sin^2\theta = \frac{1-\cos 2\theta}{2}$

$\rightarrow \sin\theta = \pm\sqrt{\frac{1-\cos 2\theta}{2}}$

Section 10.4

1. (3) **3.** (2) **5.** (2)
2. (3) **4.** (1) **6.** (3)

7. a. $\sqrt{3-2\sqrt{2}}$ b. $-\sqrt{\dfrac{2+\sqrt{2}}{4}}$ c. $-\dfrac{\sqrt{2-\sqrt{3}}}{2}$ d. $-\dfrac{\sqrt{2-\sqrt{3}}}{2}$

Section 10.5

1. $\sec A(\cot A + 1) = \csc A + \sec A$

$$\frac{1}{\cos A}\left(\frac{\cos A}{\sin A}+1\right)=\csc A+\sec A$$

$$\frac{1}{\cancel{\cos A}}\cdot\frac{\cancel{\cos A}}{\sin A}+\frac{1}{\cos A}=\csc A+\sec A$$

$$\frac{1}{\sin A}+\frac{1}{\cos A}=\csc A+\sec A$$

$$\csc A + \sec A = \csc A + \sec A$$

$$\frac{(1-\sin^2 x)(1+\sin^2 x)}{\sin^4 x}=\frac{\cos^2 x(1+\sin^2 x)}{\sin^4 x}$$

2. $\csc^4 x-1=\dfrac{\cos^2 x(1+\sin^2 x)}{\sin^4 x}$

$$\frac{1}{\sin^4 x}-1=\frac{\cos^2 x(1+\sin^2 x)}{\sin^4 x}$$

$$\frac{1}{\sin^4 x}-\frac{\sin^4 x}{\sin^4 x}=\frac{\cos^2 x(1+\sin^2 x)}{\sin^4 x}$$

$$\frac{1-\sin^4 x}{\sin^4 x}=\frac{\cos^2 x(1+\sin^2 x)}{\sin^4 x}$$

$$\frac{\cos^2 x(1+\sin^2 x)}{\sin^4 x}=\frac{\cos^2 x(1+\sin^2 x)}{\sin^4 x}$$

3. $(\sin\theta+\cos\theta)^2 = 1 + \sin 2\theta$

$\sin^2\theta + 2\sin\theta\cos\theta + \cos^2\theta = 1 + \sin 2\theta$

$1 + 2\sin\theta\cos\theta = 1 + \sin 2\theta$

$1 + \sin 2\theta = 1 + \sin 2\theta$

4. $$\frac{\tan x + \cot x}{2} = \csc 2x$$

$$\frac{\frac{\sin x}{\cos x} + \frac{\cos x}{\sin x}}{2} = \csc 2x$$

$$\frac{\frac{\sin^2 x}{\sin x \cos x} + \frac{\cos^2 x}{\sin x \cos x}}{2} = \csc 2x$$

$$\frac{\frac{\sin^2 x + \cos^2 x}{\sin x \cos x}}{2} = \csc 2x$$

$$\frac{\frac{1}{\sin x \cos x}}{2} = \csc 2x$$

$$\frac{1}{2 \sin x \cos x} = \csc 2x$$

$$\frac{1}{\sin 2x} = \csc 2x$$

$$\csc 2x = \csc 2x$$

5. $$\frac{\cos(\alpha - \beta)}{\sin \alpha \cos \beta} = \cot \alpha + \tan \beta$$

$$\frac{\cos \alpha \cos \beta + \sin \alpha \sin \beta}{\sin \alpha \cos \beta} = \cot \alpha + \tan \beta$$

$$\frac{\cos \alpha \cos \beta}{\sin \alpha \cos \beta} + \frac{\sin \alpha \sin \beta}{\sin \alpha \cos \beta} = \cot \alpha + \tan \beta$$

$$\frac{\cos \alpha \cancel{\cos \beta}}{\sin \alpha \cancel{\cos \beta}} + \frac{\cancel{\sin \alpha} \sin \beta}{\cancel{\sin \alpha} \cos \beta} = \cot \alpha + \tan \beta$$

$$\frac{\cos \alpha}{\sin \alpha} + \frac{\sin \beta}{\cos \beta} = \cot \alpha + \tan \beta$$

$$\cot \alpha + \tan \beta = \cot \alpha + \tan \beta$$

6. $$\frac{\sin 2x}{\cos 2x} = \frac{2 \tan x}{1 - \tan^2 x}$$

$$\frac{\sin 2x}{\cos 2x} = \frac{2 \frac{\sin x}{\cos x}}{1 - \frac{\sin^2 x}{\cos^2 x}}$$

$$\frac{\sin 2x}{\cos 2x} = \frac{2 \frac{\sin x}{\cos x} \cdot \cos^2 x}{\cos^2 x - \frac{\sin^2 x}{\cos^2 x} \cdot \cos^2 x}$$

$$\frac{\sin 2x}{\cos 2x} = \frac{2 \frac{\sin x}{\cancel{\cos x}} \cdot \cos^{\cancel{2}} x}{\cos^2 x - \frac{\sin^2 x}{\cancel{\cos^2 x}} \cdot \cancel{\cos^2 x}}$$

$$\frac{\sin 2x}{\cos 2x} = \frac{2 \sin x \cos x}{\cos^2 x - \sin^2 x}$$

$$\frac{\sin 2x}{\cos 2x} = \frac{\sin 2x}{\cos 2x}$$

Section 10.6

1. (3)
2. (1)
3. (4)
4. (2)
5. (4)
6. (3)
7. a. 0°, 180°, 360°
b. 90°, 270°
c. 45°, 135°, 225°, 315°
d. 0°, 60°, 180°, 300°
e. 30°, 150°, 270°
8. a. 56.3°, 104.0°, 236.3°, 256.0°
b. 70.5°, 75.5°, 284.5°, 289.5°
9. 36°

CHAPTER 11

Section 11.1

1. (2)
2. (4)
3. (2)
4. (3)
5. (4)
6. a. $123.5\sqrt{3}$ square inches
b. $26\sqrt{2}$ cm^2
7. 3943.6 square miles
8. 101 plots
9. 73.80 square units

Section 11.2

1. (3)
2. (1)
3. (4)
4. (3)
5. (2)
6. $a < c \sin A$, so there are no triangle solutions
7. 10.3 feet
8. $m\angle P = 53°50'$, $m\angle R = 85°10'$, and $r = 19.7$ or $m\angle P = 126°10'$, $m\angle R = 12°50'$, and $r = 4.4$
9. 24 feet
10. a. 33°0′
b. 8 feet
11. a. 16.9 miles
b. 9.1 miles

Section 11.3

1. (3)
2. (3)
3. (2)
4. (1)
5. (4)
6. a. 91°47′
b. 88°13′
c. 12.6

7. $x^2 = 70^2 + 90^2 - 2 \cdot 70 \cdot 90 \cdot \cos 28 = 1874.86033$ and $\sqrt{1874.86033} = 43.29965739$ or 43.3 feet

8. 10.7 units
9. a. $\sqrt{19}$
b. 23.41°
10. $a \approx 8.24$, $m\angle B = 40.4°$, and $m\angle C = 76.6°$
11. 6.9 miles
12. 88°

Section 11.4

1. 97° **2.** 15 feet per second and 36°52′

CHAPTER 12

Section 12.1

1. (3) **2.** (2) **3.** (3) **4.** $\frac{1}{3}$

5. a. $\frac{85}{452}$ b. $\frac{59}{113}$ c. $\frac{23}{108}$ d. $\frac{4}{143}$ e. $\frac{141}{226}$

6. a. $\frac{1}{4}$ b. $\frac{3}{4}$

7. a. $\frac{1}{13}$ b. $\frac{1}{4}$ c. $\frac{1}{52}$ d. $\frac{4}{13}$ e. $\frac{1}{26}$

Section 12.2

1. a. 360 b. 4 c. 120 d. 60,480
2. a. 64 b. 24 c. 6 d. 4 e. 6
3. (3)
4. (1)
5. (2)
6. 117,600
7. 3,780
8. 336
9. 42,840
10. 360,360

Section 12.3

1. a. 35
b. 5
c. 66
d. 126
2. (1)
3. (3)
4. (2)
5. (4)
6. (2)
7. a. 2,598,960
b. 1287
c. 22,308
d. 211,296
e. 0.08
8. a. 2,002 b. 56 c. $\frac{28}{1001}$
9. a. 5 b. 14 c. 35
10. 56

Section 12.4

1. (3) **2.** (1) **3.** (3) **4.** (2) **5.** 29/128

6. a. 0.050 b. 0.594 c. 0.010

7. a. $\frac{263}{16384}$ b. $\frac{567}{8192}$ c. $\frac{135}{512}$

8. a. $\frac{1}{8}$ b. $\frac{1}{8}$ c. $\frac{1}{2}$ d. $\frac{7}{8}$

9. a. 0.722 b. 0.950

Section 12.5

1. (3)
2. (1)
3. (2)
4. (2)
5. (4)
6. $x^7 - 7x^6y + 21x^5y^2 - 35x^4y^3 + 35x^3y^4 - 21x^2y^5 + 7xy^6 - y^7$
7. $81a^4 - 432a^3 + 864a^2 - 768a + 256$
8. $x^{10} + 5x^7 + 10x^4 + 10x + \frac{5}{x^2} + \frac{1}{x^5}$
9. $-448x$
10. 1.004

Section 12.6

1. Answers may vary
  a. Weigh every fifth hamburger made over a 3-day period.
  b. Ask 50 students from Binghampton High School selected randomly by their student number how many hours of homework they do each night.
  c. Examine the results on the English Regents Examination by all students who took this examination.

2. (4)

Section 12.7

1. (2) **2.** (2) **3.** (1) **4.** (1) **5.** (3)

6. mean = 3.1875, median = 3, mode = 2
7. a. mean = 7.62727, median = 7.5, mode = no mode
  b. The mean is affected the most, new mean = 7.5 with a difference of 0.12727; new median is 7.4 with a difference of 0.1
8. Janice is not correct because the data has not yet been arranged in numerical order.
9. 82
10. mean = 3, median = 3, mode = 4

11.

Number of Transactions	x_m	Frequency, f	$f \cdot x_m$
45–59	52	5	260
60–74	67	12	804
75–89	82	13	1066
90–104	97	2	194
Sum		32	2324

mean $= \frac{2324}{32} = 72.625$, 16th term ≈ 73.75 and 17th term = 74 and median ≈ 73.875, mode is in the interval 75–89

Section 12.8

1. (4) **2.** (3) **3.** (2) **4.** (2) **5.** (2)

6. $\mu \approx 23.8$, range = 5, variance ≈ 2.96, $\sigma \approx 1.72$

7. $\bar{x} \approx 3.07$, $s \approx 1.87$

8. $Q_1 = 80$, $Q_2 = 113$, $Q_3 = 122$, interquartile range = 42

9. $Q_1 = 89.5$, $Q_2 = 114$, $Q_3 = 122.5$, interquartile range = 33

10. $s \approx 26.06$, $s \approx 17.9$

Section 12.9

1. (3) **2.** (1) **3.** a. 47.7%
b. 68.2%
c. 13.6%
d. 13.6%
e. 2.2%

4. $\bar{x} = 6$, $s \approx 1.732$, and

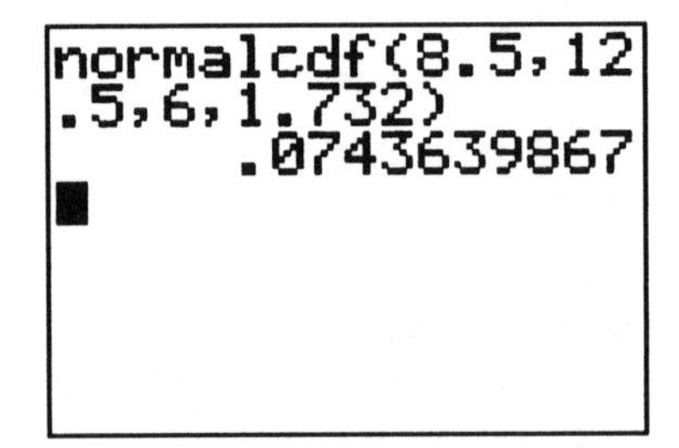

5. $\bar{x} = 8.75$, $s \approx 2.562$, and

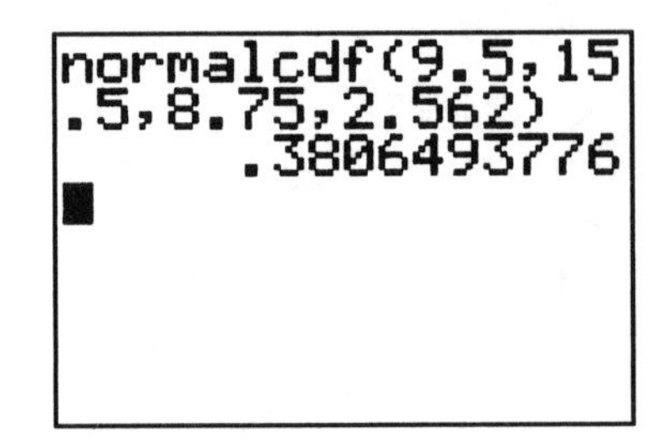

CHAPTER 13

Section 13.1

1. (2) **2.** $r \approx 0.985$

3. a.

b.
```
LinReg
 y=ax+b
 a=3.06
 b=-.0777777778
 r²=.9925103543
 r=.9962481389
```

c.

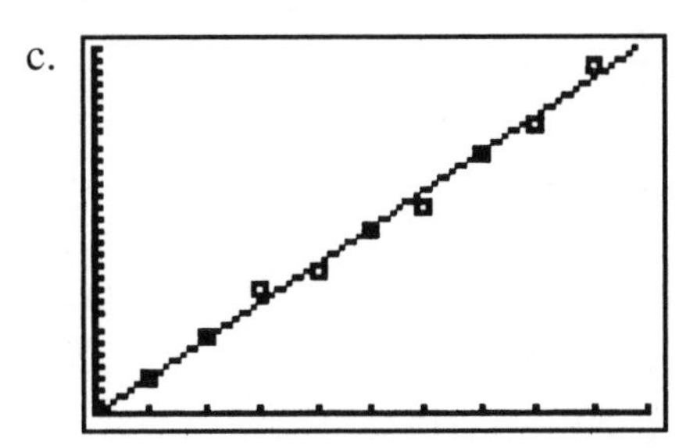

d. Correlation coefficient ≈ 0.996

4. a.

b.
```
ExpReg
 y=a*b^x
 a=3.021289959
 b=1.564994654
 r²=.9964020344
 r=.9981993961
```

c.

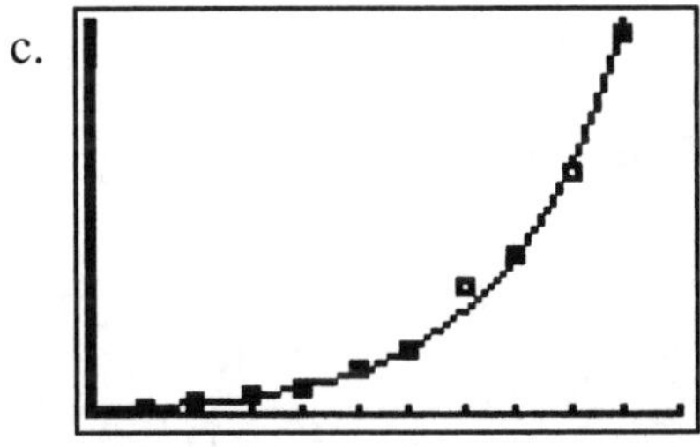

d. Correlation coefficient ≈ 0.998

5. a.

```
QuadReg
 y=ax²+bx+c
 a=1.59469697
 b=8.640151515
 c=⁻7.516666667
 R²=.9961297947
```

b.

```
ExpReg
 y=a*b^x
 a=7.055489562
 b=1.484912422
 r²=.8973430061
 r=.9472819042
```

c.

```
CubicReg
 y=ax³+bx²+cx+d
 a=⁻.0631313131
 b=2.636363636
 c=3.835858586
 d=⁻2.1
 R²=.9963416069
```

d. Cubic regression because r^2 is closer to 1.

Section 13.2

1. $y \approx -5x + 10066$, 16 students

2. $y \approx 1753.2x + 14478.5$, population in 2025 is approximately 67,076

3. $y \approx 999.89*(0.730)^x$
  a. 207 units b. 81 units c. 9 units

4. $y \approx -0.621x^3 + 17.459x^2 - 79.639x + 97.586$
  a. \$18.64 b. \$259.96 c. \$184.52

5. $y \approx 5.792x^{1.198}$
  a. 30 b. 102 c. 210

6. $y \approx 855.81(1.05)^x$, \$1,538.40

7. $y \approx 2.857x + 25.143$, 59.4 inches

CHAPTER 14

Section 14.1

1. 4, 9, 14, 19

2. 2, 10, 30, 68

3. 2, 0, –2, –4

4. 2, 4, 8, 16

5. 1, 2, 6, 24

6. $\frac{1}{2}, \frac{2}{3}, \frac{3}{4}, \frac{4}{5}$

7. 99

8. (3)

9. (3)

10. (4)

11. (2)

12. (1)

13. 18, 22, 26

14. 48, –96, 192

15. $\frac{7}{10}, \frac{2}{3}, \frac{9}{14}$

16. $a_1 = 3$, $a_2 = 4(3) - 7 = 5$, $a_3 = 4(5) - 7 = 13$, $a_4 = 4(13) - 7 = 45$
$a_5 = 4(45) - 7 = 173$

17. The sequence generator is $a_n = (0.1)^n$ and recursively defined $a_1 = 0.1$ and $a_n = (0.1) * a_{n-1}$

Section 14.2

1. a. 7 b. −12 c. 4 d. 28
2. (3)
3. (1)
4. (4)
5. (2)
6. (4)
7. $a_1 = 21$ and $a_n = a_{n-1} + 4$
8. $S_8 = 148$
9. $S_{17} = 561$
10. (3)
11. (3)
12. 650
13. 1116
14. $a_1 = -14$
15. $a_{24} = \$5300$ and $S_{24} = \$135{,}600$

Section 14.3

1. a. $r = \frac{-1}{2}$, $a_n = -8\left(\frac{-1}{2}\right)^{n-1}$, $a_1 = -8$, and $a_n = \frac{-1}{2}(a_{n-1})$

   b. $r = 4$, $a_n = 2(4)^{n-1}$, $a_1 = 2$, and $a_n = 4(a_{n-1})$

   c. $r = 0.1$, $a_n = -100\left(\frac{1}{10}\right)^{n-1}$, $a_1 = -100$, and $a_n = \frac{1}{10}(a_{n-1})$

   d. $r = 6$, $a_n = (6)^{n-1}$, $a_1 = 1$, and $a_n = 6(a_{n-1})$

2. (2)
3. (3)
4. (2)
5. 0.0000000004
6. (1)
7. (3)
8. (4)
9. 673,684.21
10. $S_5 = 5050.50505$
11. $S_{10} = 349{,}525$
12. $r = -2$, $a_1 = \frac{-3}{2}$, and $a_n = \frac{-3}{2}(-2)^{n-1}$
13. 87,480
14. 15,625

Section 14.4

1. 91
2. 105
3. 36
4. a. 196 b. 273 c. 69 d. 392 e. 177 f. 823
5. $S_n = \sum_{n=1}^{10}\left(800(.2)^{n-1}\right) = 999.68$
6. $S_n = \sum_{n=1}^{10}(-5n - 25) = -605$
7. $S_n = \sum_{n=1}^{\infty}\left((12)\left(\frac{1}{3}\right)^{n-1}\right) = 18$
8. $S_n = \sum_{n=1}^{\infty}\left(25(0.2)^{n-1}\right) = 31.25$
9. (3)
10. $\sum_{n=1}^{25} 100 = 25 \times 100 = 2500$ and $\sum_{n=1}^{25}\left(0.01(2)^{n-1}\right) = 335{,}544.31$ Take the penny and double the amount each day.

Glossary of Algebra 2/ Trigonometry Terms

A

Absolute value: The distance of a number on a number line from 0.

Absolute value equation and/or inequality: An equation and/or inequality that contains an absolute value.

Ambiguous case: A situation in which more than one solution for a triangle can be found when using the Law of Sines.

Amplitude: How high above or below the *x*-axis the graph of a function reaches.

Angle of depression: An angle between the horizontal and the line of sight to an object beneath the horizontal.

Angle of elevation: An angle between the horizontal and the line of sight to an object above the horizontal.

Antilogarithm: The inverse of a logarithm.

Approximate value: The value of a variable rounded to a specific degree of accuracy.

Arccosine: A function mapping any real number to the measure of an angle whose cosine is equal to that number.

Arc length: The radian measure of the length of an arc of a circle. The length of the radius of the circle times the radian measure of the central angle that subtends the arc.

Arcsine: A function mapping any real number to the measure of an angle whose sine is equal to that number.

Arctangent: A function mapping any real number to the measure of an angle whose tangent is equal to that number.

Arithmetic Sequence: A sequence that results from adding the same amount between terms. The *n*th term in an arithmetic sequence is $a_n = a_1 + (n - 1)d$.

Arithmetic Series: The sum of the terms of an arithmetic sequence.

Asymptote: A line (or curve) that best describes the behavior of a function as its independent variable approaches infinity or negative infinity and/or a number that is not in the domain of that function.

B

Bernouli experiment: See *binomial experiment*.

Bimodal: A data set that has two values that are both modes of the distribution.

Binomial: A polynomial with two terms.

Binomial expansion (the binomial theorem): The product of all the factors of a single binomial term raised to the *n*th power. It is expressed in combinations based on how many times each term of the binomial is selected for each term of that product.

Binomial experiment: An experiment with a fixed number of independent trials in which the probability of each event remains the same in each trial and each outcome can be classified as either a success or a failure. The sum of the probabilities for success and failure equals 1.

Box-and-Whisker Plot: A graphical representation of the spread of data that shows a 5-number summary of the lowest data value, Q_1, the median (Q_2),Q_3, and the highest data value.

C

Census: A collection of data from the entire population.

Central angle: An angle inside a circle whose vertex is the center of the circle.

Central tendency: An average data value that represents the center of the data in a distribution.

Circle: The locus of points in a plane equidistant from a given point in that plane. The equation of a circle whose center is (h, k) with radius r is $(x - h)^2 + (y - k)^2 = r^2$.

Coefficient: The number in front of or multiplied by a variable.

Cofunction: Two functions of angles so that one function evaluated at an angle is equal to the other function evaluated at the complement of that angle.

Combination: An arrangement of objects in which order does not count.

Common difference: The difference in consecutive terms of an arithmetic sequence.

Common factor: A factor that is common to each term of a polynomial expression.

Common ratio: The ratio between consecutive terms of a geometric sequence.

Complement of an event: The set of outcomes in the sample space that are not in the event.

Completing the square: Determining the number that is added to the first two terms of a trinomial, resulting in a perfect square trinomial.

Complex numbers: Numbers that can be written in the form $a + bi$, where a and b are real numbers and b is the coefficient of an imaginary number.

Complex fractional expression: A fraction that contains at least one fraction within its numerator or denominator.

Composition of functions: The composition of functions $f(x)$ and $g(x)$ is written as $h(x) = (f \circ g)(x) = f(g(x))$ such that function f is a function of function g.

Conjugates: Two binomial expressions that have the same two terms but one is the sum of these terms and the other is their difference.

Constant of variation: A constant that defines the direct or indirect variation between two variables.

Controlled experiment: Structured studies that consist of two or more groups chosen from the population. One group utilizes the subject of the study. At least one group, called a control group, does not utilize the subject of the study.

Converges: A series whose sum levels off or approaches a certain sum.

Correlation coefficient: A single number that describes the degree of relationship between two variables.

Cosecant: A function that is the ratio between the hypotenuse of a right triangle and the side opposite the angle that serves as its independent variable.

Cosine: A function that is the ratio between the side adjacent to the angle that serves as its independent variable and the hypotenuse of the right triangle.

Cotangent: A function that is the ratio between the side adjacent to the angle that serves as its independent variable and the side opposite the angle that serves as its independent variable.

Coterminal angles: Angles whose terminal sides are the same ray.

D

Degree of a polynomial: The degree of the nonzero term of highest degree (the term with the largest exponent).

Descriptive statistics: The collection, organization, presentation, and summarization of data.

Difference of two perfect squares: A binomial each of whose terms is a perfect square and where one term is subtracted from the other.

Direct variation: A relationship between two variables such that as one variable increases, the other increases also.

Discriminant: The radicand inside the quadratic formula, $b^2 - 4ac$. This expression actually determines the nature of the roots of the quadratic equation.

Diverges: A series whose sum gets extremely larger and does not level off or approaches a certain sum.

Domain: The set defined by the numbers that are the first elements in each ordered pair of a set of ordered pairs or the set of suitable values of x in a relation mapping x-values to y-values.

Double-Angle Identities: Identities used to determine a trigonometric function of an angle whose measure is twice the measure of a given angle.

Double root: A root of a polynomial that occurs twice.

E

***e*:** An irrational number that is called the natural base because it is used to model exponential growth and decay.

Empirical probability: The frequency of an event divided by the total number of frequencies of all events in the sample space.

Empirical rule: The amount of data in a normal distribution that lies within a certain number of standard deviations on either side of the mean.

Equation: An algebraic sentence that is true for a limited number of values of the variable.

Equivalent forms: Two or more forms of an expression that have the same value.

Exact value: A numerical value of a variable that has not been rounded or estimated.

Exponent: A number (or variable) superscripted above another number or variable (base) to indicate how many factors there are for that base.

Exponential function: A function that contains exponents as variables with constants as bases.

Extraneous root: A root or solution that is a solution satisfying the revised equation but not the original equation.

Extrapolation: An estimation based on a regression of a data set for which the value of the independent variable does not fall within the range of the original data set.

Event: Any subset of the sample space.

F

Factor: A divisor of a number or an expression.

Factorial: The product of a positive integer by each of its predecessors up to and including 1.

Fibonacci Sequence: A sequence defined recursively by $a_1 = 1$, $a_2 = 1$, and $a_n = a_{n-2} + a_{n-1}$.

Finite sequence: A sequence that has its domain limited to a subset of the counting numbers, starting with 1.

Fractional exponent: An exponent that is a fraction, such that $x^{\frac{a}{b}} = \sqrt[b]{x^a}$.

Frequency (of a periodic function): The number of full cycles of the curve between 0 and 2π.

Function: A relation in which each element of the domain is mapped to one and only one member of the range.

Fundamental principle of counting: The number of ways that two independent events can occur. It is found by multiplying the number of ways each event occurs.

G

Geometric Sequence: A sequence that results from multiplying the same amount between terms. The nth term in a geometric sequence is $a_n = a_1 r^{n-1}$.

Geometric Series: The sum of the terms of a geometric sequence.

Graphical solution (of equations): Those ordered pairs that satisfy both or all the equations simultaneously. The intersection of the graphs representing those functions.

Greatest common factor: The largest divisor that is common between two or more expressions.

H

Half-Angle Identities: Identities used to determine a trigonometric function of an angle whose measure is one-half the measure of a given angle.

Half-life: The interval of time required for the quantity to decay to half of its initial value.

Horizontal compression: When $a > 1$ and the y-value of each ordered pair of $f(x)$ is to be multiplied by a.

Horizontal expansion: When $0 < a < 1$ and the y-value of each ordered pair of $f(x)$ is to be multiplied by a.

Horizontal line test: If any horizontal line passes through more than one point on the graph of a function, then that function is not one-to-one.

Horizontal shift: A shift of a units on function $f(x)$ is written as $f(x - a)$

and adds a to the x-values of the ordered pairs that satisfy $f(x)$.

I

***i*:** See *imaginary unit.*

Identity: An equation that is true for all values of a variable.

Imaginary number: A number that is any multiple of the imaginary unit. The square root of any negative number.

Imaginary unit: The number i such that $i^2 = -1$.

Indirect (or inverse) variation: A relationship between two variables such that as one variable increases, the other decreases.

Infinite Geometric Series: A geometric series that contains an infinite number of terms. The sum of an infinite geometric series is $S_n = \frac{a_1}{1-r}$ where $|r| < 1$.

Infinite sequence: A sequence whose domain is the entire set of counting numbers or positive integers.

Initial side of an angle: The positive x-axis.

Interpolation: An estimation based on a regression on a data set for which the value of the independent variable does fall within the range of the original data set.

Interquartile range: The difference between the upper quartile and the lower quartile or $Q_3 - Q_1$.

Inverse of a function: A function, $f^{-1}(x)$, that maps the range of function $f(x)$ to the domain of function $f(x)$.

Inverse variation: Two variables, x and y, that satisfy the general equation $y = \frac{k}{x}$ such that as x increases, y decreases.

Irrational number: A number that cannot be expressed as the ratio of two integers.

L

Law of Cosines: A proportion involving two sides and the included angle or the three sides of a triangle that can be used to solve for the measures of the other three parts of the triangle.

Law of Sines: A proportion involving two sides and their opposite angles that can be used to solve for one of these four measures given the other three measures.

Lead coefficient: The coefficient of the first term of a polynomial when it is put into standard form. It is also the coefficient of the term in the polynomial with highest degree.

Least squares: A method used to determine a regression for which the sum of squared residuals has its least value.

Linear equation: An equation that can be written in the form $y = mx + b$.

Logarithmic equation: An equation involving logarithms.

Logarithmic function: The inverse function of an exponential function.

M

Mean: The measure of central tendency that is most frequently associated with the word average. It is calculated by adding all the data values and dividing by the number of pieces of data in the distribution.

Measures of dispersion: How the data is spread out or dispersed from the center.

Median: The middle number when the data in a distribution is arranged in numerical order.

Midline: A horizontal line midway between the smallest and largest *y*-values for a trigonometric function.

Midrange: The average of the highest data value and the lowest data value.

Mode: The data value that occurs most often in the distribution.

Monomial: A polynomial with one term.

N

Nature of the roots: Classifying the roots of an equation as being equal, rational, irrational, imaginary, and so on.

Negative exponent: An exponent that is negative such that $x^{-a} = \frac{1}{x^a}$.

Normal distribution: A distribution in which the data are symmetric to the mean.

***n*th term of a sequence:** The term defined by a_n that corresponds to the function value $f(n)$ when the sequence is thought of in functional notation. a_n is also called the sequence generator.

O

Observational study: A study in which a researcher observes the consequences for portions of a population who have already been exposed to the treatment being studied.

One-to-one function: A function such that each *y*-value comes from one and only one *x*-value.

Onto function: A function such that each member of the outcome set or range has a preimage or is actually mapped to by a member of the domain or input set.

Outcome: One of the possible occurrences in a single trial in a probability experiment.

Outlier: An observed data value that is numerically distant from the rest of the data.

P

Parabola: The graph of a quadratic function.

Parameter: A quantity that defines certain characteristics of a function.

Parent function: A basic function whose graph is easily recognizable and whose properties can be followed through various transformations.

Pascal's Triangle: A triangle representing the coefficients of the binomial $(a + b)^n$. Each row represents these coefficients for different values of n.

Percentiles: A data value in a distribution for which a portion of the data fall below.

Periodic functions: A function where, for all x in the domain of f, $f(x) = f(x + p)$ for some positive number p, called the period of that function.

Permutation: An arrangement of objects in which order counts.

Phase shift: The fraction of a complete cycle corresponding to an offset in the displacement from a specified reference point.

Polynomial: Algebraic expressions that are the sum or difference of expressions that are the product of constants and variables.

Population: All subjects that are being studied.

Principal square root: The positive number that can be squared, resulting in the radicand.

Principal value of an inverse trigonometric function: The value of the angle closest to 0 satisfying the function.

Pythagorean identity: An identity based on the Pythagorean relationship in a right triangle, $c^2 = a^2 + b^2$.

Q

Quadrant: The four regions of the plane separated by the two axes on a coordinate grid.

Quadrantal angle: An angle whose terminal side lies on one of the coordinate axes.

Quadratic equation: An equation of the form $ax^2 + bx + c = 0$.

Quadratic inequality: An inequalitiy of the form $ax^2 + bx + c < 0$ or $ax^2 + bx + c > 0$.

Quartiles: The 25th, 50th, and 75th percentiles of data values in a distribution.

R

Radian: The measure of an angle that corresponds to a central angle to a circle that subtends an arc length equal to the radius.

Radian measure of an angle: A central angle of a circle that intercepts an arc equal in length to the radius of the circle.

Radical equation: An equation that contains a radical expression.

Radical expression: An expression that is asking to find a number of equal factors of an expression. The number of factors is called the index of the radical, and the expression inside the radical is called the radicand.

Radicand: The number inside a radical.

Range (of a data set): The difference between the highest piece of data and the lowest piece of data.

Range (of a function): The set defined by the numbers that are the second elements in each ordered pair of a set of ordered pairs or the set of suitable values of y in a relation mapping x-values to y-values.

Rational equation and/or inequality: A equation and/or an inequality that contains a rational expression.

Rationalizing a denominator: Multiplication by a number or expression that makes the denominator of a fraction rational.

Reciprocal identities: Identities based on the reciprocal relationships between the six trigonometric functions.

Recursive definition: A definition in which the first term or terms of a sequence are assigned specific values and the rest of the terms are defined as a function of previous terms in the sequence.

Recursively defined sequence: A sequence in which the first term or terms are assigned specific values and the rest of the terms are defined as a function of previous terms in the sequence.

Reference angle: The acute angle formed by terminal side of the angle and the x-axis.

Reflection: A transformation of a point over a line or a point where the new point is the same distance from that line or that point as the original point.

Regression: A technique for modeling data and determining an equation that the data closely fit.

Relation: A set of ordered pairs.

Resultant: The diagonal of a parallelogram whose sides are formed by and are parallel to the initial vectors representing forces.

Root of an equation: The solution or x-intercept.

Rotation: A composite of two line reflections over intersecting lines. $R_{C,\theta}$ rotates a point about the center of rotation, C, counterclockwise through an angle of θ.

S

Sample: A subset of the population that is being studied.

Sample space: The set of all possible outcomes.

Scatter plot: A graphical representation of discrete data values.

Secant: A function that is the ratio between the hypotenuse of a right triangle and the side adjacent to the angle that serves as its independent variable.

Sector of a circle: A region enclosed by a central angle and the arc it intercepts.

Sequence: A one-to-one function whose domain is the set of counting numbers or a subset of the counting numbers.

Sequence generator: The nth term of a sequence is a function that defines each term of the sequence.

Series: The sum of the terms of a sequence.

Sigma Notation: A Greek letter, Σ, which is the symbol used to represent the sum of consecutive terms of a function.

Sine: A function that is the ratio between the side opposite the angle that serves as its independent variable and the hypotenuse of the right triangle.

Slope: The change in y-values divided by the change in x-values.

Standard deviation: The square root of the variance.

Standard form of a polynomial: $a_nx^n + a_{n-1}x^{n-1} + a_{n-2}x^{n-2} + \ldots + a_2x^2 + a_1x + a_0$.

Standard position: An angle placed so that its vertex is at the origin and its initial side is the positive x-axis.

Statistics: To collect, organize, summarize, and analyze data and draw conclusions from that data.

Sum and difference identities: Identities used to determine a trigonometric function of the sum or difference of two different angles.

Sum and product of the roots (of a quadratic equation): If $ax^2 + bx + c = 0$, then

$$r_1 + r_2 = -\frac{b}{a} \text{ and } r_1 \cdot r_2 = \frac{c}{a}.$$

Sum of a finite Arithmetic Sequence: In an arithmetic sequence, either $S_n = \frac{n}{2}(a_1 + a_n)$ or $S_n = \frac{n}{2}$

$(2a_1 + (n-1)d)$.

Sum of a Finite Geometric Sequence: In a geometric sequence, $S_n = \frac{a_1 - a_1r^n}{1-r}$ or $\frac{a_1(1-r^n)}{1-r}$ where $r \neq 1$.

Survey: Data collected in the form of a written questionnaire, a telephone survey or questions over the telephone, and personal interviews.

T

Tangent: A function that is the ratio between the side opposite the angle that serves as its independent variable and the side adjacent to that same angle.

Terminal side of an angle: The side of an angle in standard position not on the positive *x*-axis.

Theoretical probability: The ratio between the number of outcomes that satisfy an event and the total number of outcomes in the sample space.

Transformation: A mapping taking points on a plane to other points on the plane.

Translation: A composite of two line reflections over parallel lines. $T_{a,b}$: $(x, y) \rightarrow (x + a, y + b)$.

Trigonometric equation: An equation that contains trigonometric functions of an angle and whose variable is the measure of that angle.

Trigonometry: The study of the properties of triangles and the relationship between the ratio of sides of a triangle and the angles of the triangle.

Trinomial: A polynomial with three terms.

U

Unit circle: A circle whose radius is equal to 1.

V

Variable: A symbol that represents an unspecified member of a set.

Variance: The average of the squared difference of each data value from the mean of the distribution.

Vector: A ray with magnitude and direction, often representing a force.

Vertical compression: When $0 < a < 1$ and the *y*-value of each ordered pair of $f(x)$ is to be multiplied by a.

Vertical expansion: When $a > 1$ and the *y*-value of each ordered pair of $f(x)$ is to be multiplied by a.

Vertical line test: If any vertical line passes through more than one point on the graph of a relation, then that relation is not a function.

Vertical shift: A vertical shift of a units on function $f(x)$ is written as $f(x) + a$ and adds a to the *y*-values of the ordered pairs that satisfy $f(x)$.

X

***x*-intercept:** Where the graph of a function crosses the *x*-axis or the *x*-value associated with $y = 0$.

Y

***y*-intercept:** Where the graph of a function crosses the *y*-axis or the *y*-value associated with $x = 0$.

Z

Zero Product Property: If $a \cdot b = 0$, then either $a = 0$ or $b = 0$ or both.

Zero (of a function): The solution, root, or x-intercept.

***z*-score:** The number of standard deviations that a data value is from the mean.

ALGEBRA 2/TRIGONOMETRY REGENTS EXAMINATIONS

The Algebra 2/Trigonometry Examination is a three-hour examination that consists of four parts and a total of 39 questions. All questions must be answered, and there will be no opportunity to skip any questions asked.

Part I consists of multiple-choice questions each worth 2 points. The answers to these questions must be entered in the answer spaces provided in the examination booklet. The other three parts of the test consists of open-ended or constructed-response questions. Answers must be written in the space provided under the question inside the examination booklet. Part II questions are worth 2 points each. Part III questions are worth 4 points each. Part IV consists of question worth 6 points. The table below shows the breakdown of questions for this examination, worth a total of 88 credit points.

Question Type	Number of Questions
Multiple choice	27
2-credit open ended	8
4-credit open ended	3
6-credit open ended	1
Total credits	88

The percentage of questions from each of the Content Strands in the Mathematics Learning Standards for Algebra 2/Trigonometry set by the New York State Education Department are listed in the table below.

Content Strand	% of Total Credits
1) Number sense and operations	6–10%
2) Algebra	70–75%
3) Measurement	2–5%
4) Probability and statistics	13–17%

How Is the Examination Scored?

- All regents examinations are scored by a committee of qualified teachers for that subject area.
- Multiple-choice questions are scored as either right or wrong—no partial credit is awarded.
- The open-ended questions in Parts II, III, and IV may receive partial credit based on a rating guide provided by the New York State Education Department.
- Full credit for a constructed-response question is awarded only if complete and correct answer to all parts of the question are provided. Sufficient work must be shown to enable the rater to determine how the student arrived at the correct answer.
- Errors in computation, graphing, and rounding will receive a 1-credit deduction. A combination of two such errors receives a 2-credit deduction. No more than 2 credits may be deducted for such mechanical errors in any one question.
- Conceptual errors such as using the wrong formula or performing the wrong operation result in a credit reduction of no more than half the credit value of the specific question. Repeated occurrences of the same conceptual error do not result in a deduction of credit more than once in the same question. The same repeated conceptual error in multiple questions is deducted at each occurrence. Responses with two or more conceptual errors are considered completely incorrect and receive no credit.

Graphing Calculators

Graphing calculators are *required* for the Algebra 2/Trigonometry Regents Examination. This book shows you typical keystrokes that can be used in the solution to problems. A calculator is needed to work with trigonometric functions, logarithms, and many calculations that are beyond basic arithmetic operations.

The use of a graphing calculator provides you with more options when working with questions. Many questions can be solved graphically or numerically as well as with a standard algebraic approach. Be careful to indicate clearly how the calculator was used in questions that require that all work be shown! In other words, you may have to provide such things as the graphing window that was used to examine a function or some of the values from a table that was used to determine a solution to a problem.

Are Any Formulas Provided?

The Algebra 2/Trigonometry Regents Examination test booklet includes a reference sheet containing many of the formulas referred to in this book. The actual formula sheet that is provided on the Algebra 2/Trigonometry Regents Examination is included at the end of this section. Keep in mind that these formulas are not all the formulas needed to perform well on the Algebra 2/ Trigonomtry Regents Examination.

Other Important Information

- All questions must be answered on this examination. No questions can be skipped. For Part I, make sure the answer for each question is recorded in the answer blanks provided.
- Remember that unless a question specifies an approach that must be used, such as "Solve algebraically," choose any of the methods that seems the most appropriate and easiest to use to determine the solution: graphical, numerical, or algebraic.
- In the constructed-response questions of Parts II, III, or IV, if a trial-and-error approach is used, you must show the work for at least three guesses with their checks. If the first guess yields the correct answer, it is still necessary to show two guesses that do not work.
- In the constructed-response questions of Parts II, III, or IV, if a graphing calculator is used, you must explain how the calculator was used to obtain the answer. When copying graphs and/or tables, be careful to label all graphs with the appropriate equation(s) and provide the parameters used for the viewing window, coordinates of intercepts, points of intersection, and so on. It is also helpful to indicate the rationale for the approach used. Also be careful when working with trigonometry to check to see if your calculator is set for radian or degree mode.
- Remember that a formula sheet is provided, but not all formulas found in the course are provided. When using a formula to solve a problem, be careful to write out the entire formula as the first step in the solution.
- Be careful to provide your answer in the correct format. For instance, if the question asks for the measure of an angle, make sure whether degree or radian measure is required. If a specific form is not specified, an answer such as $\frac{\sqrt{2}}{2}$ can be expressed in equivalent forms such as $\frac{1}{\sqrt{2}}$ or as the full calculator display of its decimal form, .7071067812.
- Rounding errors can be easily made when using a graphing calculator. When a solution requires multiple steps and the answer from one step is used to calculate a second or multiple other steps, use the actual number displayed on the calculator rather than rounding that answer. This can often be accomplished by using **2ND** **(–)** which is **ANS** or by saving the intermediate answers before performing the next operation on the calculator. Remember, unless a problem specifies

differently, rounding should be done only when the final answer is calculated. Use the calculator display for the number π unless a problem specifies the use of an approximation for π, such as 3.14 or $\frac{22}{7}$.

- When working with multiple-choice questions, remember that the correct answer is in one of the choices and sometimes working backward from the answer is the most expedient approach. Also recognize that common errors are often represented in the incorrect answer choices. So be careful not to fall for traps in the answer choices. Do not make random guesses for multiple-choice questions. Try to eliminate some choices if the question is not on a familiar topic.
- Questions do not have to be answered in order. First work on the easy questions.

Sample Regents Examinations

Important notice: The first test that follows is *not* an actual Regents Examination. It is a practice test that has the same format and similar topic coverage of an actual Regents Examination in Algebra 2/ Trigonometry. The second test beginning on page 466 was administered to students in June 2010. This test and answers will be very valuable to you as you prepare for the Regents exam.

Algebra 2/Trigonometry Reference Sheet

Area of a Triangle

$$K = \frac{1}{2}ab\sin C$$

Law of Cosines

$$a^2 = b^2 + c^2 - 2bc\cos A$$

Functions of the Sum of Two Angles

$$\sin(A + B) = \sin A\cos B + \cos A\sin B$$
$$\cos(A + B) = \cos A\cos B - \sin A\sin B$$
$$\tan(A + B) = \frac{\tan A + \tan B}{1 - \tan A\tan B}$$

Functions of the Double Angle

$$\sin 2a = 2\sin A\cos A$$
$$\cos 2A = \cos^2 A - \sin^2 A$$
$$\cos 2A = 2\cos^2 A - 1$$
$$\cos 2A = 1 - 2\sin^2 A$$
$$\tan 2A = \frac{2\tan A}{1 - \tan^2 A}$$

Functions of the Difference of Two Angles

$$\sin(A - B) = \sin A\cos B - \cos A\sin B$$
$$\cos(A - B) = \cos A\cos B + \sin A\sin B$$

Functions of the Half Angle

$$\tan(A - B) = \frac{\tan A - \tan B}{1 + \tan A\tan B}$$

$$\sin\frac{1}{2}A = \pm\sqrt{\frac{1 - \cos A}{2}}$$

Law of Sines

$$\cos\frac{1}{2}A = \pm\sqrt{\frac{1+\cos A}{2}}$$

$$\frac{a}{\sin A} = \frac{b}{\sin B} = \frac{c}{\sin C}$$

$$\tan\frac{1}{2}A = \pm\sqrt{\frac{1-\cos A}{1+\cos A}}$$

Sum of a Finite Arithmetic Series

$$S_n = \frac{n(a_1 + a_n)}{2}$$

Sum of a Finite Geometric Series

$$S_n = \frac{a_1(1-r^n)}{1-r}$$

Binomial Theorem

$$(a + b)^n = {}_nC_0a^nb^0 + {}_nC_1a^{n-1}b^1 + {}_nC_2a^{n-2}b^2 + \ldots + {}_nC_na^0b^n$$

$$(a+b)^n = \sum_{r=0}^{n} {}_nC_ra^{n-r}b^r$$

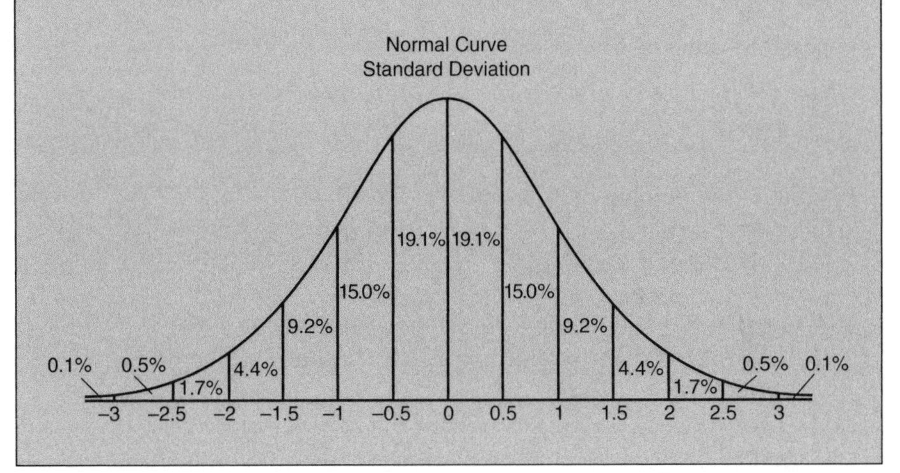

PART I

Answer all questions in this part. Each correct answer will receive 2 credits. No partial credit will be allowed. For each question, write on the separate answer sheet the numeral preceding the word or expression that best completes the statement or answers the question. [54]

1 The expression $\frac{11}{6+\sqrt{3}}$ is equivalent to

(1) $\frac{6+\sqrt{3}}{3}$ (3) $\frac{11(6+\sqrt{3})}{9}$

(2) $\frac{6-\sqrt{3}}{3}$ (4) $\frac{11(6-\sqrt{3})}{9}$ 1 _____

2 If θ is an angle in standard position and its terminal side passes through the point $(-\frac{\sqrt{3}}{2}, \frac{1}{2})$ on a unit circle, a possible value of θ is

(1) 30° (3) 120°
(2) 60° (4) 150° 2 _____

3 What is the domain of $g(x) = \sqrt{x^2 + 2x - 8}$?
(1) $\{x|x \geq 4 \text{ or } x \leq -2\}$ (3) $\{x|-4 \leq x \leq 2\}$
(2) $\{x|x \geq 2 \text{ or } x \leq -4\}$ (4) $\{x|-2 \leq x \leq 4\}$ 3 _____

4 A function, f, is defined by the set {(5, 1), (3, 2), (–4, 6)}. If f is reflected over the line $y = x$, which point will be in the reflection?
(1) (5, –1) (3) (1, 5)
(2) (–5, 1) (4) (–1, 5) 4 _____

5 $\sum_{i=1}^{8}(3i - i^2) =$
(1) –96 (2) 42 (3) –100 (4) –42 5 _____

6 The expression i^{42} is equivalent to
(1) 1 (3) i
(2) –1 (4) $-i$ 6 _____

7 From the sum of $\frac{2}{3}s^3 - 4s^2 + \frac{1}{5}s + \frac{5}{8}$ and $\frac{3}{4}s^2 - \frac{3}{5}s + \frac{1}{6}$ subtract $\frac{2}{9}s^3 - \frac{1}{10}s + \frac{3}{8}$.

(1) $\frac{4}{9}s^3 - \frac{13}{4}s^2 - \frac{1}{2}s - \frac{7}{6}$
(2) $\frac{4}{9}s^3 - \frac{19}{4}s^2 - \frac{3}{10}s - \frac{7}{6}$
(3) $\frac{4}{9}s^3 - \frac{13}{4}s^2 - \frac{3}{10}s + \frac{5}{12}$
(4) $\frac{4}{9}s^3 - \frac{13}{4}s^2 - \frac{1}{2}s - \frac{5}{12}$

7 _____

8 What is the amplitude of the function shown in the accompanying graph?

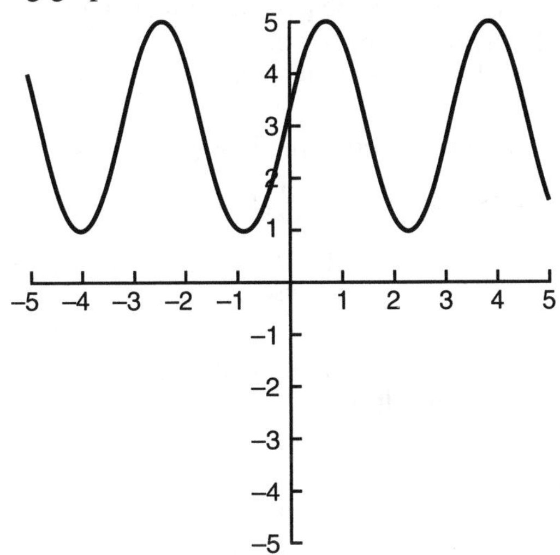

(1) 1.5 (2) 2 (3) 4 (4) 5

8 _____

9 If $\log_b 2 = r$ and $\log_b 3 = s$, then express $\log_b 9\sqrt{2}$ in terms of r and s.

(1) $2r + s$
(2) $2s + \frac{1}{2}r$
(3) $2r + \frac{1}{2}s$
(4) rs

9 _____

10 What is the solution set of the equation $x = \sqrt{3x + 4}$?

(1) $\{-1\}$
(2) $\{2\}$
(3) $\{4\}$
(4) $\{4, -1\}$

10 _____

11 The roots of the equation $3x^2 - 2x = 6$ are

(1) real, irrational, and unequal
(2) real, rational, and equal
(3) real, rational, and unequal
(4) imaginary

11 _____

12 A function is defined by the equation $y = 3x + 4$. Which equation defines the inverse of this function?

(1) $y = \frac{1}{3x+4}$

(2) $y = \frac{1}{3}x - \frac{4}{3}$

(3) $x = \frac{1}{3y+4}$

(4) $x = 3y + 4$

12 _____

13 Bill is planning a trip. He knows that if he can drive the route he plans at 55 miles per hour, the trip will take 4 hours. Instead, once he reaches his destination, he found that the trip took 5 hours because of heavy traffic. If the rate of travel is inversely proportional to the number of hours, at what rate did he travel for this trip?

(1) 44 mph (2) 45 mph (3) 50 mph (4) 54 mph

13 _____

14 If ${}_nC_r$ represents the number of combinations of n items taken r at a time, what is the value of $\sum_{r=2}^{4} {}_6C_r$?

(1) 3

(2) 20

(3) 30

(4) 50

14 _____

15 If the sum of the roots of $2x^2 - 6x + 7 = 0$ is multiplied by the product of its roots, the result is

(1) 3

(2) $\frac{21}{2}$

(3) 21

(4) $\frac{7}{2}$

15 _____

16 If $\sec(4\theta) = \csc(6\theta)$ and θ is an angle in radian measure, determine $m\angle\theta$.

(1) 9 (2) $\frac{\pi}{20}$ (3) $\frac{\pi}{10}$ (4) 10

16 _____

17 The fraction $\dfrac{x - \frac{y^2}{x}}{\frac{y}{x} + 1}$ is equivalent to

(1) $\frac{x^2 - xy^2}{y+x}$

(2) $\frac{x^2 - y^2}{y+1}$

(3) $x + y$

(4) $x - y$

17 _____

18 In $\triangle ABC$, if $AC = 10$, $BC = 8$, and $m\angle A = 40°$, angle B could be
(1) an obtuse angle, only
(2) a right angle, only
(3) an acute angle, only
(4) either an obtuse angle or an acute angle 18 ______

19 Which equation represents the parabola shown in the accompanying graph?

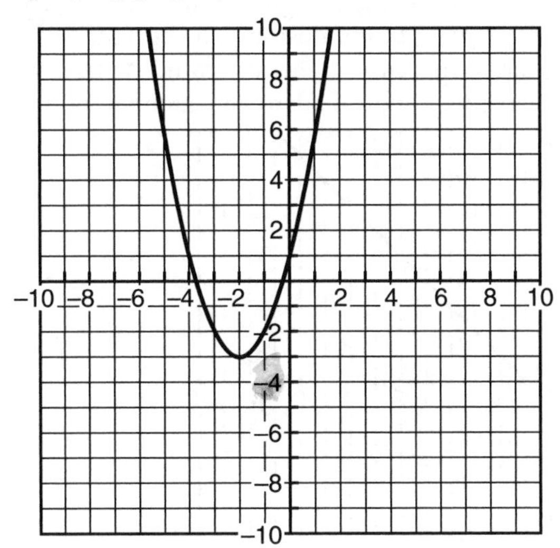

(1) $f(x) = (x + 2)^2 - 3$
(2) $f(x) = (x - 2)^2 - 3$
(3) $f(x) = -(x - 2)^2 - 3$
(4) $f(x) = (x - 2)^2 + 3$ 19 ______

20 If $f(x) = x^2 - 10$ and $g(x) = \sqrt{x+4}$, what is the value of $(f \circ g)(12)$?

(1) $\sqrt{138}$
(2) -4
(3) 6
(4) 36 20 ______

21 Define the sequence, 64, 16, 4, 1, . . . through a recursive formula.
(1) $a_1 = 64$ and $a_n = a_{n-1}(0.3)^n$
(2) $a_1 = 64$ and $a_n = a_{n-1}(0.25)$
(3) $a_1 = 64$ and $a_n = a_{n-1}(0.4)$
(4) $a_1 = 64$ and $a_n = a_{n-1}(0.25)^{n-1}$ 21 ______

22 The radian measure of an angle is $\frac{23\pi}{12}$. What is the degree measure of the angle?
(1) 165° (2) 15° (3) 690° (4) 345° 22 ______

23 The expression $\frac{\cos 2\theta}{\cos\theta + \sin\theta}$ is equivalent to
(1) $\cot 2\theta$ (3) $\cos\theta - \sin\theta$
(2) $\sec 2\theta$ (4) $\sin\theta - \cos\theta$ 23 ______

24 Which graph is *not* a function?

(1)

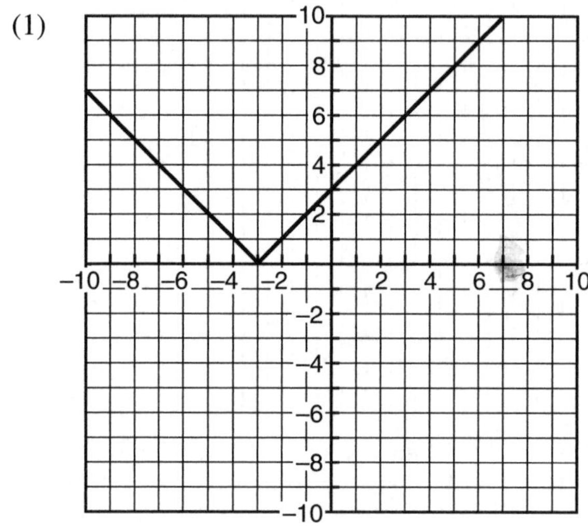

(2)

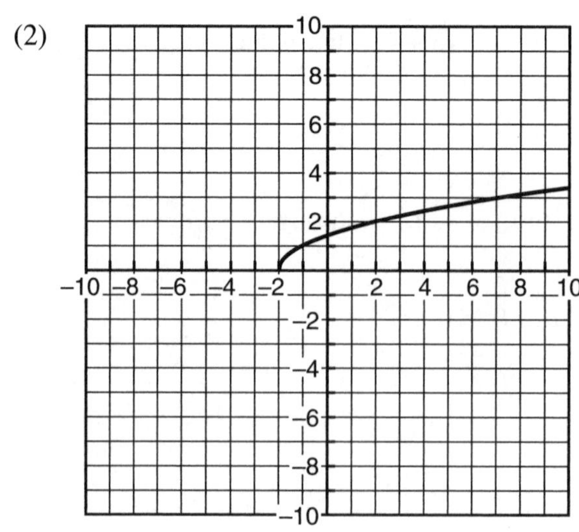

(3)

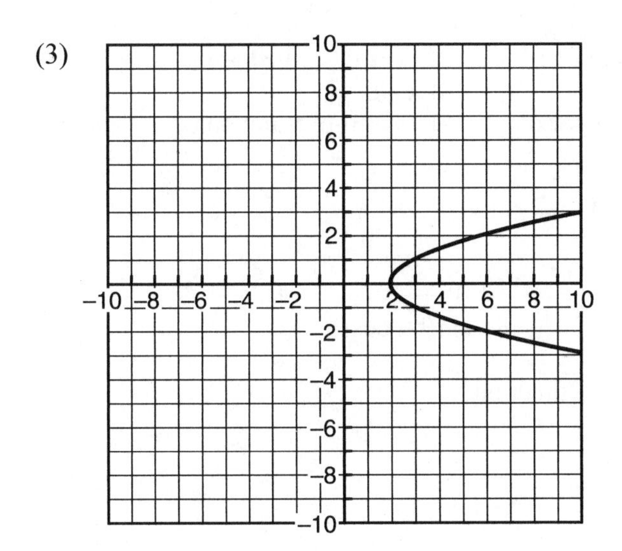

(4)

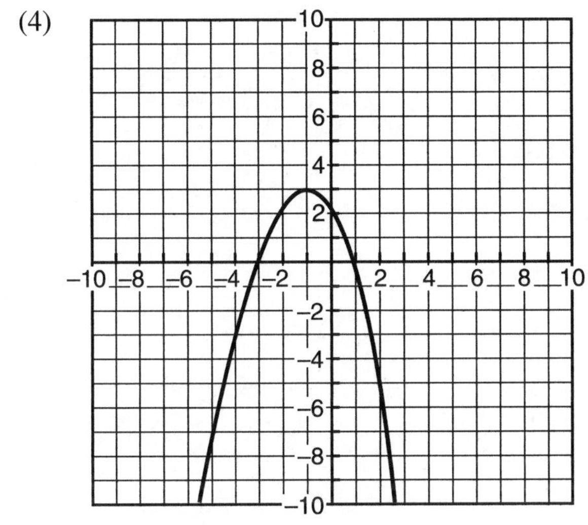

24 ______

25 Jana puts 25¢ into her piggy bank one day. The next day, she puts in 40¢. On the third day, she puts in 55¢. If she continues increasing the amount she puts into her piggy bank the same way each day, how much will she put into her piggy bank on the twelfth day?

(1) \$1.90 (2) \$2.05 (3) 70¢ (4) \$12.90

25 ______

26 The function $g(x) = \sqrt[3]{x-3} + 2$ is best described as
(1) one-to-one
(2) one-to-one and not onto
(3) onto
(4) one-to-one and onto
26 ______

27 Evaluate: $\cos\left(\frac{5\pi}{4}\right)$
(1) $-\frac{\sqrt{2}}{2}$
(2) $-\frac{\sqrt{3}}{3}$
(3) $\frac{\sqrt{2}}{2}$
(4) $-\frac{1}{2}$
27 ______

PART II

Answer all questions in this part. Each correct answer will receive 2 credits. Clearly indicate the necessary steps, including appropriate formula substitutions, diagrams, graphs, charts, etc. For all questions in this part, a correct numerical answer with no work shown will receive only 1 credit. [16]

28 A child walks around the edge of a circular water fountain in a park. If the radius of the water fountain is 15 feet, what distance does the child walk when the subtended arc is 72°? Express your answer to the *nearest hundredth of a foot.*

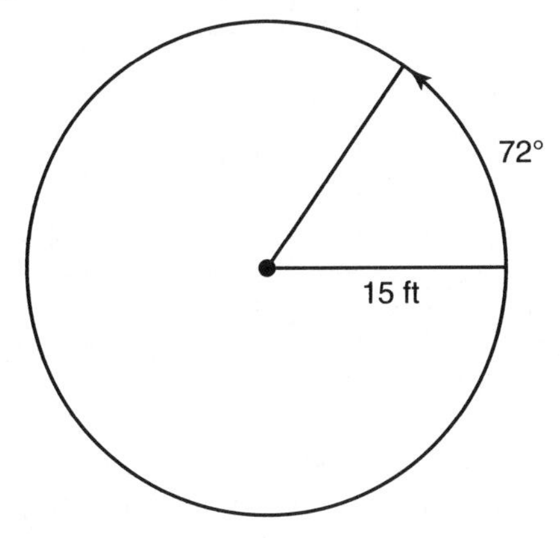

29 Write the equation of a circle whose center is at (2, –1) and passes through (–3, 5).

30 What is the fifth term in the expansion of $(3a - b)^7$?

31 In an electrical circuit, the voltage, E, in volts, the current, I, in amps, and the opposition to the flow of current, called impedance, Z, in ohms, are related by the equation $E = IZ$. A circuit has a current of $(5 - i)$ amps and an impedance of $(-4 + 2i)$ ohms. Determine the voltage in $a + bi$ form.

32 In triangle ABC, $m \angle A = 47°$, $AB = 8$ cm, and $AC = 7$ cm. Determine the area of triangle ABC to the *nearest hundredth.*

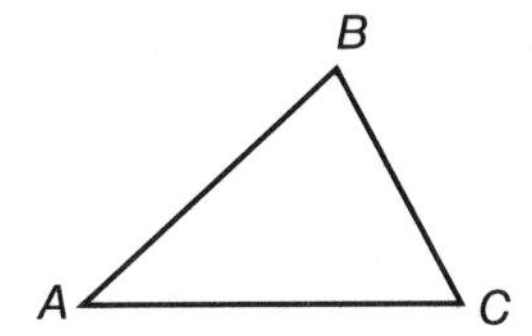

33 On the accompanying set of axes, graph the equation $y = 3\sin 2x$ in the domain $-\pi \leq x \leq \pi$.

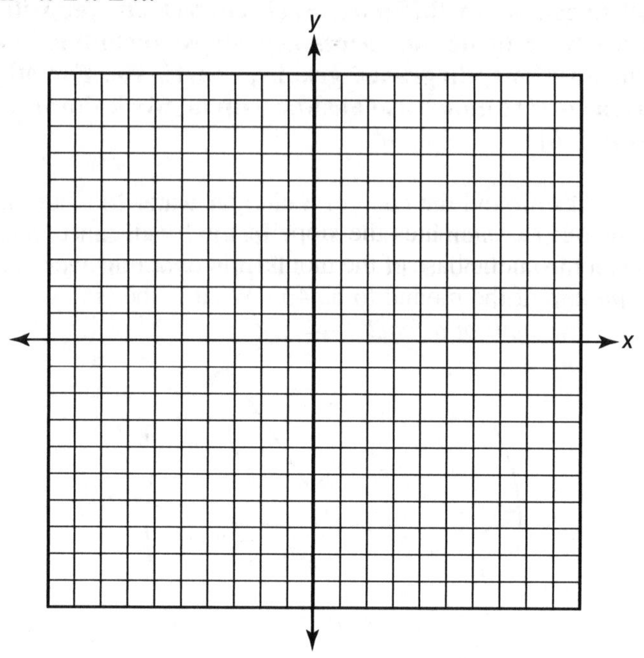

34 Using only positive exponents, rewrite and simplify $\dfrac{y^{-1}-2}{2y-1}$.

35 Evaluate in radical form: $\cos(\sec^{-1}(-2) + \tan(-1))$

PART III

Answer all questions in this part. Each correct answer will receive 4 credits. Clearly indicate the necessary steps, including appropriate formula substitutions, diagrams, graphs, charts, etc. For all questions in this part, a correct numerical answer with no work shown will receive only 1 credit. [12]

36 Phil wants to determine the height of a mound of earth that is on a construction site. He estimates the slope of the mound to be 62°. Then he walks 5 feet from the base of the mound and measures the angle of elevation to the top of the mound to be 43°. What is the height of the mound to the *nearest tenth of a foot*?

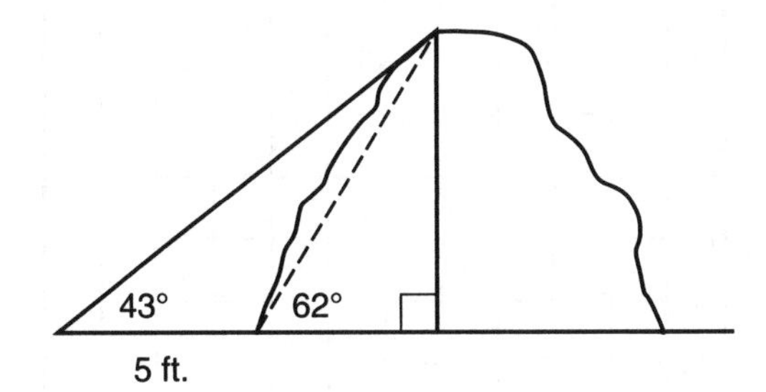

37 The probability that Frank will bowl a strike anytime he bowls a frame is $\frac{5}{8}$. Use a normal approximation to estimate the probability to the *nearest thousandth* that if Frank bowls 20 frames, he will bowl a strike at least 12 times.

38 The demand function for a new video game is given by the function $p(x) = -x^2 + 5x + 39$. Correspondingly, the supply function for that same video game is given by $p(x) = 7x - 9$, where p represents the price per video game in dollars and x represents the number of video games sold in thousands of games. Determine the equilibrium price (where demand equals supply) and the number of video games sold to achieve equilibrium. The accompanying graph grid is provided to assist in the solution but does not have to be used in the solution.

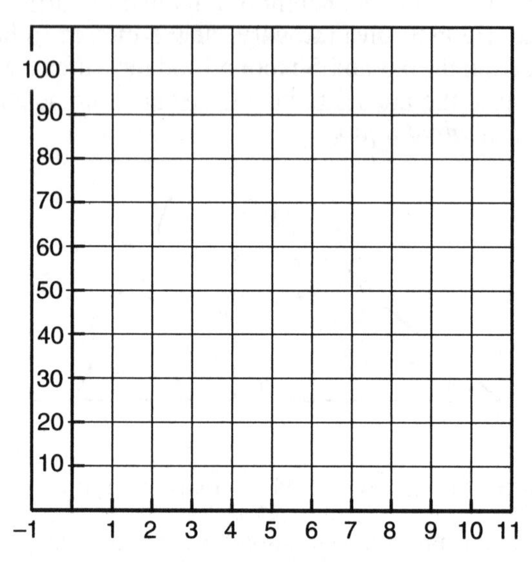

PART IV

Answer all questions in this part. A correct answer will receive 6 credits. Clearly indicate the necessary steps, including appropriate formula substitutions, diagrams, graphs, charts, etc. For all questions in this part, a correct numerical answer with no work shown will receive only 1 credit. [6]

39 The table below shows the average growth rate of Scotch pine trees planted on a farm in Sullivan County, New York. At the time of planting, all trees were 1 year old and 5 feet high.

Age of Tree in Years	Height in Feet
1	5
2	9.5
3	13
4	14.5
5	16
6	17.5
7	18.5
8	19.5

Using the data in the table, create a scatter plot on the grid and state the logarithmic regression equation with the coefficient and base rounded to the *nearest hundredth*.

Using your written regression equation, estimate the height to the *nearest tenth* of a Scotch pine tree on the farm after 12 years.

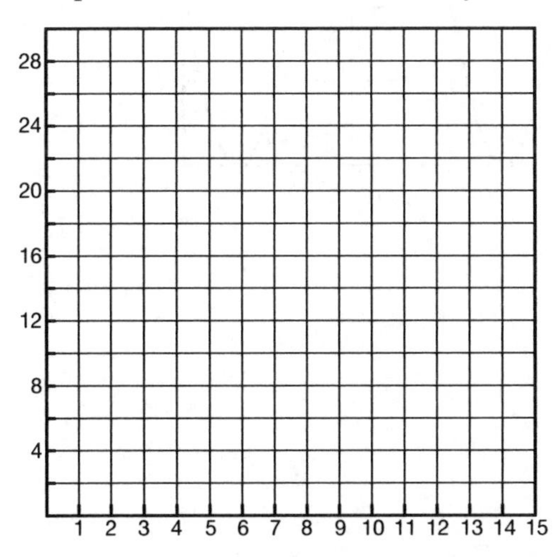

Answers to Sample Regents Examination

PART I

1. (2)	**6.** (2)	**11.** (1)	**16.** (2)	**21.** (2)	**26.** (4)
2. (4)	**7.** (3)	**12.** (2)	**17.** (4)	**22.** (4)	**27.** (1)
3. (2)	**8.** (2)	**13.** (1)	**18.** (4)	**23.** (3)	
4. (3)	**9.** (2)	**14.** (4)	**19.** (1)	**24.** (3)	
5. (1)	**10.** (3)	**15.** (2)	**20.** (3)	**25.** (1)	

PART II

28. 18.85 feet
29. $(x - 2)^2 + (y + 1)^2 = 61$
30. $945a^3b^4$
31. $-18 + 14i$
32. 20.48 cm^2

33.

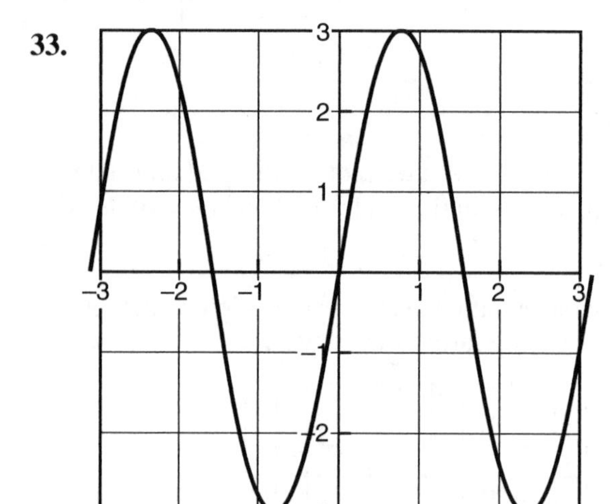

34. $\frac{-1}{y}$

35. $\frac{\sqrt{6} - \sqrt{2}}{4}$

PART III

36. 9.2 feet

37. 0.678

38.

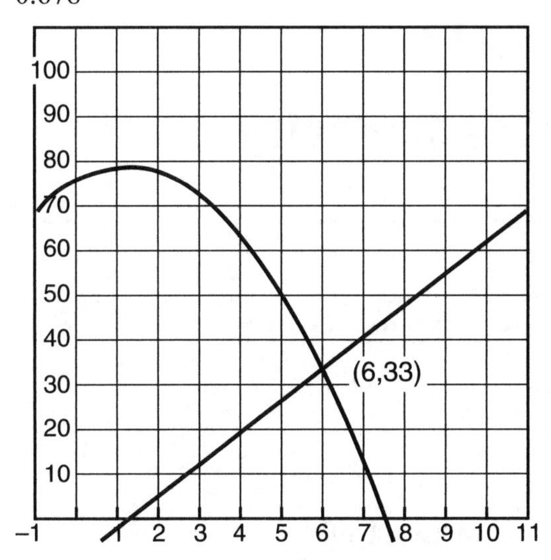

The equilibrium price is \$33 for 6000 video games

PART IV

39. $y = 4.94 + 6.98 \ln x$, 22.3 feet

In PARTS II–IV you are required to show how you arrived at your answers.

Examination June 2010

Algebra 2/Trigonometry

REFERENCE SHEET

Area of a Triangle

$$K = \frac{1}{2}ab \sin C$$

Law of Cosines

$$a^2 = b^2 + c^2 - 2bc \cos A$$

Functions of the Sum of Two Angles

$$\sin(A + B) = \sin A \cos B + \cos A \sin B$$

$$\cos(A + B) = \cos A \cos B - \sin A \sin B$$

$$\tan(A + B) = \frac{\tan A + \tan B}{1 - \tan A \tan B}$$

Functions of the Double Angle

$$\sin 2A = 2 \sin A \cos A$$

$$\cos 2A = \cos^2 A - \sin^2 A$$

$$\cos 2A = 2 \cos^2 A - 1$$

$$\cos 2A = 1 - 2 \sin^2 A$$

$$\tan 2A = \frac{2 \tan A}{1 - \tan^2 A}$$

Functions of the Difference of Two Angles

$$\sin(A - B) = \sin A \cos B - \cos A \sin B$$

$$\cos(A - B) = \cos A \cos B + \sin A \sin B$$

$$\tan(A - B) = \frac{\tan A - \tan B}{1 + \tan A \tan B}$$

Functions of the Half Angle

$$\sin\frac{1}{2}A = \pm\sqrt{\frac{1-\cos A}{2}}$$

$$\cos\frac{1}{2}A = \pm\sqrt{\frac{1+\cos A}{2}}$$

$$\tan\frac{1}{2}A = \pm\sqrt{\frac{1-\cos A}{1+\cos A}}$$

Law of Sines

$$\frac{a}{\sin A} = \frac{b}{\sin B} = \frac{c}{\sin C}$$

Sum of a Finite Arithmetic Sequence

$$S_n = \frac{n(a_1 + a_n)}{2}$$

Sum of a Finite Geometric Sequence

$$S_n = \frac{a_1(1-r^n)}{1-r}$$

Binomial Theorem

$$(a+b)^n = {}_nC_0a^nb^0 + {}_nC_1a^{n-1}b^1 + {}_nC_2a^{n-2}b^2 + \cdots + {}_nC_na^0b^n$$

$$(a+b)^n = \sum_{r=0}^{n} {}_nC_r a^{n-r}b^r$$

Normal Curve
Standard Deviation

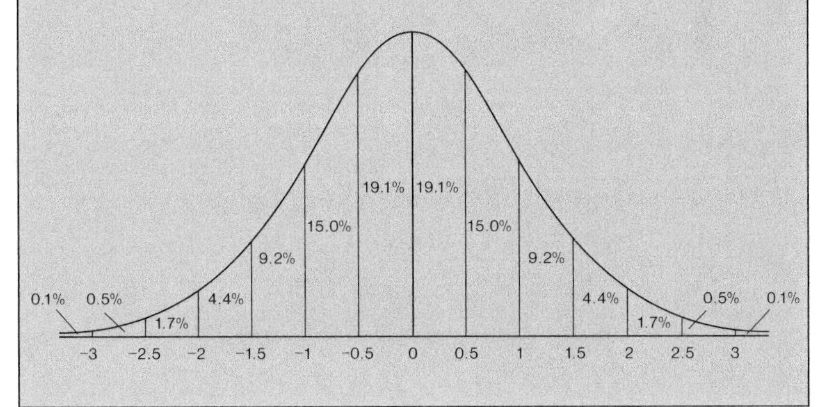

PART I

Answer all 27 questions in this part. Each correct answer will receive 2 credits. No partial credit will be allowed. For each question, write in the space provided the numeral preceding the word or expression that best completes the statement or answers the question. [54 credits]

1 What is the common difference of the arithmetic sequence 5, 8, 11, 14?

(1) $\frac{8}{5}$ (3) 3

(2) -3 (4) 9

1 _____

2 What is the number of degrees in an angle whose radian measure is $\frac{11\pi}{12}$?

(1) 150 (3) 330

(2) 165 (4) 518

2 _____

3 If $a = 3$ and $b = -2$, what is the value of the expression $\frac{a^{-2}}{b^{-3}}$?

(1) $-\frac{9}{8}$ (3) $-\frac{8}{9}$

(2) -1 (4) $\frac{8}{9}$

3 _____

4 Four points on the graph of the function f(x) are shown below.

$$\{(0,1), (1,2), (2,4), (3,8)\}$$

Which equation represents f(x)?

(1) $f(x) = 2^x$
(2) $f(x) = 2x$
(3) $f(x) = x + 1$
(4) $f(x) = \log_2 x$

4 _____

5 The graph of $y = f(x)$ is shown below.

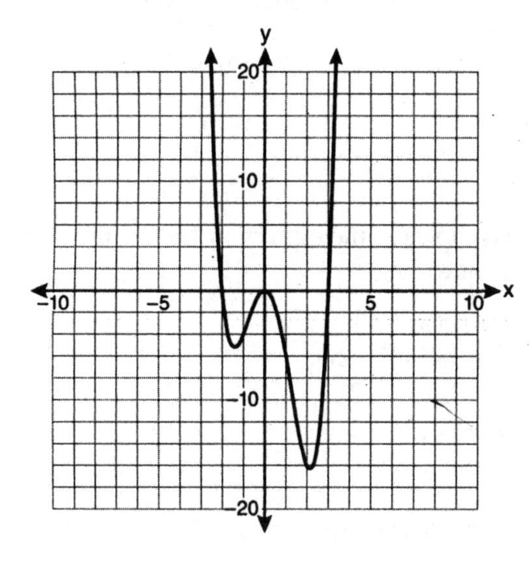

Which set lists all the real solutions of $f(x) = 0$?

(1) {–3, 2}
(2) {–2, 3}
(3) {–3, 0, 2}
(4) {–2, 0, 3}

5 _____

6 In simplest form, $\sqrt{-300}$ is equivalent to

(1) $3i\sqrt{10}$
(2) $5i\sqrt{12}$
(3) $10i\sqrt{3}$
(4) $12i\sqrt{5}$

6 _____

7 Twenty different cameras will be assigned to several boxes. Three cameras will be randomly selected and assigned to box *A*. Which expression can be used to calculate the number of ways that three cameras can be assigned to box *A*?

(1) $20!$
(2) $\frac{20!}{3!}$
(3) ${}_{20}C_3$
(4) ${}_{20}P_3$

7 _____

8 Factored completely, the expression $12x^4 + 10x^3 - 12x^2$ is equivalent to

(1) $x^2(4x + 6)(3x - 2)$
(2) $2(2x^2 + 3x)(3x^2 - 2x)$
(3) $2x^2(2x - 3)(3x + 2)$
(4) $2x^2(2x + 3)(3x - 2)$

8 _____

9 The solutions of the equation $y^2 - 3y = 9$ are

(1) $\frac{3 \pm 3i\sqrt{3}}{2}$
(2) $\frac{3 \pm 3i\sqrt{5}}{2}$
(3) $\frac{-3 \pm 3\sqrt{5}}{2}$
(4) $\frac{3 \pm 3\sqrt{5}}{2}$

9 _____

10 The expression $2 \log x - (3 \log y + \log z)$ is equivalent to

(1) $\log \frac{x^2}{y^3 z}$ (3) $\log \frac{2x}{3yz}$

(2) $\log \frac{x^2 z}{y^3}$ (4) $\log \frac{2xz}{3y}$ 10 ______

11 The expression $(x^2 - 1)^{-\frac{2}{3}}$ is equivalent to

(1) $\sqrt[3]{(x^2 - 1)^2}$ (3) $\sqrt{(x^2 - 1)^3}$

(2) $\frac{1}{\sqrt[3]{(x^2 - 1)^2}}$ (4) $\frac{1}{\sqrt{(x^2 - 1)^3}}$ 11 ______

12 Which expression is equivalent to $\frac{\sqrt{3}+5}{\sqrt{3}-5}$?

(1) $-\frac{14+5\sqrt{3}}{11}$ (3) $\frac{14+5\sqrt{3}}{14}$

(2) $-\frac{17+5\sqrt{3}}{11}$ (4) $\frac{17+5\sqrt{3}}{14}$ 12 ______

13 Which relation is *not* a function?

(1) $(x - 2)^2 + y^2 = 4$ (3) $x + y = 4$
(2) $x^2 + 4x + y = 4$ (4) $xy = 4$ 13 _____

14 If $\angle A$ is acute and $\tan A = \frac{2}{3}$, then

(1) $\cot A = \frac{2}{3}$ (3) $\cot(90° - A) = \frac{2}{3}$

(2) $\cot A = \frac{1}{3}$ (4) $\cot(90° - A) = \frac{1}{3}$ 14 _____

15 The solution set of $4^{x^2 + 4x} = 2^{-6}$ is

(1) $\{1, 3\}$ (3) $\{-1, -3\}$
(2) $\{-1, 3\}$ (4) $\{1, -3\}$ 15 _____

16 The equation $x^2 + y^2 - 2x + 6y + 3 = 0$ is equivalent to

(1) $(x - 1)^2 + (y + 3)^2 = -3$
(2) $(x - 1)^2 + (y + 3)^2 = 7$
(3) $(x + 1)^2 + (y + 3)^2 = 7$
(4) $(x + 1)^2 + (y + 3)^2 = 10$ 16 _____

17 Which graph best represents the inequality $y + 6 \geq x^2 - x$?

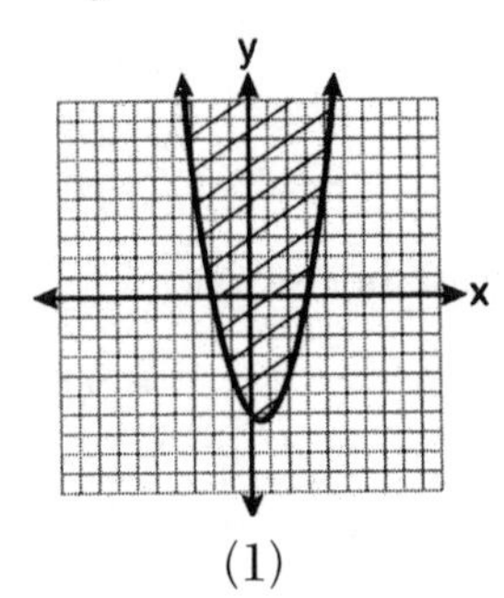

(1)

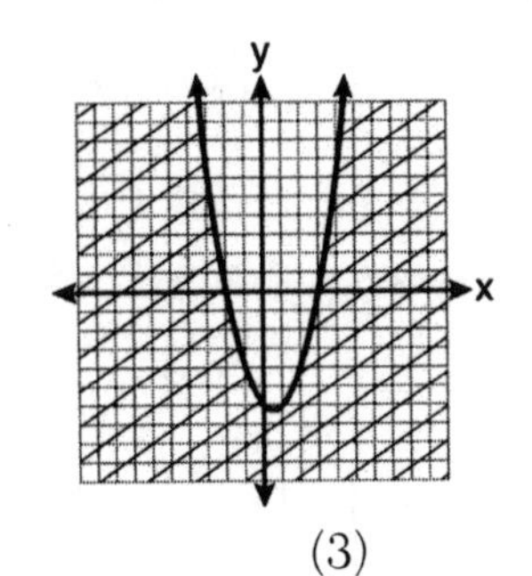

(3)

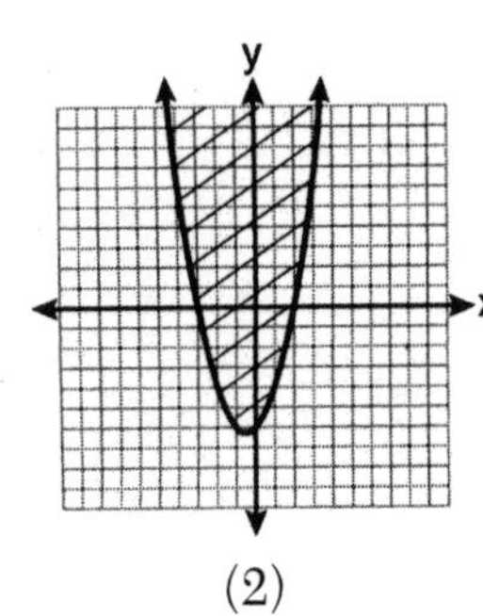

(2)

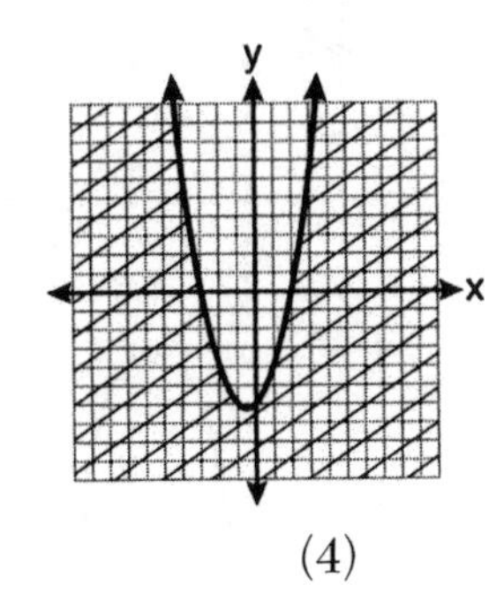

(4)

17 _____

18 The solution set of the equation $\sqrt{x+3} = 3 - x$ is

(1) {1}
(2) {0}
(3) {1, 6}
(4) {2, 3}

18 _____

19 The product of i^7 and i^5 is equivalent to

(1) 1
(2) −1
(3) i
(4) $-i$

19 _____

20 Which equation is represented by the graph below?

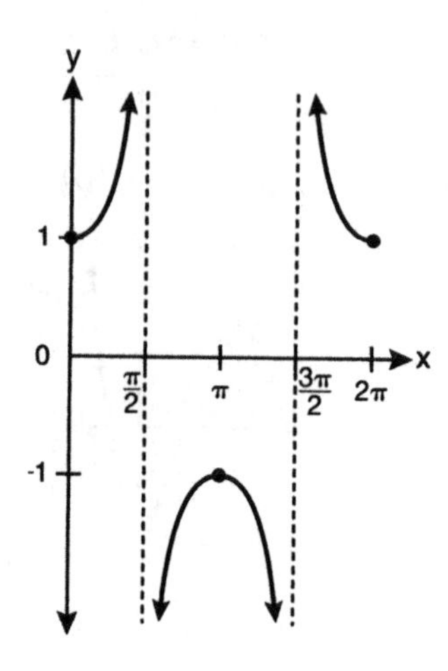

(1) $y = \cot x$
(2) $y = \csc x$
(3) $y = \sec x$
(4) $y = \tan x$

20 _____

21 Which value of r represents data with a strong negative linear correlation between two variables?

(1) −1.07
(2) −0.89
(3) −0.14
(4) 0.92

21 _____

22 The function $f(x) = \tan x$ is defined in such a way that $f^{-1}(x)$ is a function. What can be the domain of $f(x)$?

(1) $\{x \mid 0 \le x \le \pi\}$
(2) $\{x \mid 0 \le x \le 2\pi\}$
(3) $\left\{x \middle| -\frac{\pi}{2} < x < \frac{\pi}{2}\right\}$
(4) $\left\{x \middle| -\frac{\pi}{2} < x < \frac{3\pi}{2}\right\}$

22 ______

23 In the diagram below of right triangle KTW, $KW = 6$, $KT = 5$, and $m\angle KTW = 90$.

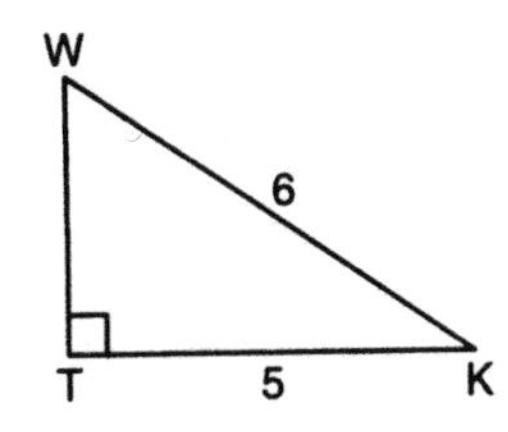

What is the measure of $\angle K$, to the *nearest minute*?

(1) 33°33'
(2) 33°34'
(3) 33°55'
(4) 33°56'

23 ______

24 The expression $\cos^2 \theta - \cos 2\theta$ is equivalent to

(1) $\sin^2 \theta$
(2) $-\sin^2 \theta$
(3) $\cos^2 \theta + 1$
(4) $-\cos^2 \theta - 1$

24 ______

25 Mrs. Hill asked her students to express the sum $1 + 3 + 5 + 7 + 9 + \ldots + 39$ using sigma notation. Four different student answers were given. Which student answer is correct?

(1) $\sum_{k=1}^{20} (2k-1)$

(2) $\sum_{k=2}^{40} (k-1)$

(3) $\sum_{k=-1}^{37} (k+2)$

(4) $\sum_{k=1}^{39} (2k-1)$

25 ______

26 What is the formula for the nth term of the sequence 54, 18, 6, . . . ?

(1) $a_n = 6\left(\frac{1}{3}\right)^n$

(2) $a_n = 6\left(\frac{1}{3}\right)^{n-1}$

(3) $a_n = 54\left(\frac{1}{3}\right)^n$

(4) $a_n = 54\left(\frac{1}{3}\right)^{n-1}$

26 ______

27 What is the period of the function $y = \frac{1}{2}\sin\left(\frac{x}{3} - \pi\right)$?

(1) $\frac{1}{2}$

(2) $\frac{1}{3}$

(3) $\frac{2}{3}\pi$

(4) 6π

27 ______

PART II

Answer all 8 questions in this part. Each correct answer will receive 2 credits. Clearly indicate the necessary steps, including appropriate formula substitutions, diagrams, graphs, charts, etc. For all questions in this part, a correct numerical answer with no work shown will receive only 1 credit. [16 credits]

28 Use the discriminant to determine all values of k that would result in the equation $x^2 - kx + 4 = 0$ having equal roots.

29 The scores of one class on the Unit 2 mathematics test are shown in the table below.

Unit 2 Mathematics Test

Test Score	Frequency
96	1
92	2
84	5
80	3
76	6
72	3
68	2

Find the population standard deviation of these scores, to the *nearest tenth.*

30 Find the sum and product of the roots of the equation $5x^2 + 11x - 3 = 0$.

31 The graph of the equation $y = \left(\frac{1}{2}\right)^x$ has an asymptote. On the grid below, sketch the graph of $y = \left(\frac{1}{2}\right)^x$ and write the equation of this asymptote.

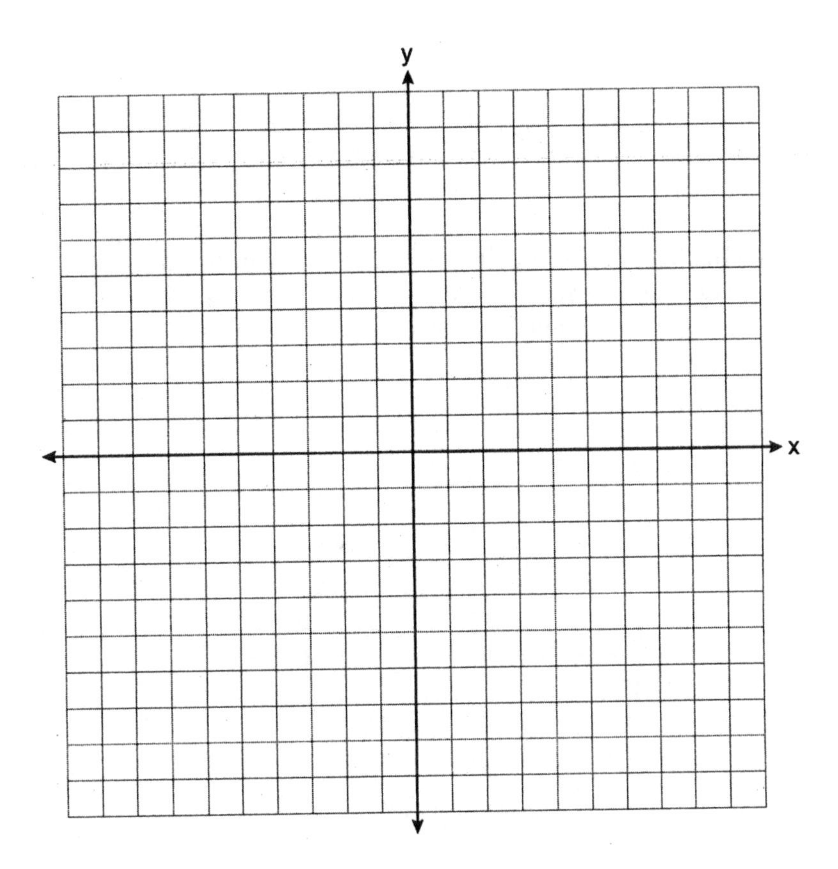

32 Express $5\sqrt{3x^3} - 2\sqrt{27x^3}$ in simplest radical form.

33 On the unit circle shown in the diagram below, sketch an angle, in standard position, whose degree measure is 240 and find the exact value of sin 240°.

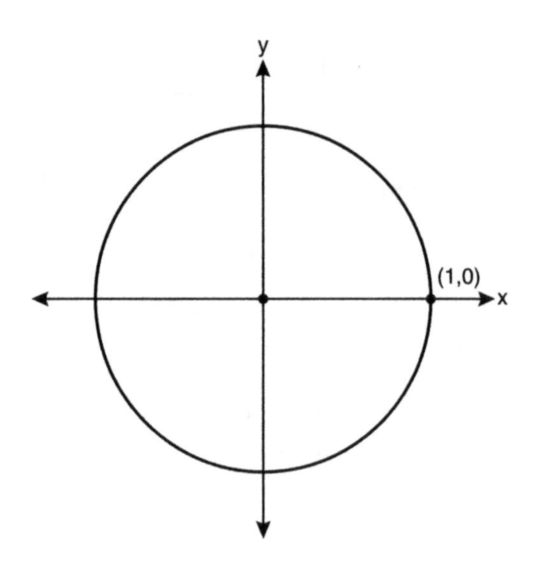

34 Two sides of a parallelogram are 24 feet and 30 feet. The measure of the angle between these sides is 57°. Find the area of the parallelogram, to the *nearest square foot.*

35 Express in simplest form: $\dfrac{\frac{1}{2}-\frac{4}{d}}{\frac{1}{d}+\frac{3}{2d}}$

PART III

Answer all 3 questions in this part. Each correct answer will receive 4 credits. Clearly indicate the necessary steps, including appropriate formula substitutions, diagrams, graphs, charts, etc. For all questions in this part, a correct numerical answer with no work shown will receive only 1 credit. [12 credits]

36 The members of a men's club have a choice of wearing black or red vests to their club meetings. A study done over a period of many years determined that the percentage of black vests worn is 60%. If there are 10 men at a club meeting on a given night, what is the probability, to the *nearest thousandth*, that *at least* 8 of the vests worn will be black?

37 Find all values of θ in the interval $0° \leq \theta < 360°$ that satisfy the equation $\sin 2\theta = \sin \theta$.

38 The letters of any word can be rearranged. Carol believes that the number of different 9-letter arrangements of the word "TENNESSEE" is greater than the number of different 7-letter arrangements of the word "VERMONT." Is she correct? Justify your answer.

PART IV

Answer the question in this part. The correct answer will receive 6 credits. Clearly indicate the necessary steps, including appropriate formula substitutions, diagrams, graphs, charts, etc. A correct numerical answer with no work shown will receive only 1 credit. [6 credits]

39 In a triangle, two sides that measure 6 cm and 10 cm form an angle that measures 80°. Find, to the *nearest degree*, the measure of the smallest angle in the triangle.

Answers June 2010

Algebra 2/Trigonometry

Answer Key

PART I

1. (3)	**6.** (3)	**11.** (2)	**16.** (2)	**21.** (2)	**26.** (4)
2. (2)	**7.** (3)	**12.** (1)	**17.** (1)	**22.** (3)	**27.** (4)
3. (3)	**8.** (4)	**13.** (1)	**18.** (1)	**23.** (1)	
4. (1)	**9.** (4)	**14.** (3)	**19.** (1)	**24.** (1)	
5. (4)	**10.** (1)	**15.** (3)	**20.** (3)	**25.** (1)	

PART II

28. 4 and –4

29. 7.4

30. sum = $-\frac{11}{5}$ and product = $-\frac{3}{5}$

31. correct graph and $y = 0$

32. $-x\sqrt{3x}$

33. angle correctly drawn and $-\frac{\sqrt{3}}{2}$

34. 604

35. $\frac{d-8}{5}$

PART III

36. 0.167

37. 0°, 60°, 180°, and 300°

38. no, with justification.

PART IV

39. 33°

In Parts II–IV, you are required to show how you arrived at your answers.